INSIDE form • Z
GUIDE TO 3D MODELING AND RENDERING

2ND EDITION

**Eden Greig Muir,
Cory Clarke, and Nam-ho Park**

Africa • Australia • Canada • Denmark • Japan • Mexico • New Zealand
Phillipines • Puerto Rico • Singapore • United Kingdom • United States

NOTICE TO THE READER

Publisher does not warrant or guarantee any of the products described herein or perform any independent analysis in connection with any of the product information contained herein. Publisher does not assume, and expressly disclaims, any obligation to obtain and include information other than that provided to it by the manufacturer.

The reader is expressly warned to consider and adopt all safety precautions that might be indicated by the activities herein and to avoid all potential hazards. By following the instructions contained herein, the reader willingly assumes all risks in connection with such instructions.

The publisher makes no representation or warranties of any kind, including but not limited to, the warranties of fitness for particular purpose or merchantability, nor are any such representations implied with respect to the material set forth herein, and the publisher takes no responsibility with respect to such material. The publisher shall not be liable for any special, consequential, or exemplary damages resulting, in whole or part, from the readers' use of, or reliance upon, this material.

Trademarks

form*Z, RenderZone, and RadioZity are trademarks of auto*des*sys, Inc. LightWorks is a registered trademark of LightWork Design Limited. Macintosh, QuickDraw 3D, QuickTime, QuickTime VR, and MacOS are registered trademarks or trademarks of Apple Computer, Inc. Microsoft, Windows, Windows NT, and Windows 95/98 are registered trademarks or trademarks of Microsoft Corporation. All other trade names mentioned in this book are registered trademarks or trademarks of their respective companies. OnWord Press and the authors make no claim to these marks.

OnWord Press Staff

Publisher: Alar Elken

Executive Editor: Sandy Clark

Managing Editor: Carol Leyba

Development Editor: Daril Bentley

Editorial Assistant: Allyson Powell

Executive Marketing Manager: Maura Theriault

Executive Production Manager: Mary Ellen Black

Production and Art & Design Coordinator: Cynthia Welch

Manufacturing Director: Andrew Crouth

Technology Project Manager: Tom Smith

Cover Design by Cammi Noah

Copyright © 2000 by Eden Greig Muir
SAN 694-0269
10 9 8 7 6 5 4 3 2
Printed in Canada

Library of Congress Cataloging-in-Publication Data
Muir, Eden.
   Inside form*Z : guide to 3D modeling and rendering / Eden Greig Muir. – 2nd ed.
   p. cm.
   ISBN 1-56690-189-8
   form*Z. 2. Computer simulation 3. Computer graphics. Title.
QA76.9.C65M86 2000
006.6'869—dc21   99-39798
                                                                                       CIP

For more information, contact

OnWord Press   An imprint of Thomson Learning

Box 15-015   Albany, New York   USA 12212-15015

All rights reserved Thomson Learning. The text of this publication, or any part thereof, may not be reproduced or transmitted in any form or by any means, electronic or mechanical, including photocopying, recording, storage in an information retrieval system, or otherwise, without prior permission of the publisher.

You can request permission to use material from this text through the following phone and fax numbers.

Phone: 1-800-730-2214; Fax: 1-800-730-2215; or visit our Web site at www.thomsonrights.com

About the Authors

Eden Greig Muir is founder and chair of CyberSites, an Internet firm that develops Web properties such as AncientSites.com, an online community dedicated to the digital reconstruction of ancient sites for entertainment and education. He also serves as Director of Computers and teaches architecture at the graduate level at Columbia University and is the author of numerous articles and books, including *Guide to Creating 3D Worlds*.

Cory Clarke is an adjunct assistant professor at Columbia University's Graduate School of Architecture, Planning, and Preservation. He also works professionally in architectural design and Web site and Internet development.

Nam-ho Park is an adjunct professor at Columbia University's Graduate School of Architecture, Planning, and Preservation and works professionally in architectural design in New York.

Acknowledgments

The authors would like to extend their thanks and appreciation to an exceptional group of Columbia University students and recent graduates who helped create this second edition of *INSIDE form•Z*. We must first mention Brandon Hicks, who developed many of the renderings and, as editorial assistant, miraculously kept the authors, the manuscript, and the hundreds of illustrations organized as the project team and the scope of the undertaking grew week by week.

Joey Kosinksi deserves special praise for his huge contribution to the new rendering and animation material. Jason Jiminez and Henry Gunawan helped bring the tutorials up to date and explored many of version 3.1's new modeling tools. Many thanks as well to Dean DiSimone, William Gatchell, Solomon Frausto, and Kyung-San Kim (who built the Villa Savoye model); Michael Hoffman, Christopher Perry, and Ben Yorker (who contributed the Construction System model in Chapter 9); and Robert Quevedo (who developed the Pedestrian Bridge with Joey Kosinski).

Special thanks must go to Bernard Tschumi, Dean of the Graduate School of Architecture, for his support of this project, as well as Professor Kenneth Frampton, who encouraged his students Dean DiSimone and Joey Kosinksi to include their images and animation of Frank Lloyd Wright's Insurance Building project. A dedicated technical support team keeps the "paperless studios" (and form•Z) humming at Columbia; this book was greatly facilitated by the technical wizardry of Soo Young Lee, Mark McNamara, Nick McNamara, Theresa Ling, Shae Russell, Ameet Doshi, Matt Martin, and Brendon O'Toole.

Many thanks to Chris Yessios of auto•des•sys, Inc., for supporting this effort and above all for creating a brilliant product, form•Z. And last but not least, we salute our patient and indefatigable editor, Daril Bentley.

Cory wishes to thank his parents and sister, without whose support he would not have had the opportunity to write this book. Nam-ho wishes to thank Joon, Eliot, Moonkyu, Sung, and Phil for their inspiration and insight, along with a special thanks to Jimin for her warm understanding and support. Above all, he would like to thank his parents, without whom none of this would have been possible.

Contents

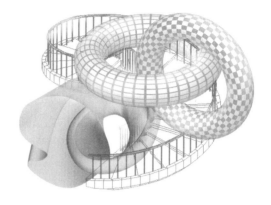

INTRODUCTION 1

PART 1: USING form•Z 1

CHAPTER 1: GETTING STARTED 3
   Introduction 3
   3D Versus 2D 4
   3D First 4
   Your 3D Modeling Playground 5
   The Structure of form•Z 8
   The form•Z Environment 8
   The form•Z Interface 8
   Menus 9
      Palettes 9
      Modeling Tools   10
      Window Tools   10
      The 3D Window   10
   Get Started! 11

CHAPTER 2: GETTING HELP 13
   Introduction 13
   Your First Stop for Help 14
   Built-in Help Menus 15
      Customizing the Modeling Tools Palette 17
      Window Tools Help 19
      Menus Help 20
      Keyboard Help 21
      Drafting Tools Help 25
      General Help 26
      Introductory Help 27
   Using the Query Tool 28
   Internet Help 28
      auto•des•sys Help 29
      CyberSites, Inc., Help 30

CHAPTER 3: MENUS 31
   Introduction 31
      Menus Interface 32
      Concepts and Terminology 32
      Shortcuts and Conventions 33
   File Menu 34
      New (Model) 35
      New (Draft) 35
      New Imager Set 36
      Open 37
      Open Recent 37
      Close 37
      Save Command 38
      Save As... 39
      Save A Copy As... 40
      Save QuickTime VR... 41
      Export Animation... 41
      Revert To Saved 41
      Import... 42
      View File... 42
      Page Setup 43
      Plot/Print Setup... 44
      Page Preview... 44
      Print... 45
      Quit 46
   Edit Menu 46
      Undo* 46
      Redo 47
      Redo All 47
      Replay 47
      Reset Undo/Redo 48
      Undo Options... 48
      Cut 49
      Copy 49
      Paste 50
      Duplicate* 50

Duplication Offset... 51	Axonometric* ... 89
Paste From Modeling*/Paste From Drafting . 51	Isometric* ... 91
Grab Image ... 52	Oblique* ... 92
Paste Image ... 53	Perspective* ... 93
form•Z Clipboard... 53	Panoramic* ... 94
Select Previous ... 54	View Parameters... 96
Select All UnGhosted ... 54	Save View... 97
Select All Ghosted ... 55	Views... 97
Select By... 55	Animation From Keyframes... 97
Deselect ... 58	Sun Position... 98
Clear ... 58	Edit Cone Of Vision... 99
Clear All Ghosted ... 59	Display Menu ... 103
Hide Ghosted ... 59	1/8" = 1' - 0" Scale, 1/16" = 1' - 0" Scale, and
Key Shortcuts... 59	1/32" = 1' - 0" Scale ... 104
Preferences... 60	Custom Display Scale... 104
Windows Menu ... 65	Wire Frame* ... 105
New Model Window ... 65	Quick Paint* ... 108
New Draft Window ... 66	Hidden Line* ... 109
Tile Windows ... 66	Surface Render* ... 110
Close ... 67	Shaded Render* ... 111
Close All ... 67	RenderZone* ... 113
Window Frames ... 68	QuickDraw 3D* ... 114
Extended Cursor* ... 69	OpenGL* ... 115
Extended Cursor Options... 71	Radiosity Options..., Initialize Radiosity*,
Auto Scroll ... 72	Generate Radiosity Solution*, and Exit
Show Plane Axes* ... 72	Radiosity ... 116
Show World Axes* ... 73	Display Options... 116
Show Grid* ... 73	Generate Animation... 117
Window Setup... 73	Play Animation... 117
Show Rulers* ... 75	Draft Layout Mode ... 117
Ruler Options... 76	Clear Rendering Memory... 117
Snap Options... 76	Always Clear Rendering Memory ... 117
Underlay... 77	Show Surfaces As Double Sided ... 117
Heights Menu ... 80	Redraw Buffers... 118
Graphic/Keyed ... 80	Image Options... 118
Custom... 81	Options Menu ... 119
Edit Menu... 81	Input Options... 119
View Menu ... 82	Pick Options... 120
z=30° x=60° ... 82	Zoom Options... 121
z=45∞ x=45∞ ... 83	Layers... 121
z=120° x=20° ... 84	Project Colors... 122
z=220° x=45° ... 84	Symbol Libraries... 123
z=60° x=30° ... 85	Working Units... 124
Custom View Angles... 85	Color Palette... 125
[+XY] : Top ... 86	Lights... 126
[–XY] : Bottom ... 86	Macro Transformations... 126
[+YZ] : Right Side ... 87	Objects... 127
[–YZ] : Left Side ... 88	Profiles... 128
[+ZX] : Back ... 88	Reference Planes... 128
[–ZX] : Front ... 88	Status Of Objects... 129
Plane Projection ... 89	Surface Styles... 130

Contents

Color Palette..., Line Styles...,
    and Line Weights... 130
2D and 3D Digitizer Options 131
Palettes Menu 131
    Customize Tools... 132
Help Menu 133
    Error Messages... 133
    Project Info... 134

CHAPTER 4: PALETTES 137
Introduction 137
Palettes and Screen Size 138
Manipulating Palettes 141
    Rearranging Palettes 141
    Collapsing Palettes 142
    Closing and Opening Palettes 143
    Hiding Palettes 144
    Expanding Palettes 144
    Transferring Palettes 144
    Ranking Palettes 144
    The Prompts Palette 145
    The Tool Options Palette 148
    The Views Palette 150
    The Colors/Surface Styles Palette 151
    The Layers Palette 154
    The Lights Palette 157
    The Coordinates Palette 158
    The Objects Palette 159
    The Planes Palette 161
    The Profiles Palette 162
    The Symbols Palette 162
    The Window Tools Palette 165
    The Animation Palette 166

CHAPTER 5: MODELING TOOLS 165
Introduction 165
    The Interface 165
    Concepts and Terminology 167
    Polygons 167
    Solids and Surfaces 167
    Parametric Objects and Wires and Facets .. 168
    Modifiers and Generators 168
    Derivative Tools 169
Using the Tools 169
    Primitives Tool Palette 170
    Balls Tool Palette 174
    Object Type Tool Palette 178
    Insertions Tool Palette 182
    Polygons and Circles Tool Palette 185
    Lines, Splines, and Arcs Tool Palette 190
    Topological Levels Tool Palette 196

Pick Tool Palette 199
Derivatives Tool Palette 204
Parametric Derivatives Tool Palette 213
Meshes and Deform Tool Palette 226
Rounding and Draft Angles Tool Palette .. 238
NURBS and Patches Tool Palette 242
Metaformz Tool Palette 254
Booleans and Intersections Tool Palette .. 256
Join and Group Tool Palette 268
Text Tool Palette 271
Symbols Tool Palette 273
Line Editing Tool Palettes 278
Topologies Tool Palette 284
Self/Copy Tool Palette 288
Query Tool Palette 291
Geometric Transformations Tool Palette .. 295
Relative Transformations Palette 303
Attributes Tool Palette 312
Ghost and Layers Tool Palette 324
Delete Objects Tool Palette 329
Delete Parts Tool Palette 330

CHAPTER 6: WINDOW TOOLS 333
Introduction 333
    The Interface 333
    Concepts and Terminology 335
Window Tools 335
    Reference Planes 336
    Perpendicular Switch 338
    Reference Planes 339
    Grid Snap Switch 341
    Direction Snaps 344
    Object Snaps 347
    Zoom and Pan 354
    View Tools 359

PART 2: BEGINNER TO ADVANCED EXERCISES 363

CHAPTER 7: EXERCISES FOR BEGINNERS 365
Introduction 365
Exercises 366
    Exercise 7-1: The Five-minute Barn 366
    Exercise 7-2: The Five-minute Goblet 370
    Exercise 7-3: The Five-minute Mace 373
    Exercise 7-4: The Five-minute City 375
    Exercise 7-5: The Five-minute Logo 378
    Exercise 7-6: The Five-minute Maze 381
    Exercise 7-7: The Five-minute Coin 383
    Exercise 7-8: The Five-minute Landscape .. 387
Further Exercises 391

CHAPTER 8: INTERMEDIATE EXERCISES 393
- Introduction 393
  - Exercise 8-1: Modeling Skyscraper Towers .. 394
  - Exercise 8-2: Modeling a Tape Dispenser ... 408
  - Exercise 8-3: Modeling a Picture Frame 418
  - Exercise 8-4: Modeling a Magnifying Glass .. 425
  - Exercise 8-5: Modeling a Screwdriver 433
  - Further Exercises 441

CHAPTER 9: ADVANCED EXERCISES 443
- Introduction 443
- The Exercises 443
  - Exercise 9-1: The Hand: Modeling.......... 444
  - Exercise 9-2: Modeling a Telephone 451
  - Exercise 9-3: An Umbrella: Modeling Surfaces Using Nurbz 465
  - Exercise 9-4: Modeling a Flashlight 474
  - Further Exercises 484

PART 3: RENDERING, DRAFTING, ANIMATION, AND TIPS ... 487

CHAPTER 10: RENDERING 489
- Introduction 489
- Rendering Display Types 490
  - Wire Frame 490
  - Quick Paint 491
  - Hidden Line 492
  - Surface Render 492
  - Shaded Render 492
  - RenderZone 493
  - QuickDraw 3D/OpenGL 494
- RenderZone Options 495
  - Rendering Types 496
  - Shading Options 499
  - Environment 508
  - Depth Effects 511
  - Illumination 514
- Lighting and Shadows 516
  - Name 517
  - Light Types 517
- Radiosity 525
  - Lighting an Interior with RadioZity 526
  - Concepts of Radiosity 526
  - Setting Up Renderings with RadioZity 527
  - Radiosity Options 529
  - Tips and Getting Best Results 536

CHAPTER 11: DRAFTING 543
- Introduction 543
- The Drafting Environment 543
- Pasting from Modeling into Drafting 546
- Preparing to Add Dimension Lines 548
- Setting Numeric Accuracy 549
- Setting Witness Lines 550
- Setting Terminators 551
- Setting Dimension Text 552
- Placing Dimension Lines 553
- Placing a Text Label 555
- Draft Layout Mode 555
- Preparing to Print 556

CHAPTER 12: ANIMATION 557
- Introduction 557
- form•Z Animation Basics 558
- Animation Quick Guide 559
- Animation Tour and Exercises 561
  - Exercise 12-1: A Casual Walk Through a City 561
  - Exporting Your File to QuickTime or AVI .. 566
  - Exporting Stills from Your Animation 566
  - Exercise 12-2: A Day in the Life of a Fly 567
  - Editing the Velocity Control Curve 572
  - Adding Motion Blur to Your Animation 576
  - Super Sampling 577
  - Adding a Background Image to Your Animation 578
  - Previewing Your Animation in OpenGL/QuickDraw 3D 578
  - Exercise 12-3: Advanced Animation Example 579
  - Using Walkthrough View to Create an Animation Path 593
  - Generating an Animation 594

CHAPTER 13: FIFTY-FIVE form•Z TIPS 595

APPENDIX A: form•Z ON MacOS AND WINDOWS 617

APPENDIX B: FILE FORMATS 619

APPENDIX C: KEYBOARD SHORTCUTS 627

SUBJECT INDEX TO LEARNING OBJECTIVES 635

SUBJECT INDEX TO CHAPTER 13 637

INDEX 639

Foreword

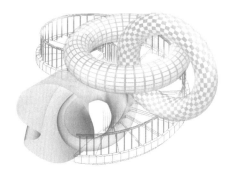

It is only a little over a year ago that I wrote the foreword to the first edition of *INSIDE form•Z*. I am now delighted to confirm that all the optimism expressed in that foreword has proven correct. This past year, *INSIDE form•Z* was a best seller and quickly sold out. In spite of this, the publisher did not take the easy way out. Rather than reprinting the first edition, the authors were asked to revise and update it to include the new features introduced in form•Z 3.0, which was released after the first edition was published. Once again, the authors did an admirably timely and effective job in developing the material for the new edition.

What I said a year ago is still valid. In the text that follows, other than some future tenses changed to past, I changed only the number 250 to 300, which is the number of universities known to be involved with the teaching of form•Z through the Joint Study program. Much of the "wishful thinking" I expressed a year ago is now "fact," and the text that follows reflects this.

Most software publishers will agree that the task of preparing exhaustive and well-written manuals is monumental. In the quest to satisfy every possible reader by being as detailed as possible, the sheer size of documentation can intimidate some users. With this in mind, I can only express admiration for the job professor Eden Muir and his coauthors have done in writing and updating *INSIDE form•Z*. Within a very manageable number of pages, they offer both a comprehensive overview of the individual features of form•Z and a collection of valuable techniques and tutorial examples that are original and clever.

Having personally written many, and closely supervised the balance, of the form•Z manuals, I can attest to how difficult it is to write concise yet pleasantly flowing presentations of technical material. The authors have succeeded in both. In addition, the tutorial techniques employed in the book are unavailable elsewhere. The exercise chapters alone are worth more than the cost of the entire publication.

At auto•des•sys, we like to remember and cherish the fact that form•Z's roots are in academia, which, without question, has impacted the philosophy of the program. Even today, the research behind the software's operations is always

Foreword

aimed toward tools that facilitate, encourage, and enhance the creative process of design, rather than just visualization. We strive to develop tools that are readily familiar to new users or that require very little adjustment. Consequently, we are always happy when instructors—from the more than 300 universities worldwide at which form•Z is taught—tell us that the software is unsurpassed as a tool for motivating students and developing their interest in the virtual arts, and that it is a highly effective teaching tool in 3D modeling and visualization.

At the same time, we also recognize that it is no accident form•Z is taught most successfully at schools where some gifted instructor makes it happen. Before *INSIDE form•Z* was first published, many dedicated teachers of form•Z had to prepare their own instructional aids, wishing that appropriate material were available so that they did not have to reinvent the wheel. They were looking for a text that would effectively take their students through the theory of 3D modeling and offer examples of and exercises on how to produce practical applications quickly and efficiently. They can find all of these in *INSIDE form•Z*.

In producing the first edition, Eden Muir collected and refined material he had been preparing for teaching form•Z for some five years at Columbia University. His proven teaching techniques at the university's highly regarded design school proved equally effective when adopted at other institutions. This material has now been expanded to include the new features of form•Z 3.0, which introduced animation. Even though these features are as new for the authors as for everyone else, Muir's team has again done an excellent job of generating examples that will make the new features easy to learn. In addition, the new edition of *INSIDE form•Z* contains a section with color images that will further facilitate learning the program. As it happened with the first edition, I again expect this book to reach beyond academia and into the professional realm, where the experienced and inexperienced alike have discovered the benefits of, and are now moving into, the world of computer-assisted 3D modeling and design.

Because *INSIDE form•Z* does significantly more than cover how the features of the program work, it has proven very effective in easing an established professional's transition from traditional methods to the new electronic media. The book does and excellent job of demonstrating not only what one can do with the various tools but what one can achieve that is not easily accomplished by hand, as well as how one can expand his or her design repertoire. At auto•des•sys, we would like to believe that to a large extent this is also a reflection of the quality and richness of form•Z.

Nonetheless, a book that exposes the program's features and makes sure they are clearly conveyed as easily usable tools—as they truly are—makes a world of difference to both the manufacturer of the software and the user. All of us at auto•des•sys feel both gratified and thankful for the excellent job this book does in presenting form•Z in general, and the new features of version 3.0 in particular. With the publication of the second edition of *INSIDE form•Z*, we finally feel that version 3.0 is undoubtedly complete.

Chris I. Yessios
President, auto•des•sys, Inc.
September 1999

Introduction

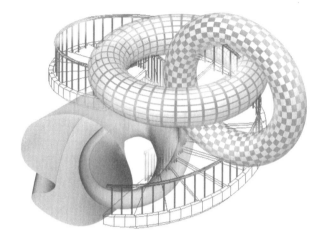

Philosophy, Approach, and Structure

The approach to form•Z taken in this book stems from the philosophy of teaching through example and application; 40 design-model exercises are used to show you how form•Z can be applied in real situations. Unlike most computer manuals, which describe tools through abstract text and diagrams, the highly visual *INSIDE form•Z* discusses each tool succinctly and critically—and whenever possible in the context of an actual application. The purpose of this method is to show you not only how the tools are used to create 3D models but when and why to use each tool. The reasoning behind this is that the best of modelers eventually gain the ability to apply learned functionality abstractly.

Editor's Note

From an editorial perspective, if you have not already done so, you are encouraged to read the Foreword by Christopher Yessios, president of auto•des•sys, Inc., manufacturer of the form•Z software. In addition to singing the book's well-deserved praises, Yessios does a nice job in the Foreword of explaining the authorial mind-set in tackling the writing of this book and the approach employed in it, as well as why you might want to use both the software and the book.

A Note on Structural Mechanics

Note that chapters 7, 8, and 9—the exercise chapters—are each continued on the companion CD-ROM. See the section on using the CD-ROM at the end of this introduction, as well as the "Exercises" section, for information on the exercise material in the book and on the CD.

Who Should Use This Book

form•Z is a powerful 3D visualization tool used by architects, engineers, industrial designers, graphic designers, animators, and many others of widely diverging disciplines. The software's broad range as a modeling tool makes it as well suited for creating buildings as it is for generating 3D graphics and interface designs for the Web.

In addition to breadth of application, this book addresses the range of user level. The book is structured around sets of exercises that not only introduce the beginner to the tools and modeling methods in form•Z but attempt to guide the more experienced user though the advanced techniques that can be employed to produce complex modeling solutions.

Beginner Through Advanced User

For the beginner, this book contains concise descriptions of all tools in form•Z, as well as a set of basic exercises that can be followed to get acquainted with the tools and modeling environment. For the more experienced user, there are intermediate and advanced exercises that demonstrate the most powerful features of form•Z, such as rendering and animation. Chapter 13, Quick Tips, also provides insight into a collection of shortcuts and techniques for optimizing form•Z performance.

The Instructor

For educators, the book provides an easily accessible and structured way of teaching form•Z through hands-on application of the tools and other features of the software. This approach centers on the exercises, facilitated by additional teaching information contained in headings such as "Learning Objectives" and "Tips and Study and Application Notes," and by features such as subject indexes—all of which are explained in detail in the material that follows. For example, the tool-description sections can also be used as a quick reference, and can be used to efficiently target—along with headings, subject indexes, and other book features—sets of tools that pertain to a specific instructional context.

Version Information

The second edition of *INSIDE form•Z* is written to form•Z 3.1, including RenderZone and RadioZity. (The first edition, released in 1998, covered form•Z 2.9.5.) The most important changes in version 3.1 include the introduction of animation, which is discussed in Chapter 12, and the revamping of the Modeling Tools interface, which is now displayed as two columns of icons at the left side of the screen. This book introduces several new tutorials in order to present the new features of form•Z 3.1. Those exercises that have been carried over from the previous edition have been carefully reconstructed in the new version of the software.

MacOS Versus Windows

Aside from a couple of minor differences between the two environments, form•Z is practically identical on both Macintosh and PC (Windows) platforms. All important differences are noted and explained in the text (see "Visual Guides" under "Book Features and Mechanics" in this introduction), and wherever applicable, both variations of the feature are provided.

Although the book was originally written and screens captured on the MacOS version, Windows users will find little difficulty in locating all of the features

Hardware Requirements

illustrated in the figures. See also Appendix A, form•Z on PC and Macintosh, for a discussion of the differences between the two platforms.

The recommended hardware requirements for the CD-ROM that accompanies this book differ slightly from the minimum system requirements for form•Z. This is due to the fact that form•Z, like most modeling/rendering software, tends to put high demands on the processor and memory of your computer. These requirements reflect the authors' views on what hardware configurations would be necessary to achieve a satisfactory result without wasting too much time.

Processor Speed and RAM

The recommended processor speed is based on the requirements for the advanced exercises. Users who do not meet these requirements should have no problem modeling and rendering the beginner and intermediate exercises. If you do not have these minimum requirements, you will probably not be able to render the models you make in the advanced exercises. The amount of RAM in your computer becomes an issue when rendering large files, especially when your model contains many light sources that cast shadows.

Hardware for Macintosh

The recommended hardware requirements for the MacOS version of form•Z are a Power Macintosh with a 120-MHz processor or better, and with 64 Mb or more of RAM.

Hardware for Windows

The recommended hardware requirements for the PC (Windows) version of form•Z is a Pentium 133-MHz processor or better with 64 Mb or more of RAM.

How This Book Is Organized

The book consists of 13 chapters divided into three parts. Part I consists of an overview of the software methodology, interface, and functionality, as well as chapters containing detailed descriptions of the form•Z tools. Part II consists of the beginner through advanced user exercises. Part III covers rendering, drafting, and animation, and ends with a chapter that provides "quick tips" on the range of the software's use.

Chapter Structure and Content

The first six chapters, constituting Part I, cover the form•Z interface, including all of the menus and tools available in the software. Each function is discussed briefly to give you a quick grasp of what each menu or tool can do and what effect it produces. The following three chapters—7, 8, and 9, which constitute Part II—teach you how to apply the form•Z modeling tools in progressively more difficult sets of modeling exercises.

Chapters 10, 11, and 12 introduce the extra modules in form•Z: RenderZone, Drafting, and Animation. These chapters cover enough of the options and tools in each module to get you working in no time. The final chapter, Chapter 13, is a set of more than 50 tips for everyone from the beginner to the advanced form•Z user. The following is a brief description of the book by chapter.

Chapter 1: Getting Started

Chapter 1 covers the fundamental concepts behind the software and explains some of the basic ideas of 3D computer modeling. The chapter explains how the software is organized, as well as the geometrical concepts behind how it represents 3D space.

Chapter 2: Getting Help

Chapter 2 explores all of the resources available for getting help in form•Z. The resources listed in this chapter are useful for those who want to continue teaching themselves after completing this book, as well as for those who need a little more help getting started.

Chapter 3: Menus

Chapter 3 explains all of the pull-down menu bars that run across the top of the screen. The function of each menu bar selection is explained briefly. The menu bars discussed in this chapter are File, Edit, Windows, Heights, View, Display, Options, Palettes, and Help.

Chapter 4: Palettes

Chapter 4 covers all of the form•Z palettes, the small floating windows on the screen. This chapter explains how to arrange the palettes to customize your interface and presents an overview of each palette's function. The palettes covered in this chapter include Coordinates, Layers, Lights, Objects, Planes, Profiles, Prompts, Surface Styles, Symbols, Views, and Window Tools. Two important palettes new in version 3.1—Animation and Tool Options—are discussed in detail.

Chapter 5: Modeling Tools

Chapter 5 covers the form•Z modeling tools. The modeling tools are the backbone of the form•Z software. These tools are used to model all of the objects in form•Z. This chapter briefly explains what each tool does, but more importantly explains how and why you would apply each tool when creating a model. The arrangement of the version 3.1 modeling tools is completely different from version 2.9.5. Chapter 5 is an important chapter for anyone making the transition to the recent upgrade.

Chapter 6: Window Tools

Chapter 6 covers all of the window tools in form•Z, the small palette of tools in the bottom of the modeling window. The window tools control the viewing of your model and the behavior of the cursor. The tools covered in this section include the Reference Plane tools, Perpendicular Snap, Grid Snap, Direction Snaps, Object Snaps, Zoom & Pan, and the View tools.

Chapter 7: Exercises for Beginners

Chapter 7 begins Part II. This chapter incorporates a series of five-minute models and easy exercises to introduce basic modeling tools and techniques. The five-minute models take advantage of some of the built-in generative tools in form•Z that can be used to produce complex models in five minutes. The five-minute models included in the chapter are Barn, Goblet, Mace, City Block, Logo, Maze, Coin, and Landscape. Additional five-minute models—Bolt, Geodesic Dome, and Pear—can be found on the companion CD-ROM.

The easy exercises are a little more difficult than the five-minute models. These exercises apply some of the basic tools in form•Z to make simple architectural and household objects. The easy exercises included as part of the chapter's continuation on companion CD-ROM are Paper Clip, Swiss Cheese, Ice Cream Cone, Monument, Portal on a Podium, and Staircase.

Chapter 8: Intermediate Exercises

Chapter 8 builds on the basic tools you learned in the previous chapter and shows you how to combine them to produce more involved and complex models. The intermediate exercises are Skyscrapers, Key, Picture Frame, Magnifying Glass, Screwdriver. Exercises for Chair, Knife, Fork, Spoon, Soda Can, Hammer, and Teapot can be found on the companion CD-ROM.

Chapter 9: Advanced Exercises

Chapter 9 assumes you have a basic understanding of the modeling tools and concepts of form•Z. This chapter deals less with instruction in specific modeling tools and places an emphasis on technique and application. The exercises in this chapter show you how to combine tools in form•Z to produce complicated models and effects. The advanced exercises are Hand, Telephone, Umbrella, and Flashlight. Advanced exercises continued on the CD-ROM include Fruit Bowl, Rowboat, Toothpaste Tube, Sandals, Space Frame, Construction System, the Villa Savoye, Complicated Model, and Office Interior.

Chapter 10: Rendering

Chapter 10 covers the basics of rendering and producing images with form•Z RenderZone. The chapter discusses the RenderZone options and how to get the best effects and images for your model. This chapter has been significantly expanded, including more material on advanced rendering techniques and more examples of the RadioZity renderer. Some of these examples are reproduced in the color insert at the center of this book.

Chapter 11: Drafting

Chapter 11 discusses form•Z's 2D drafting module and how to import and export data to be used in the drafting module. The chapter also discusses some of the tools specific to the drafting module and focuses on applying dimension lines and simple annotation to your drawing.

Chapter 12: Animation

Chapter 12 presents the animation tools of form•Z, which were introduced with version 3.0. Three step-by-step examples take you through the production of simple fly-bys and walk-throughs up to advanced animation techniques that depend on the manipulation of complex function curves.

Chapter 13: 55 form•Z Tips

Chapter 13 presents more than 50 tips on everything from rendering and modeling to saving and printing. These tips are useful for beginners who want to improve their skills and work on technique, as well as for the regular users of form•Z who want to speed up the modeling process and pick up some fancy new tricks.

⇢ **NOTE:** *To locate tips topically, see the Subject Index to Chapter 13 at the back of the book.*

Color Insert

New in the second edition is a color insert that demonstrates the increased sophistication and power of the form•Z renderers RenderZone and RadioZity in version 3.1. Some of these remarkably photorealistic images are renderings of simple five-minute models presented in Chapter 7. Most of the other renderings show objects and buildings that can be constructed by means of the step-by-step tutorials included in chapters 8 and 9.

Appendices

You will find three appendices at the back of the book. These are Appendix A, form•Z on PC and Macintosh; Appendix B, File Formats; and Appendix C, Keyboard Shortcuts. The following sections discuss the content and use of these appendices.

Appendix A: form•Z on PC and Macintosh

The PC appendix is a reference section for PC users reading this book. The PC version and Macintosh version of form•Z are almost identical. Therefore, there should be no difficulty using this book no matter what platform you are using. However, this section is included to cover those minor differences between the Macintosh and PC versions of the software.

Appendix B: File Formats

The File Formats appendix discusses all of the file formats available in form•Z for export and import of data. This section discusses file formats for images, 3D/2D data, and lighting data. The file formats covered are EPS, JPG, PICT, TIFF, Targa, PNG, DWG, DXF, FACT, form•Z, IGES, Illustrator, Lightscape, OBJ, RIB, SAT, STL, 3DGF, 3DMF, VRML, CIE, CIBSE, and IES.

Appendix C: Keyboard Shortcuts

Appendix C, Keyboard Shortcuts, lists all of the keyboard shortcuts available in form•Z. Note also that keyboard shortcuts applicable to text discussion are called out in page margins (see the "Book Features and Mechanics" section of this introduction for details).

Indexes

In addition to a complete general index, this book contains a Subject Index to Learning Objectives and a Subject Index to Chapter 13. As noted in the "Book Features and Mechanics" section, which follows, these indexes augment the general index and make it possible to easily locate particular types of information. These indexes are also meant to allow the user and the teacher to build tailored self-study and instructional outlines.

Book Features and Mechanics

Knowing how a book is intended to be used, as well as an awareness of the useful features it provides, helps you get the most out of it. The following sections discuss the features of this book and the mechanics of how to use them individually and together to best advantage.

Visual Guides

The editorial and graphic design of this book incorporates visual cues that will help you note, keep track of, and guide you to classes or particular topics of information. These include icons, use of headings, notes, tips, warnings, tables, margin material, and so on. The following sections discuss these visual features.

MacOS/PC Icon

The MacOS/PC icon (shown at left) signals either a MacOS/PC Note in the margin (discussed in the next section) or that the text to the right of the icon contains material that discusses a "MacOS versus PC use" issue. This icon appears wherever MacOS/PC is discussed. Note also that Appendix A specifically deals with this topic. If you were to collect the relevant text wherever the icon appears and combine it with the appendix, you would have all of the information this book offers on the differences and similarities between the two environments.

MacOS/PC NOTE:
The Macintosh command key (with the cloverleaf or apple icon) is referred to as Cmd. Cmd-S means press the Command key and type S. On a Windows machine, keyboard shortcuts use the Control key. Ctrl + S means press the Control key and type S.

MacOS/PC NOTE

Wherever possible, MacOS/PC information is set off from the text as in-margin notes so that you can easily locate this material. They appear with the MacOS/PC icon. Where this material runs more naturally in-text, the MacOS/PC icon is placed in the margin as a "sign post" to its text location. An example of the MacOS/PC NOTE appears at left.

Keyboard Shortcut Notes

In the margin you will find notes titled "Keyboard Shortcut." These also appear with the MacOS/PC icon. These notes indicate either a Mac or a PC keyboard alternative, or both, to on-screen functionality. An example of a Keyboard Shortcut appears at left.

Notes, Tips, and Warnings

- **NOTE:** *Notes highlight important information and are used for all cross references to other parts of the book.*

- **TIP:** *Tips offer helpful suggestions for using the software more efficiently or with better results.*

- **WARNING:** *Warnings alert you to the consequences of a particular action, preventing major mistakes or the loss of work.*

Special Headings

Several repeat headings are intended to facilitate both the learner and the instructor in terms of form•Z functionality. These headings appear in chapters 5 through 9, which cover modeling tools, and beginner to advanced exercises—the core of the "know-how" portion of the book. The following sections describe these headings and their intended uses.

Learning Objective

This heading appears in chapters 5 through 9. The heading provides you with a succinct synopsis of the substance of the tool, menu, modeling exercise, or modeling exercise subpart under discussion. Note that all learning objectives are accessible topically by referring to the Subject Index to Learning Objectives at the back of the book. The objectives headings are numbered to correspond to the locators in the index. (See the "Indexes" section of this introduction for details.)

Tools Covered

This heading is specific to chapters 5 and 6, which cover the form•Z modeling and window tools. The list that appears under this heading informs you of the individual tools, discussed in depth in the respective section.

Screen Location

This heading is also specific to chapters 5 and 6, the tools chapters. In Chapter 5, a small grid diagram, corresponding to the two-columned layout of the modeling tools, shows you the location of the tool under discussion. In Chapter 6, the heading tells you the tool's location in relation to the form•Z modeling window.

Tips and Study and Application Notes

This heading appears in chapters 5 through 7 and 9. The content of these tips and notes either summarizes or adds to the content of the text. These headings are numbered consecutively throughout chapters for ease of instructional setting reference.

Tools

This heading is specific to chapters 5 and 6, the tools chapters. These headings tell you the nature of the tool or group of tools discussed and are numbered consecutively throughout the chapters for ease of instructional setting reference.

Supporting and Further Reference

This heading appears in Chapter 7, "Exercises for Beginners." The heading cross-references other parts of the book that discuss the topic of the exercise just completed.

Tips

Tips run throughout the text, as described in the previous section, "Notes, Tips, and Warnings." These work together with Chapter 13, to present a wealth of helpful suggestions that will save you time, make you more efficient, and allow you to learn more quickly and produce better models. Note that the tips within Chapter 13 are numbered consecutively to correspond to the Subject Index to Chapter 13, which allows you to locate tips topically.

Exercises Continued on Companion CD-ROM

As previously noted in the descriptions of chapters 7, 8, and 9 in the "Chapter Structure and Content" section, these three exercise chapters are each continued on the companion CD-ROM at the back of the book. For information on the CD-ROM, see "The Companion CD-ROM" section at the end of this introduction. Note that the color insert includes models from chapters 7 through 9. In some cases, the step-by-step instructions for constructing these models are not found in the book but on the companion CD-ROM.

Exercise Methodology and Mechanics

Cross-referencing via the "Supporting and Further Reference" heading is done in Chapter 7 to keep the new learner from having to absorb too much material when he or she is trying to learn the interface, learn basic functionality, and follow the process of a given exercise all at once.

Cross-referencing in chapters 8 and 9 is relegated to Notes that run with the text because the intermediate user has a grasp of basic functionality and needs the "why" as well as the "what" immediately accessible in the context of a particular exercise—before moving on to the advanced stage. The "Analysis and Method" and "Process" headings in these chapters supply the approach and the "why" that make the "what" (skills learned) of the exercises translatable to new modeling challenges and situations.

Exercise Content and Format

Chapters 7, 8, and 9 (continued on the companion CD-ROM) contain all of the modeling exercises in the book, which begin with the "Exercise" heading and are enclosed by a shaded background for "flip-through" identification of where they appear in the book, as well as where individual exercises begin and end. The exercises increase in difficulty within chapters and from chapter to chapter. Animation examples are included in Chapter 12.

The Companion CD-ROM

Included on the CD-ROM that accompanies this book is a demo version of form•Z 3.1, including RenderZone and RadioZity. The CD also contains form•Z files of the models used in the exercise chapters 7, 8, and 9. The demo version of form•Z 3.1 RenderZone and RadioZity is a fully capable version of the form•Z modeling environment, with only the Save functions disabled. The form•Z Drafting module is not available with the form•Z demo version. The demo version of the form•Z modeler also has a time limitation: after one hour, the program will force you to quit and restart the software.

All of the models supplied on the CD-ROM can be opened, viewed, and rendered using this demo version of the software. In addition, barring time limitations, all of the exercises shown in this book can be completed using the demo version of form•Z. The CD-ROM also includes the form•Z files of the exercise models in chapters 7, 8, and 9. These files are organized in folders by chapter.

Also included with each exercise are high-resolution, full color renderings, as well as a copy of all of the images used for surface style texture maps in the exercises. The models are all set to the same screen size and rendering settings. Examples of movie files from the animation chapter are included, as well as samples of QuickTime VR panoramic movies—all of which were created in form•Z.

These settings are selected with the minimum hardware requirements in mind. However, even with the rendering settings set at a minimum, if you are using the minimum hardware, you may have some difficulty rendering some of the advanced exercise models with Shadows turned on in the RenderZone options. If you have a more powerful computer than described in the "Hardware Requirements" section of this introduction, you may want to adjust the RenderZone options.

part 1

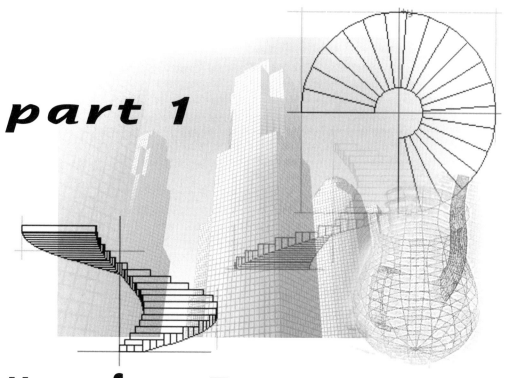

Using form•Z

chapter 1

Getting Started

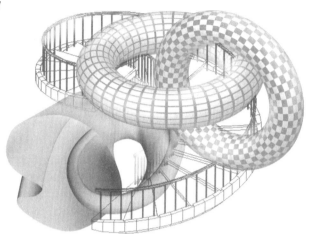

Introduction

This book is a guide to an exceptional computer-aided design (CAD) product, form•Z, which will change the way you approach the design of 3D objects, buildings, and spaces. form•Z was the first moderately priced CAD software to bring the most advanced 3D modeling tools to the average PC or Macintosh computer. These tools can make you a sculptor of virtual form, a molder of digital clay, and an architect of real or virtual structures. The most recent additions to form•Z will take you even further; you will now be able to add animator and moviemaker to your resume!

The goal of this book is to facilitate your mastery of form•Z so that you can experience the full power of its extraordinary 3D modeling, rendering, and animation capabilities. To be sure, computer-aided 3D design is an intellectual challenge; however, with the right approach and the right software, it can be a joy, and the rewards can be great.

The step-by-step tutorials in this book have grown out of the authors' experience in the computer labs at the Graduate School of Architecture at Columbia University, where we have taught form•Z since 1993. When the drafting tables were thrown out in favor of our "paperless" design studios, form•Z became a cornerstone of the new design curriculum. form•Z is now a far more powerful and comprehensive package, and is more popular than ever at the school.

With form•Z, students can be building meaningful and complex 3D models in just a few weeks if they systematically work through a series of modeling exercises that introduce them to the major features of the software. Our students are usually creating high-quality renderings and animated sequences within a few weeks. By following the tutorials in this book and on the companion CD-ROM, you will be able to do the same.

3D Versus 2D

form•Z has set itself apart from many other software products for designers by focusing on the development of a superior interface and a collection of tools and procedures for modeling 3D objects and spaces. The earliest generation of software products for designers favored 2D drafting because it was an easier problem to solve than 3D modeling; that is, 2D graphics are less computationally intensive, and are easier to think about and program. In addition, a 2D drafting emphasis solved the immediate problem of designers who needed an efficient system for editing what was typically 2D documentation.

For many years, high-end computer workstations, costing tens of thousands of dollars each, provided the only hardware platform for a few equally rare and expensive 3D software packages. These systems were used mainly by engineers, industrial designers, and special effects experts in environments where cost was not a major concern.

It is a fairly recent development that moderately priced software running on average personal computers has been able to handle complex 3D modeling, rendering, and animation tasks. form•Z running on a PC is an example of how, in recent years, state-of-the-art 3D design tools became accessible and affordable to a much larger segment of the design profession. Now, architects, engineers, illustrators, interior designers, animators, game designers, and any other creative amateur or professional can benefit from these developments.

3D First

The affordability and power of 3D software such as form•Z has far-reaching implications for the design profession. A new 3D methodology can now be implemented in all design environments—an approach summed up by the phrase "3D first." 3D first means that 3D modeling is the primary activity of the designer, with 2D documentation produced afterward by transporting images of the 3D model into the drafting module.

This may sound perfectly logical and not at all radical to designers who are already using form•Z. However, in many design environments today, both professional and academic, it is still common practice to design and document 3D projects using 2D methods and to limit 3D work—if any is done at all—to a final perspective rendering taken from a single viewpoint.

This book, like form•Z itself, is an expression of the 3D-first design philosophy. In all of the exercises, you will be building a model that is not described as a 2D drawing, but is built as a 3D model that occupies virtual 3D space. Although it is possible to use form•Z as a 2D drafting product by transferring images from the modeling to the drafting module, the 3D capabilities of the software are the primary focus of this book.

Other benefits of using form•Z for design will quickly become clear as you read this book and master the program. In the course of a form•Z design project, you will have instant access to presentation-quality perspectives at any time. In fact, you will be able to produce many types of views—from panoramic to axonometric—which once were time-consuming presentation tasks reserved for the final stages of a project and required special skills. You will be able to grab images from the screen to insert directly into Web pages, send as e-mail attachments, or place on CD-ROMs. Perhaps the most exciting news is that you will be able to produce movies!

The animation features of form•Z are described in detail in Chapter 12. Architectural walk-throughs and fly-bys are now easy to produce from any form•Z model, but you can go even further. If you obtain video editing software such as Adobe Premiere, you will be able to combine your form•Z sequences with titles and other images and add a sound track to produce a presentation movie.

Perhaps the most startling benefit of a 3D-first strategy is the fact that you can export your form•Z models to a rapid prototyping system to have them fabricated as real 3D objects you can hold in your hand. For the engineer, architect, or product designer, this is surely one of the most tangible benefits of moving toward a digital design methodology with a powerful 3D tool such as form•Z.

Your 3D Modeling Playground

If you are not familiar with the basics of 3D coordinate geometry (figure 1-1), a brief review is in order. In figure 1-1, the three dark intersecting lines are the x, y, and z axes, which have been identified with large text labels consisting of 3D letters. Measuring along these three axes allows the program to use just three numbers to represent any point in space, which is what makes 3D computer graphics possible.

Fig. 1-1. 3D coordinate geometry provides a 3D environment for designers.

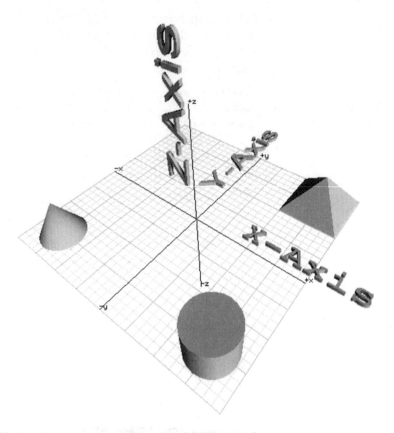

form•Z encourages an intuitive approach to 3D modeling. That is, you will be able to ignore the underlying mathematics, if you wish, and use the visual feedback supplied by the various types of views (perspective, isometrics, and so on) to understand the 3D forms you create.

Note that in figure 1-1, the cone, cylinder, and pyramid are resting on the xy plane, defined by the x and y axes. This is the reference plane, or the active drawing surface. In general, items are described on this surface, which is represented by a light blue grid. The z axis is the one that brings this virtual "playground" into the 3D realm. The positive z axis, rising up from the reference plane, is indicated with a plus sign (+), whereas the negative z axis goes "underground." By specifying a z-axis value, flat shapes can be extruded into 3D objects.

Keep in mind that you can look into the virtual space of this 3D world from any viewpoint or angle. The "camera," or cone of vision, which creates the views, is itself bound to the laws of coordinate geometry.

That is, it is merely another object in the 3D scene and can be positioned anywhere and can look in any direction. With the introduction of animation in form•Z, the camera can be considered a moving object that can glide smoothly from one viewpoint to another, producing hundreds of individual views or "frames." The perspective shown in figure 1-1 is just one of an infinite number of views that could be generated of that 3D model.

For many designers, especially those who have toiled with 2D systems for most of their careers, it is a startling revelation to realize that they can model almost any 3D shape that can be imagined, and that they can view it from any vantage point. Once designers master the tools and methods of form•Z, their productivity and creativity can truly soar.

In contrast, a 2D drafting system is a primitive and limited environment in which to conduct 3D design (figure 1-2). A 2D computer-aided drafting environment is certainly more efficient than manual drafting for documenting a project, but it lacks the 3D information needed to effectively communicate space and form. It does not supply the designer with the instant 3D visual feedback that can inform the design process. With a 3D-first approach, drafting assumes its proper place as a final documentation phase based on 2D images that can be easily and cost-effectively extracted from the 3D model.

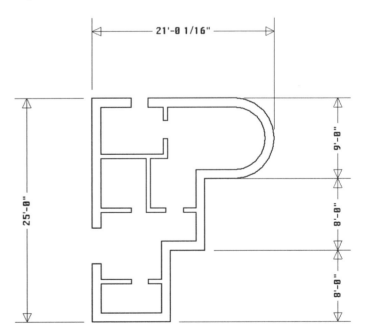

Fig. 1-2. A 2D drafting environment is limited by the lack of the third dimension.

> **NOTE:** *Drafting is discussed in Chapter 11.*

The Structure of form•Z

form•Z software is offered in a regular version that includes modeling and drafting modules, and in more expensive versions that add the RenderZone and RadioZity rendering modules. The latter versions are highly recommended, due to their impressive texture mapping and rendering capabilities, which are demonstrated in many of the exercises in this book. If you do not have the form•Z RenderZone or RadioZity versions, you will notice that you can work with colors but not textures, and that some of the rendering choices on your Display menu are grayed out.

The 3D modeling and 2D drafting modules exist side by side, allowing you to easily move entities between modules by means of cut-and-paste operations. RenderZone is seamlessly integrated into form•Z; that is, it appears as additional choices in the Display menu and requires no special saving or transferring of files. The RadioZity renderer has been handled the same way; it appears in a special section under the Display menu.

> **NOTE:** *Moving back and forth between the 3D modeling and 2D drafting modules is discussed in Chapter 11.*

The form•Z Environment

The form•Z environment consists primarily of the following components: (1) 3D modeling, for form generation and editing, (2) 2D drafting, for documentation, (3) Rendering in RenderZone, with or without the RadioZity module, and (4) the animation feature. The Menus, Palettes, Command Tools, and Window Tools that control the modeling environment are discussed in chapters 3, 4, 5, and 6 of this book. Step-by-step examples are provided in chapters 7, 8, and 9.

The state-of-the-art renderers RenderZone and RadioZity—whose many uses include creating presentation drawings and photorealistic images—are discussed in Chapter 10. The 2D drafting module (which offers dimensioning, line weights, hatching, and so on) is discussed in Chapter 11. Chapter 12 is devoted entirely to animation.

The form•Z Interface

Before you jump into form•Z, take a moment to acquaint yourself with the structure of the interface. Figure 1-3 shows the form•Z screen as it will appear during a typical session of 3D modeling. The five major components of the form•Z modeling interface are the menus, modeling tools, palettes, window tools, and the 3D window, which provides a 3D view of your project as you construct it.

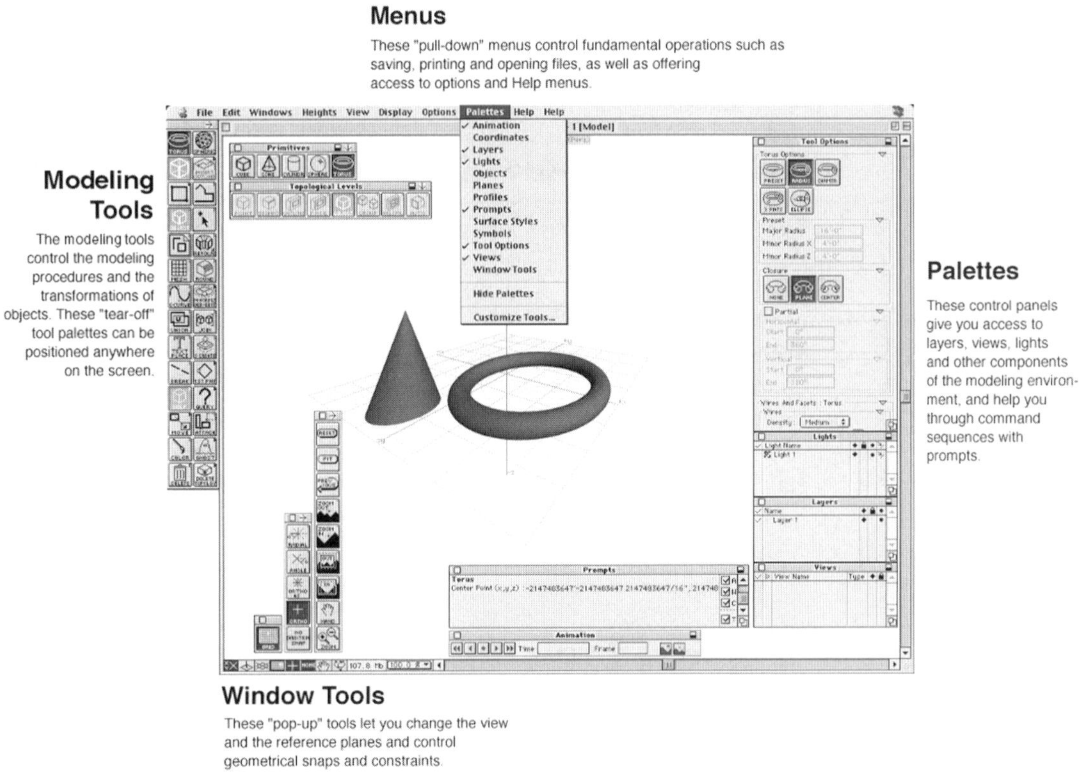

Menus

Modeling Tools
The modeling tools control the modeling procedures and the transformations of objects. These "tear-off" tool palettes can be positioned anywhere on the screen.

Menus
These "pull-down" menus control fundamental operations such as saving, printing and opening files, as well as offering access to options and Help menus.

Palettes
These control panels give you access to layers, views, lights and other components of the modeling environment, and help you through command sequences with prompts.

Window Tools
These "pop-up" tools let you change the view and the reference planes and control geometrical snaps and constraints.

Fig. 1-3. The major components of the form•Z modeling environment.

Menus

The form•Z pull-down menus, which always appear at the top of your screen, provide access to fundamental system operations such as opening, closing, saving, and printing files. Other menus control heights, views, and display styles. An extremely useful Help feature is located here as well. Figure 1-3 shows the Palettes menu selected and extended downward.

⇨ **NOTE:** *The form•Z menus are discussed in detail in Chapter 3.*

Palettes

The Palettes are control and information panels that sit above the modeling window and act as shortcuts to various aspects of the modeling environment. Mastering the palettes will dramatically increase your speed and efficiency in form•Z. In figure 1-3, the Coordinates, Tool

Options, Lights, Objects, Layers, Views, Prompts, and Animation palettes are shown at the right side of the modeling window.

➤ **NOTE:** *Palettes are discussed in detail in Chapter 4.*

Modeling Tools

The modeling tools are presented in two columns of square icons at the left side of the screen. Each icon expands into a horizontal tear-off palette that can be dragged to any location on the screen. The icons also act as shortcuts to dialog boxes that modify each tool with controls that are also displayed in the Tool Options palette. The modeling tools are the heart of the form•Z interface; almost nothing can be accomplished in form•Z without them. Figure 1-3 shows the Polygons & Circles/Ellipses tool palette with the Circle tool highlighted. Below that is the Topological Level modifier palette. The modifier icons, which appear in turquoise, affect the drawing and modeling tools, which appear in gray.

➤ **NOTE:** *Command tools are discussed in detail in Chapter 5.*

Window Tools

The window tools are convenient controls that allow you to switch reference planes, geometrical constraints, and views. These tools become more essential as the objects to be modeled increase in sophistication or complexity. In figure 1-3, the Grid Snap, Direction Snap, and View tools are shown extended upward from their position at the bottom left corner of the modeling window.

➤ **NOTE:** *Window tools are discussed in detail in Chapter 6.*

The 3D Window

The 3D window is your viewing portal into the 3D workspace of form•Z. You can open several windows at once, you can resize the windows, and you can use both 2D and 3D views. The types of 3D views available include panoramic, perspective, isometric, oblique, and axonometric.

➤ **NOTE:** *Views and window setup issues are discussed in detail in Chapter 3.*

Now that you have some understanding of the basics of the form•Z environment and interface, you should consider how to approach learning the software. The following section offers some suggestions on getting started.

Get Started!

How you begin your form•Z adventure depends on your personal learning style, and on your previous computer experience. The book is organized both as a reference guide and a tutorial based on step-by-step examples. Those who want to jump right in and start modeling immediately can proceed to Chapter 7 and try some of the Five-minute Models. Those who are more cautious and methodical will want to completely explore the form•Z interface, examining every command, menu item, and palette while reading about them.

A third approach may be right for you: Try a few of the simplest examples in Chapter 7, and then study each of the commands they use. After having read more about each tool used in the easy exercise, see if you can repeat the exercise without following the step-by-step instructions.

When you find you can repeat an exercise from memory, you can be sure that you have learned the necessary commands. At that point it might be useful to repeat the example one more time, this time customizing it to your personal taste. For example, instead of modeling a wine goblet, create a plate or a classical urn. In this way you can practice being creative and experimental within a familiar structure so that you do not get lost. If you do get lost, the next chapter, on the form•Z Help function, is a good place to turn.

chapter 2

Getting Help

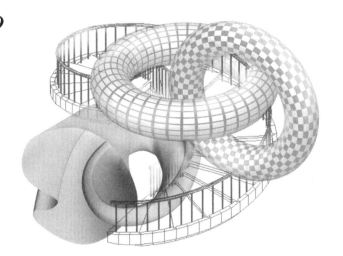

Introduction

Learning form•Z can be approached in a variety of ways. The right approach to mastering the software, and the best strategies for getting help when you need it, will depend on your learning style and your personal circumstances. For example, if a form•Z course is available in your area, and you respond well to a structured learning experience, you would be well advised to enroll. Getting help directly from an expert instructor and meeting students who are on the same learning curve can be beneficial.

For the majority of users, however, a more practical strategy is to purchase this book and work systematically through the exercises. The book's structure as well as its table of contents, in-text cross-referencing, indexes, and appendices is designed to help you quickly find answers to particular problems. This chapter focuses on getting answers from the help features built into form•Z. For this reason, the Help pull-down menus are discussed in greater detail here than in Chapter 5, where they are also discussed.

In this chapter, the Help menus are presented in the order of their importance and frequency of use for the beginner. This structure differs from the order in which the menus appear in the interface so that you will more efficiently be able to obtain the information you need at a given stage of the learning process.

Chapter 2: Getting Help

If you need or have advice on a form•Z issue or have knowledge to share with others, you can join an Internet forum or chat room. Several of these are reviewed in the last section of this chapter.

➥ **NOTE:** *auto•des•sys, the producer of form•Z, also provides help by telephone, as described in the last section of this chapter, "Internet Help."*

Your First Stop for Help

In most cases, the first stop in your search for help should be the built-in help functions of form•Z. These are available to you as soon as you launch the program via the Help menu in the row of Menu Commands at the top of the screen, shown in figure 2-1.

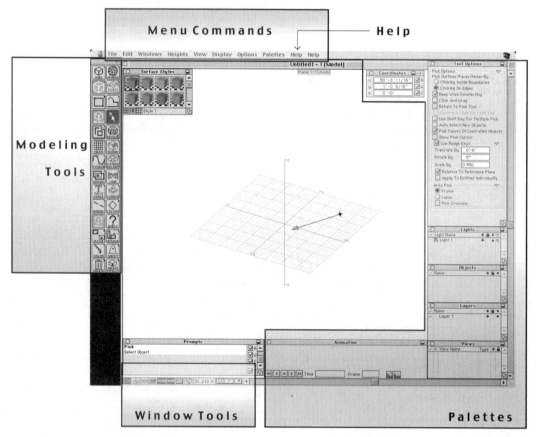

Fig. 2-1. form•Z screen with labels added to show the main components of the interface: Menu Commands (including Help), Modeling Tools, "floating" Palettes, and Window Tools.

Built-in Help Menus

> **NOTE:** *You cannot leave the Help menu open while you work in the program; you have to close the Help menu in order to return to your form•Z work.*

The built-in form•Z Help function is a great tool for the beginner. In fact, it is so convenient and well designed that even experienced form•Z users access it to remind themselves about particular commands and features they use infrequently.

The form•Z Help function is always available in the pull-down menus at the top right-hand corner of the screen, shown in figure 2-1. This menu lists the Main Help dialog box, as well as an Introduction and sections on Modeling, Menus, Keyboard Shortcuts, Drafting, and Window Tools (figure 2-2). The two final items, Error Messages and Project Info, are discussed at the end of Chapter 3. The seven major items in the Help menu are discussed in the material that follows, in order of their usefulness to the form•Z beginner.

Fig. 2-2. Pull-down Help menu.

The Modeling Tools Help section of the Help interface is by far the most frequently used. This is a great place to familiarize yourself with the essence of the form•Z program's powerful 3D modeling capabilities.

Modeling tools are represented as a column of square icons at the left-hand side of the screen, as shown in figure 2-1. Use the scroll bar to view the complete list.

> **NOTE 1:** *The Modeling Tools Help section mimics the "tear-off" functionality of the tool palettes by presenting all tool icons simultaneously, as if you had held the mouse down on all of the tools at the same time (figure 2-3).*

Fig. 2-3. Normal state of the Modeling Tools Help window.

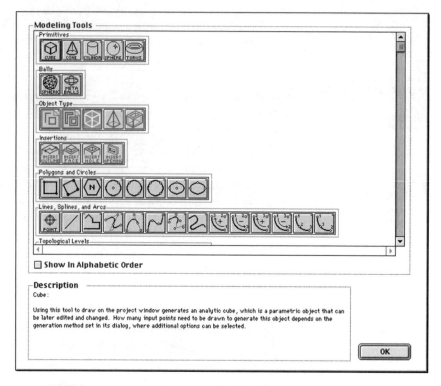

➻ **NOTE 2:** *The Project Info command is useful for getting helpful information and stats on your model. This menu item is discussed in Chapter 3.*

Dragging the mouse over the Help sections activates a text box that provides a brief description of each command. Note that the Description box also provides the official name of the modeling tool, which is sometimes longer than the icon label in the Tool palettes. For example, the Modeling Tools Help window informs you that the Revolve tool is actually named Revolved Object. When there are discrepancies, this book refers to the tool's full name rather than its icon label.

➻ **NOTE:** *The Modeling Tools Help descriptions are brief and may not adequately explain certain 3D concepts. For this reason, Chapter 5 provides a comprehensive listing and detailed explanation of all modeling tools, including supporting 3D illustrations.*

Customizing the Modeling Tools Palette

Under the Palettes pull-down menu, you can use the Customize Tools options to customize the display of the modeling tools (figure 2-4). The three distinctive areas—Tool Bars, Tool Palettes, and Tool Set—let you modify your Modeling Tools palette. Tool Bar simulates the actual tool bar as it appears in figure 2-1, and Tool Palettes illustrates the tools as if they were floating palettes on your screen. By clicking your mouse on one of the icons, either inside Tool Bar or Tool Palettes, you can remove and/or add tools according to your needs. Forming new palettes can be easily achieved by dragging the tool from Tool Set to Tool Bars or Tool Palettes.

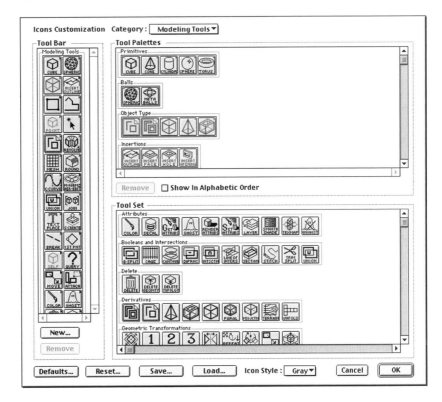

Fig. 2-4. Icons Customization dialog box.

Eliminating certain modeling icons can be especially useful in a controlled classroom teaching situation in which a teacher wants the students to concentrate on mastering certain modeling tools in a particular order. Commands that are not needed can be made inactive and invisible. You may find this convenient if you are performing specialized repetitive tasks in form•Z and do not need all modeling tools visible.

✓ **TIP:** *Eliminating tool icons unlikely to be used is a way of streamlining the modeling interface, which will increase your productivity.*

Once you have customized the Modeling Tool interface in this way, you can save this setting by clicking on the Save button. You will then be prompted to provide a name for your latest selection of visible and invisible modeling tools, and you will be able to return to this custom configuration at a later date.

Figure 2-5 shows the same Icons Customization window, with several rows of tools and several individual tools removed. Figure 2-6 shows what the form•Z toolbar would look like if these tools were suppressed. Note that the column of icons at the left-hand side of the screen is much shorter than it appeared in figure 2-1.

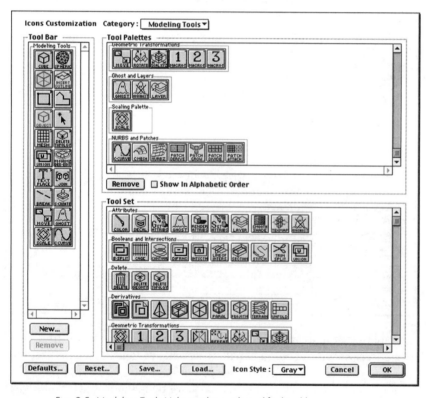

Fig. 2-6. Modeling Tools toolbar reduced from 28 to 20 icons.

Fig. 2-5. Modeling Tools Help window with modified tool bars.

Built-in Help Menus

➥ **NOTE:** *If you quit the program after customizing the modeling tools, you will find that they are still customized when you return. You will need to return to the Icon Customization menu to restore them.*

Window Tools Help

Window Tools appear at the bottom of each modeling or drafting window you open. Each tool is a pop-up menu that expands upward into a column of rectangular icons. These tools are associated with reference planes, snaps, geometric constraints, and the setting of views.

The Window Tools Help window, shown in figure 2-7, helps to familiarize you with Window Tools by displaying all eight rows of icons. As you move the cursor over an icon, you get an instant readout of that tool's function.

➥ **NOTE:** *See Chapter 6 for a comprehensive listing and detailed explanation of the Window Tools.*

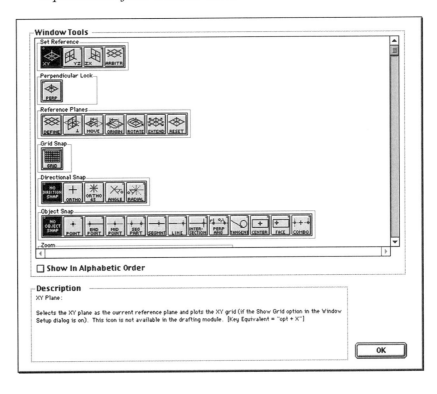

Fig. 2-7. Window Tools window.

Menus Help

The form•Z menus are displayed at the top of the screen. The menus control fundamental aspects of the system, such as opening files and printing, as well as particular modeling issues such as height and view settings.

∞ **NOTE:** *Pull-down menus are discussed in detail in Chapter 3.*

The Menu Help window displays the content of all menus in a series of lists that will not fit on your screen. Use the scroll bars to navigate and view the entire content, as indicated in figures 2-8 and 2-9. These lists are displayed as you would see them if you scrolled across the actual pull-down menus at the top of the form•Z screen.

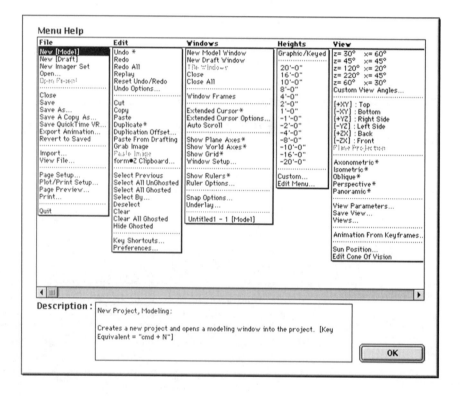

Fig. 2-8. Menu Help window, left side.

Built-in Help Menus

Fig. 2-9. Menu Help window, right side.

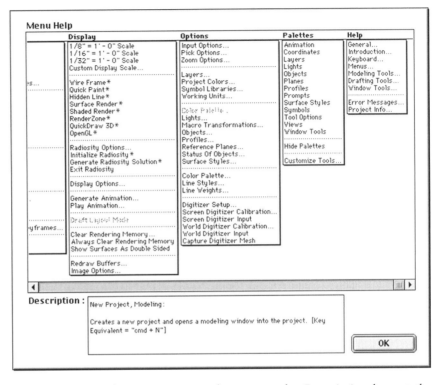

As you move the cursor across the screen, the Description box at the bottom of the screen provides a brief explanation of each command, along with keyboard shortcuts, if any. It also informs you if a particular command is not available in the Drafting or Modeling module.

➥ **NOTE:** *See Chapter 3 for a comprehensive list of all menu items, including detailed explanations and illustrations of the dialog boxes and preview windows associated with each menu item.*

Keyboard Help

Keyboard shortcuts are essential if you want to improve your speed and efficiency in form•Z. The mouse is not as effective an input device as the keyboard for certain operations, especially repetitive ones.

➥ **NOTE:** *Keyboard shortcuts are listed in Appendix C.*

The most convenient source of information on keyboard shortcuts is only a mouse click away once you have launched the form•Z program. There are two ways to access keyboard help: (1) in the Keyboard Layout mode (figure 2-10) in the Help menu, you can hold down various keys and see what, if any, shortcut they correspond to, and (2) in Key Shortcuts Manager window mode (figure 2-11), accessed through Key Shortcuts in the Edit menu, you will see an alphabetical list of keyboard commands and summaries of their functions.

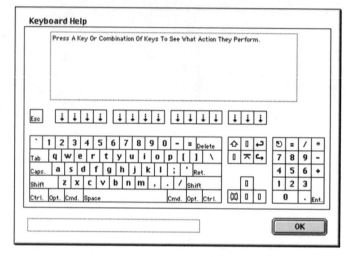

Fig. 2-10. Keyboard Help window in Keyboard Layout mode.

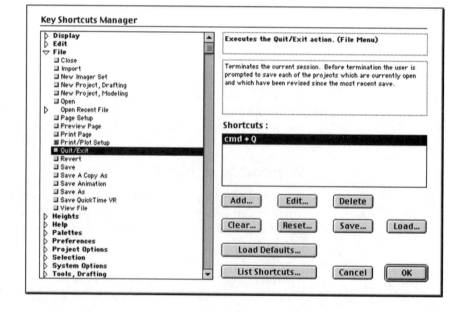

Fig. 2-11. Key Shortcuts Manager window, accessed through Key Shortcuts under the Edit menu.

On the left side of the Key Shortcuts Manager window is a list that comprises virtually every aspect of the form•Z environment. These items range from Heights and Help to System Options and Preferences.

The Key Shortcuts Manager also allows you to customize keyboard shortcuts. This is not as valuable to the beginner, who is learning basics, as it is to the expert trying to increase productivity, but the ability exists for the beginner nonetheless.

To customize a shortcut, select a command and then click on the Add button. You will then be prompted to define the keystrokes and mouse clicks that will be recognized as a shortcut for that command. On the right side of the Key Shortcuts Manager window is a short explanation of the command, which you will recognize as the same text displayed in the Description box of one of the other Help windows.

Below the explanation of the command is a box that displays the shortcut. A series of buttons here allows you to Add, Clear, Edit, Reset, and Delete shortcut definitions. The Save button allows you to write your shortcuts to a file. The Load button allows you to use a set of shortcuts previously saved. The Load Defaults button restores the standard settings.

Note that shortcuts are color coded. Blue squares represent selection tools, red represents actions that bring up dialog boxes, and yellow represents actions that execute operations.

The Shortcuts List

The most helpful option in the Key Shortcuts Manager window is the List Shortcuts button, which brings up a complete list of current shortcuts (figure 2-12). Access this list to familiarize yourself with the keyboard shortcuts available in form•Z.

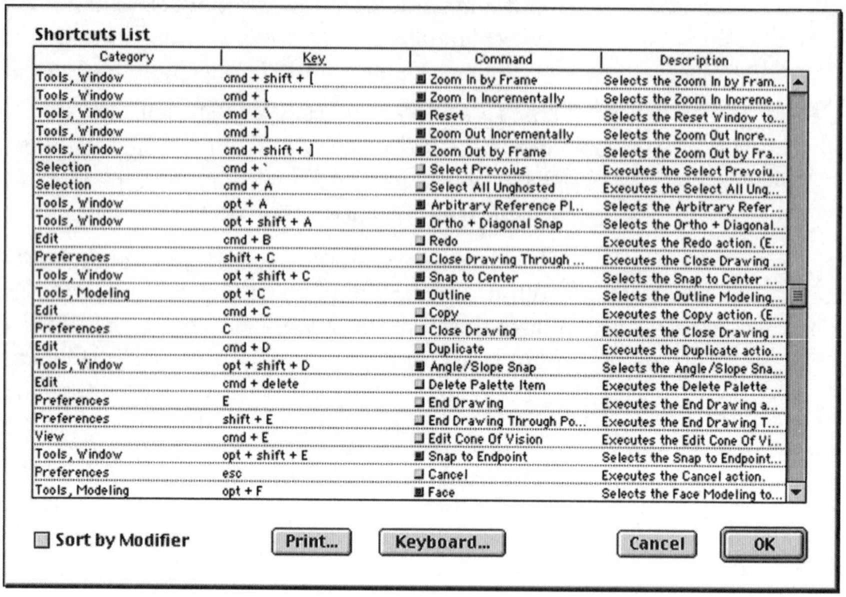

Shortcuts List

Category	Key	Command	Description
Tools, Window	cmd + shift + [	Zoom In by Frame	Selects the Zoom In by Fram...
Tools, Window	cmd + [	Zoom In Incrementally	Selects the Zoom In Increme...
Tools, Window	cmd + \	Reset	Selects the Reset Window to...
Tools, Window	cmd +]	Zoom Out Incrementally	Selects the Zoom Out Incre...
Tools, Window	cmd + shift +]	Zoom Out by Frame	Selects the Zoom Out by Fra...
Selection	cmd + `	Select Prevoius	Executes the Select Prevoiu...
Selection	cmd + A	Select All Unghosted	Executes the Select All Ung...
Tools, Window	opt + A	Arbitrary Reference Pl...	Selects the Arbitrary Refer...
Tools, Window	opt + shift + A	Ortho + Diagonal Snap	Selects the Ortho + Diagonal...
Edit	cmd + B	Redo	Executes the Redo action. (E...
Preferences	shift + C	Close Drawing Through ...	Executes the Close Drawing ...
Tools, Window	opt + shift + C	Snap to Center	Selects the Snap to Center ...
Tools, Modeling	opt + C	Outline	Selects the Outline Modeling...
Edit	cmd + C	Copy	Executes the Copy action. (E...
Preferences	C	Close Drawing	Executes the Close Drawing ...
Edit	cmd + D	Duplicate	Executes the Duplicate actio...
Tools, Window	opt + shift + D	Angle/Slope Snap	Selects the Angle/Slope Sna...
Edit	cmd + delete	Delete Palette Item	Executes the Delete Palette ...
Preferences	E	End Drawing	Executes the End Drawing a...
Preferences	shift + E	End Drawing Through Po...	Executes the End Drawing T...
View	cmd + E	Edit Cone Of Vision	Executes the Edit Cone Of Vi...
Tools, Window	opt + shift + E	Snap to Endpoint	Selects the Snap to Endpoint...
Preferences	esc	Cancel	Executes the Cancel action.
Tools, Modeling	opt + F	Face	Selects the Face Modeling to...

Fig. 2-12. Shortcuts List, accessed through the Key Shortcuts Manager window.

The Sort by Modifier option (figure 2-13) is valuable if you are searching for a particular shortcut you remember began with a particular key, such as a Control or Option.

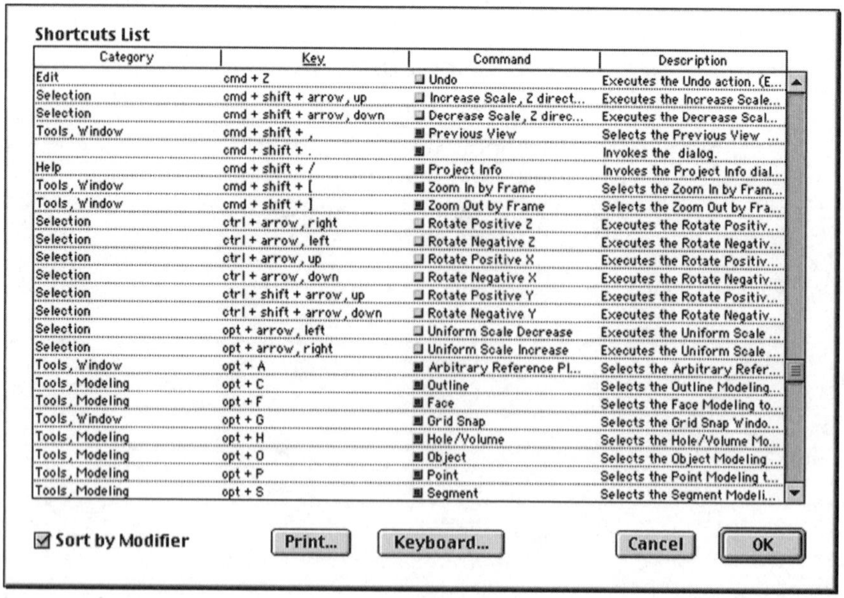

Shortcuts List

Category	Key	Command	Description
Edit	cmd + Z	Undo	Executes the Undo action. (E...
Selection	cmd + shift + arrow, up	Increase Scale, Z direct...	Executes the Increase Scale...
Selection	cmd + shift + arrow, down	Decrease Scale, Z direc...	Executes the Decrease Scal...
Tools, Window	cmd + shift + ,	Previous View	Selects the Previous View ...
	cmd + shift + .		Invokes the dialog.
Help	cmd + shift + /	Project Info	Invokes the Project Info dial...
Tools, Window	cmd + shift + [	Zoom In by Frame	Selects the Zoom In by Fram...
Tools, Window	cmd + shift +]	Zoom Out by Frame	Selects the Zoom Out by Fra...
Selection	ctrl + arrow, right	Rotate Positive Z	Executes the Rotate Positiv...
Selection	ctrl + arrow, left	Rotate Negative Z	Executes the Rotate Negativ...
Selection	ctrl + arrow, up	Rotate Positive X	Executes the Rotate Positiv...
Selection	ctrl + arrow, down	Rotate Negative X	Executes the Rotate Negativ...
Selection	ctrl + shift + arrow, up	Rotate Positive Y	Executes the Rotate Positiv...
Selection	ctrl + shift + arrow, down	Rotate Negative Y	Executes the Rotate Negativ...
Selection	opt + arrow, left	Uniform Scale Decrease	Executes the Uniform Scale ...
Selection	opt + arrow, right	Uniform Scale Increase	Executes the Uniform Scale ...
Tools, Window	opt + A	Arbitrary Reference Pl...	Selects the Arbitrary Refer...
Tools, Modeling	opt + C	Outline	Selects the Outline Modeling...
Tools, Modeling	opt + F	Face	Selects the Face Modeling to...
Tools, Window	opt + G	Grid Snap	Selects the Grid Snap Windo...
Tools, Modeling	opt + H	Hole/Volume	Selects the Hole/Volume Mo...
Tools, Modeling	opt + O	Object	Selects the Object Modeling ...
Tools, Modeling	opt + P	Point	Selects the Point Modeling t...
Tools, Modeling	opt + S	Segment	Selects the Segment Modeli...

Fig. 2-13. Shortcuts List sorted by modifier.

➥ **NOTE:** *See Appendix C for a complete list of standard keyboard shortcuts.*

The Shortcuts list is not directly accessible through the pull-down Help menus. You must remember to use the Edit menu to access the Key Shortcuts Manager and then click on the List Shortcuts button.

Essential Shortcuts

There are several standard keyboard shortcuts that should be immediately memorized by the beginner. The shortcuts that follow are extremely helpful in arranging the view in the modeling window.

Cmd + F	Selects the Fit All window tool
Cmd + [	Selects the Zoom In Incrementally tool
Cmd +]	Selects the Zoom Out Incrementally tool
Cmd + 2	Selects a preset 3D view
Cmd + 6	Selects the Top view
Cmd + 8	Selects the Right view

Drafting Tools Help

The drafting tools appear at the left-hand side of the screen when you enter the form•Z Drafting module. They are similar to the modeling tools in layout and functionality, except that they are 2D.

➥ **NOTE:** *Drafting tools are discussed in Chapter 11.*

The Drafting Tools Help window, shown in figure 2-14, is a convenient guide to the 2D capabilities of form•Z. This help window is similar in functionality to the Modeling Tools Help window previously discussed.

Fig. 2-14. Drafting Tools Help window.

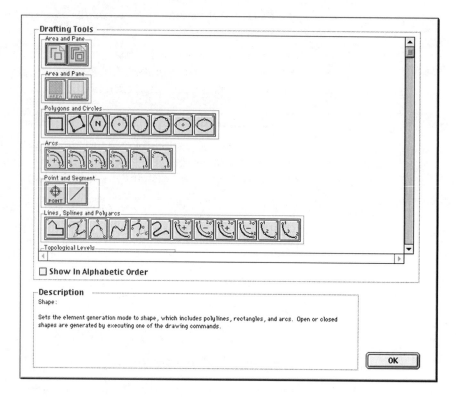

General Help

The General Help dialog box (figure 2-15) displays the icon and name of the currently active command, along with a brief description. This dialog box functions primarily as an interface to the other help windows, which can just as easily be accessed directly via the Help pull-down menu.

Built-in Help Menus

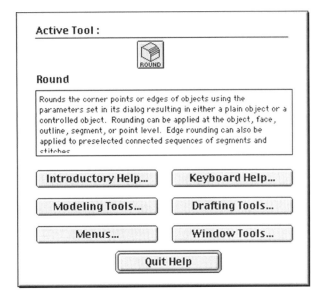

Fig. 2-15. General Help dialog box.

Introductory Help

This section of the Help function is a review of some of the basic features and principles of form•Z. The Introductory Help window (figure 2-16) provides a good overview of the product. In particular, read the paragraphs on the difference between solid and surface objects, and the comparisons between the 3D modeling and 2D drafting modules.

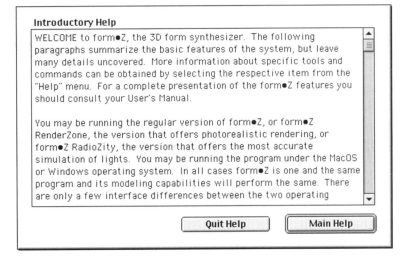

Fig. 2-16. Introductory Help window.

Using the Query Tool

Another helpful form•Z function is the Query tool, located in the Query tool palette (figure 2-17). This tool brings up the Query Object dialog box, which displays information about a selected object: the object's name, layer, color, object type, and other geometric properties (figure 2-18).

Fig. 2-17. Query tool highlighted in the Query tool palette.

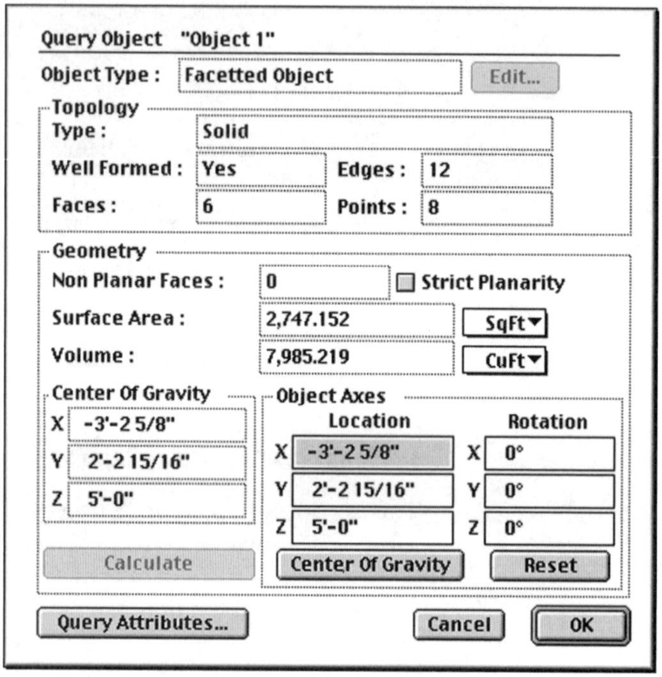

Fig. 2-18. Query Object dialog box used to calculate the surface area and volume of a sphere.

The Query tool is particularly helpful if you are confused as to whether you have created a solid or a surface, or whether the object you have created is well formed. The Query tool is more than a help function; it allows you to change the attributes of a selected object.

> **NOTE:** *The object-editing aspect of the Query tool is discussed in Chapter 5.*

Internet Help

There are several resources available to form•Z users on the Internet. auto•des•sys is the manufacturer of the form•Z product and offers many resources for help and assistance on its Web page. Another resource is *3DNation.com*, an Internet community dedicated to 3D modeling, rendering, and animation, which devotes online bulletin boards to form•Z.

auto•des•sys Help

If the built-in help features of form•Z and other resources at your immediate disposal do not provide the answers you need, you can telephone experts at auto•des•sys in Columbus, Ohio.

➥ **NOTE:** *The auto•des•sys product information number is 614-488-9777. auto•des•sys can also be contacted by e-mail at formz@autodessys.com.*

Another strategy is to access the Internet and visit some of the Web sites devoted to computer-aided design issues. The following is the official form•Z web site, produced by auto•des•sys. In addition to a forum for questions and answers on the product, the site contains a collection of excellent tutorials that can be downloaded and printed (figure 2-19).

www.formz.com

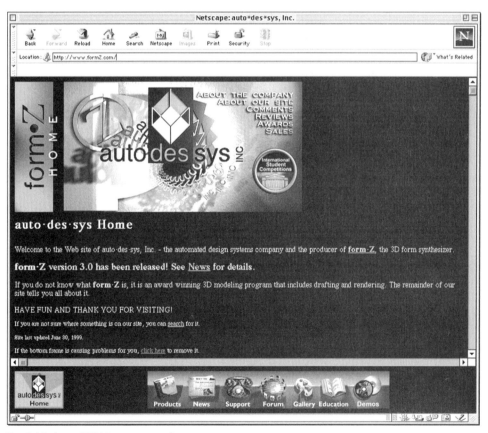

Fig. 2-19. form•Z Web site at www.formz.com.

CyberSites, Inc., Help

CyberSites, Inc., created the following site, which features bulletin boards devoted to form•Z, as well as other 3D packages. The features of this site include discussion boards, personal and public chat rooms, and a Who's Online panel with instant messaging, which allows you to easily find and communicate with other members who use form•Z.

www.3DNation.com

Members of this site select nicknames based on form•Z commands, such as George Boolean, Re Union, and The Shadow. They publish their computer graphics on personal home pages, and the best work is featured on the site's main page. The site includes a Who's Online panel with instant messaging, which allows you to easily find and communicate with other members. Figure 2-20 shows the front page of the site.

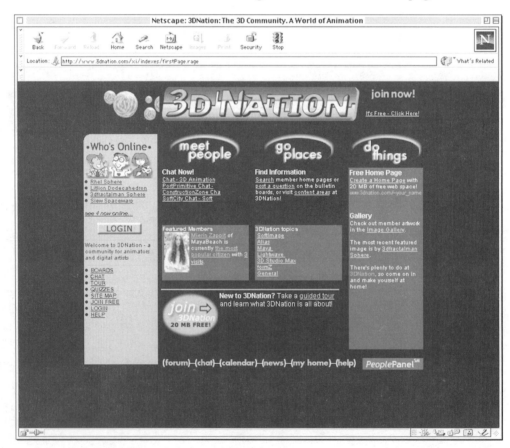

Fig. 2-20. 3DNation virtual community at www.3dnation.com.

chapter 3

Menus

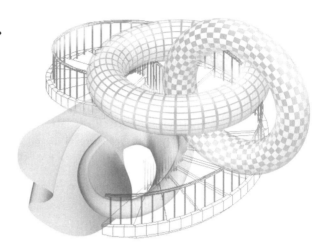

Introduction

form•Z's menus are the most fundamental of the program's four main interface systems; that is, the menu bar always appears in its entirety and is always in the same place at the top of the screen. In contrast, the appearance of the three other main interface components is less predictable and may depend on the preferences of the last person to use the program.

For example, the palettes may appear to float anywhere on the screen, and they may be partially or totally hidden by the user; the Modeling Tools may appear to lack certain icons that have been made inactive by the user; and the Window Tools are duplicated for each modeling window opened on the screen. You may see several Window Tools control bars on the screen at one time, but you will never see more than one menu bar at the top of the screen.

The reliable, ever-present form•Z menus control fundamental aspects of the system, such as printing, as well as particular modeling issues such as height. The menus also provide means of controlling the three other main interface systems. The Palettes menu, for example, controls all palettes and provides access to the Window Tools as a palette, and the Help menu can be used to modify the appearance of the Command Tools.

Menus Interface

As previously mentioned, menus are permanently displayed at the top of the form•Z screen. Included are fundamental commands such as Open, Save, and Print (common to most other software programs), as well as form•Z-specific menus such as Heights and View. For beginners, one of the most important of the menus is Help. The row of menus at the top of the interface screen is highlighted in figure 3-1.

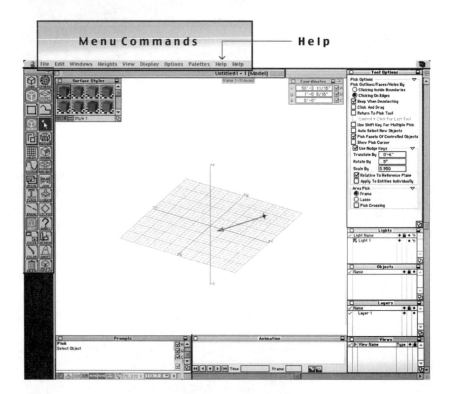

Fig. 3-1. Menus row at the top of the form•Z screen.

Concepts and Terminology

form•Z menus are referred to as "pull-down" menus; in fact, on a small screen, some of them may pull down below the bottom of the screen, at which point they begin to scroll. Note that the Macintosh menu has a small apple symbol at the left end of the menu row. This is part of the Macintosh operating system and is not present in the Windows version of form•Z. On the Macintosh, one of the items appearing under the apple is About form•Z, which contains software-version information.

Introduction

The same information can be found in the Windows version at the bottom of the Options menu. The menus are discussed in the order in which they appear on the screen, starting from File on the left to Help on the right. The following are the form•Z main menus.

- File
- Windows
- View
- Options
- Help
- Edit
- Heights
- Display
- Palettes

If a menu item is discussed in detail elsewhere, which is common due to the built-in redundancies of the form•Z interface, there will be a note to that effect, and the entry will be kept brief. In each command grouping—such as File, Edit, or Window—the individual commands are discussed in the order in which they are listed, from top to bottom.

➥ **NOTE:** *Subsequent discussion of commands is up to date for form•Z version 3.1. There were significant differences in the menus between versions 2.8 and 2.9.5, and commands were in some cases moved from one menu to another. For example, in version 2.8, the Window Setup command was in the Options menu; in version 2.9.5 and 3.0, it is found in the Windows menu.*

MacOS/PC NOTE: *The Macintosh command key (with the cloverleaf or apple icon) is referred to as Cmd. Cmd-S means press the Command key and type S. On a Windows machine, keyboard shortcuts use the Control key. Ctrl + S means press the Control key and type S.*

Shortcuts and Conventions

The ellipsis (...) after a command name is used to indicate that the command will immediately bring up a dialog box with various option settings. Keyboard equivalents are often shown on the pull-down menus and are listed here after the commands for which shortcuts are provided.

➥ **NOTE:** *The Key Shortcuts item in the Edit menu allows you to create keyboard equivalents for any form•Z command.*

The asterisk (*) means that there are optional settings for that menu item. Press the Option key (Ctrl + Shift in Windows) and select the menu item to bring up the options dialog box. After selecting the desired options, click on the OK button to proceed, or on Cancel to abort the process.

Note that the families of related commands are grouped and separated by a thin gray line in each pull-down menu. Some menu items will appear to be dimmed, or grayed out, if they are not available for the current mode or command. For example, if you have not yet saved your file, Revert to Saved (in the File menu) will be dimmed.

File Menu

The File menu controls basic operations such as creating, opening, saving, and exporting files. Figure 3-2 shows the Macintosh version of this menu. The Windows version is virtually identical, with the exception that it uses the word *Exit* instead of *Quit*.

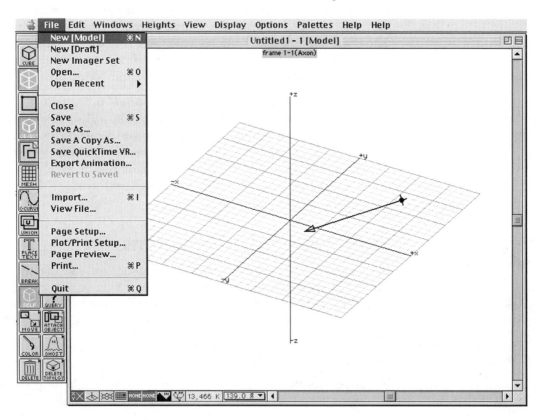

Fig. 3-2. The New (Model) command in the File menu opens a modeling window with a 3D axonometric view.

Keyboard Shortcut:
Mac (Cmd-N), Windows (Ctrl + N)

New (Model)

This command creates a new project and opens a new window into the project's 3D modeling space. The project will be named Untitled1 (or New1 in Windows 3.1) until you provide another name, which you should do immediately with the Save command. The new 3D modeling window will appear on top of any other window and will be active, and thus ready for you to start modeling in it. The window will display a 3D axonometric view of the reference plane, as shown in figure 3-2.

The project is named Untitled1 (in Windows, *New1.fmz*). Note the 3D view of the reference plane and a black dot and arrow, which represent the "sun" light source. You can hide the light source symbol by clicking on the black diamond beside the light's name in the Lights palette. Note in figure 3-2 that all palettes have been hidden.

New (Draft)

The New (Draft) command creates a new project and opens a new drafting window for 2D drawing and dimensioning. It is similar to New (Model), except that it is used for 2D work. The window will display a 2D view of the XY drafting plane. Note that the default grid extends to the edges of the window. This is in contrast to the modeling window, which displays an "aerial" view of a limited square grid contained within the boundaries of the window. Figure 3-3 shows the 2D drafting screen. Note that all palettes have been hidden except for the Prompts window.

Chapter 3: Menus

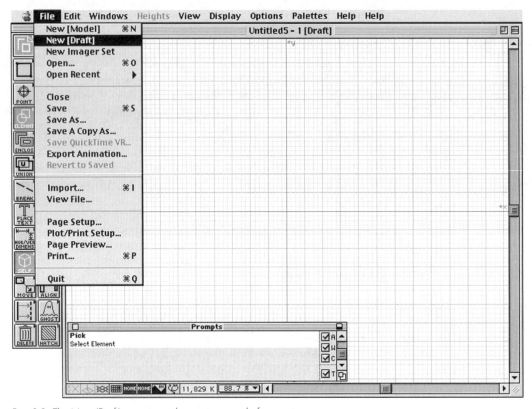

Fig. 3-3. The New (Draft) menu is used to start a new drafting project.

New Imager Set

The New Imager Set command opens up the Imager Set window (figure 3-4), which allows you to prepare batch renderings to be executed later in form•Z Imager, a separate but related program.

Fig. 3-4. Imager Set window.

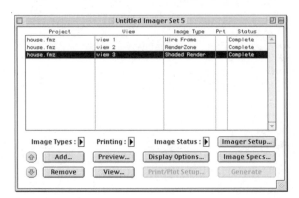

File Menu

➪ **NOTE:** *Imager is discussed in Chapter 10.*

Keyboard Shortcut:
Mac (Cmd-O), Windows (Ctrl + O)

Open

The Open command is used to open a form•Z project file that already exists, or another type of file that form•Z can import, such as DXF or IGES. The Open command will access the system's Open File dialog box (figure 3-5), which allows you to browse through your folder structure and specify the file you want to open. The File Format selection lets you browse and choose special types of file formats.

➪ **NOTE:** *The standard file formats are discussed in Appendix B.*

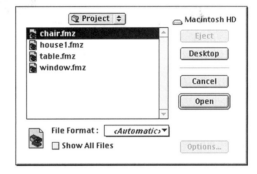

Fig. 3-5. Macintosh version of the form•Z Open File dialog box.

When you open a preexisting form•Z file, it appears on your screen exactly as it was the last time you saved it. The same window will be active, and the same palettes will be visible. Note in the example shown in figure 3-5 that the folder is named Project and resides on a disk named Macintosh HD. The file selected to be opened is *chair.fmz*.

Open Recent

The Open Recent command is new for version 3.0. It conveniently expands the File/Open menu to the side and displays the names of the form•Z projects you have most recently worked on. In the Edit menu, under Preferences, you will find a Recent Files option that controls the number of files displayed by the Open Recent command.

Close

The Close command closes the current project. All windows connected with that project are closed immediately. If you do not remember to save your file before closing, the system will prompt you to do so.

Chapter 3: Menus

Macintosh users will note that they cannot use the familiar Cmd-W keyboard shortcut to close windows in form•Z because this key sequence is reserved for the Wire Frame view. A custom keyboard shortcut, such as Option-W, can be set with the Key Shortcuts command in the Edit menu.

Save Command

This command writes the current form•Z file to the disk, overwriting the previous version saved with the same name. You should save your work regularly. However, remember that when you save your file you will lose the ability to use the Undo command to back up to a previous command, unless the Reset After Saving Project option has been deselected in Undo Options within the Edit menu.

The first time you save a project using the Save command, the Save As dialog box will appear. To save your file under a new name, you must use the Save As command (figure 3-6).

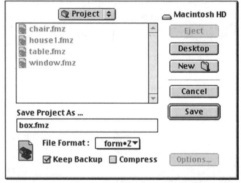

Fig. 3-6. Saving a file for the first time accesses the Save As dialog box.

The Save As dialog box displays a few useful option buttons. The Desktop button allows you to save directly to the desktop of your computer. The New button lets you create a new folder before saving your file. Cancel lets you abort the Save operation. The Keep Backup option is on by default; this will determine whether a backup file, named with the file suffix .bak, will be generated for that file. The Compress button will cause your files to be smaller, but the system will then require a bit more time to open them. The use of this button is not recommended unless you are running out of space on your hard drive, or need to send the files over the Internet.

Note that in the Edit/Preferences menu there is a general control for specifying that form•Z always save a backup file automatically. This

File Menu

Auto Save feature is turned on by default, and it is recommended that you keep it on.

Save As...

The Save As command accesses the Save Project As dialog box (figure 3-7), which includes a pull-down menu with various standard file types in which you may save your project. This is also what you would see if you were saving your file for the first time. You must enter a name in the Save Project As field.

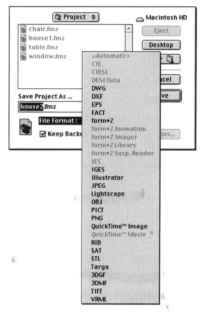

Fig. 3-7. Save Project As dialog box with its File Format pop-up menu.

A file name suffix is provided automatically in order to indicate the type of file you are saving. Normally, the suffix is .fmz, to indicate a standard form•Z file. You would change the file type from the .fmz suffix only if you did not want to save your file as a regular form•Z file.

For example, you would select PICT from the File Format menu if you wanted to save a 2D picture file for further work in a photo manipulation software such as Adobe Photoshop. You would select DXF or IGES if you wanted to save the 3D file for export to another CAD program. Depending on which file format you select, a special dialog box will appear with options pertaining to that particular format.

∞ **NOTE:** *File formats and transferring files to and from other CAD programs are discussed in Appendix B.*

Save A Copy As...

The Save A Copy As command saves a copy of the current project, which is different from saving the project itself. Save A Copy As does not affect the current project. This command can save files in form•Z format only; therefore, the Save A Copy As dialog box (figure 3-8) it invokes is simpler than that of the Save As command, which offers multiple file format options.

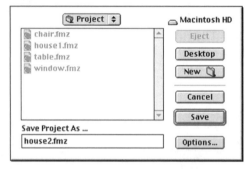

Fig. 3-8. Save A Copy As dialog box.

The Save A Copy As dialog box contains additional options that affect how a project is written to the hard drive. These include the ability to save a project as a different form•Z version number. This may be useful if you have to transfer form•Z files to another system running an older version of the software. Note also the Visible Entities Only option. When this option is activated, you can save, via the Project Options dialog box (figure 3-9), a new version of a project that excludes any currently hidden parts of the model.

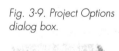

Fig. 3-9. Project Options dialog box.

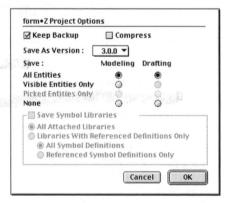

File Menu

Save QuickTime VR...

The Save QuickTime VR command will save a file in QuickTime Virtual Reality (QTVR) movie format. This panoramic presentation format can be played with software that supports QTVR, such as Apple's MoviePlayer. The Panoramic setting in the Views menu will create a 360-degree view of the movie imagery created, which you can pan the full 360 degrees. The program actually renders a rectangular image, which the QTVR engine converts to a dynamic, user-controlled perspective image with pan and zoom controls.

Save QuickTime VR can also be used to generate Object movies, which create the illusion of the user being able to rotate a 3D object and see it from all sides. Note that QTVR is often generated from photographs, which involves the multiple problems of lenses, tripods, light meters, leveling devices, and focusing. By contrast, with form•Z's Panoramic setting you are virtually guaranteed a picture-perfect QTVR panorama of any well-constructed and rendered form•Z model. This is because the virtual camera is easy to level, the lighting is strictly controlled, and a single "snapshot" generates the panorama.

Export Animation...

The Export Animation command is a new command for version 3.0, and allows you to save a form•Z animation in various formats, such as QuickTime Movie.

➥ *NOTE: Animation is discussed in detail in Chapter 12.*

Revert To Saved

The Revert To Saved command will erase any memory of the changes you have made to your project since the last time you saved a file. Note that Revert To Saved will be grayed out and inactive until you save the project and make subsequent additions or revisions.

➥ *NOTE: Use Revert To Saved with caution! Revert To Saved is useful when you get into a real mess and want to start again from a point just before things started to go wrong. However, this tactic works only if you have been regularly saving your work.*

Chapter 3: Menus

Keyboard Shortcut:
Mac (Cmd-I), Windows (Ctrl + I)

Import...

The Import command is similar to the Open command except that the content of the file opened is placed into the active project rather than in a new window. This is the recommended method of combining two projects you have worked on separately, such as a model of a house and a model of a garden (figure 3-10).

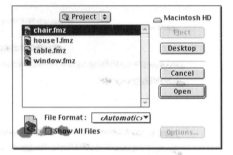

Fig. 3-10. Open dialog box accessed from Import command but no new window is opened.

View File...

The View File command lets you see what image files look like without having to leave the form•Z environment. Image files in several formats—including JPG, TIFF, and Targa—are visible in this manner, as well as PICT on the Macintosh and BMP or Metafile on Windows. This command is especially convenient if your machine does not have enough memory to run a digital image processing software (such as Adobe Photoshop) at the same time as form•Z. You can use View File to refer to previous renderings as you set up new perspectives, or to preview texture maps you plan to use in RenderZone. View File will also let you view form•Z Animations, QuickTime movies, and 3DMF files.

Figure 3-11 shows the file type options available with this command. Figure 3-12 shows the View Image File window, which opens up to display the image and provides information such as its size in pixels. This window also provides a magnifying glass tool for zooming in and out, as well as a Pan tool for scrolling sideways if the image is larger than the screen.

File Menu

Fig. 3-11. View File command accesses the Open dialog box and allows picture, animation, and 3DMF formats.

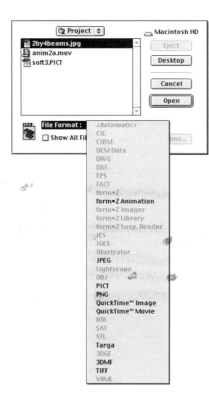

Fig. 3-12. View Image File window shows a JPG-format file named 2by4beams.jpg, 280 by 220 pixels in size.

Page Setup

Page Setup displays the standard printing dialog box for the current printer. For example, figure 3-13 shows the Macintosh Page Setup dialog box for an Epson 740 color printer. Refer to your printer's manual for details.

Fig. 3-13. Macintosh Page Setup dialog box.

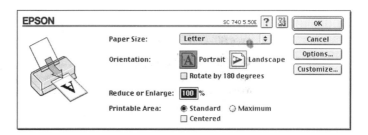

Plot/Print Setup...

The Plot/Print Setup command provides various ways of adjusting the plotter or printer output from your form•Z project. You can set the scale directly in the Plot Scale field, or specify options under the Scale To Fit Media command, which is usually a more convenient way of getting the image to fit on the page. In either case, you should always click on the Page Preview button to verify the results before clicking on Print. The X and Y Justification options let you position the image on the page. The Plot/Print Type section gives you several choices. In this section, Extents prints the entire model, whereas Window Contents prints just what you see.

With the Dump Window or Dump Screen options in the Plot/Print Setup dialog box (figure 3-14), you can also send the image of the window, or entire screen, to print. Plot Grid, Plot Axis, Background, Crop Marks, File Name & Date, and Frame determine whether these elements are printed as part of the image.

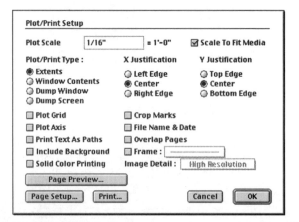

Fig. 3-14. Plot/Print Setup dialog box.

Print Text As Paths will use the character outlines of TrueType or PostScript fonts, which is important if using a plotter. Solid Color Printing applies solid fill, instead of pixel patterns, on color polygons. Overlap Pages makes it easier to assemble multiple printed pages by providing an overlap in the printed areas at the edges of the page.

Page Preview...

The Page Preview command, which accesses the Page Preview window (figure 3-15), provides you with a glimpse of exactly what the printer or plotter is about to produce. Use this command every time you print,

File Menu 45

especially if the print job involves more than one page. Use the Close icon at the upper left corner to close the window.

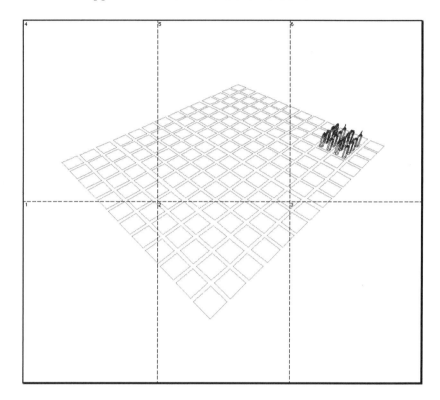

Fig. 3-15. The Page Preview window shows how an image may be spread over several pages, depending on settings in the Plot/Print Setup dialog box.

Print...

The Print command accesses your system's Print dialog box (figure 3-16), which controls the number of pages and copies printed. The Page Preview window will be shown as the document prints.

Keyboard Shortcut:
Mac (Cmd-P), Windows (Ctrl + P)

Fig. 3-16. The Print dialog box.

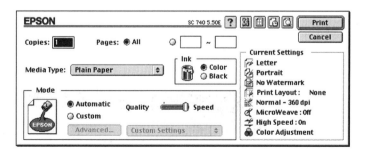

Quit

Keyboard Shortcut:
Mac (Cmd-Q), Windows (Ctrl + Q)

The Quit command terminates a form•Z session. Note that the Windows version of form•Z uses the word *Exit* instead of *Quit*. If you have made changes since the last time you saved your project, the system will prompt you to save.

Edit Menu

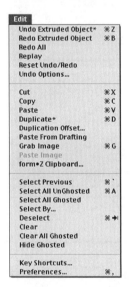

Fig. 3-17. The pull-down Edit menu.

The Edit menu (figure 3-17) contains several commonly used commands that will be familiar from other applications, such as Cut and Paste, as well as a variety of Selection and Undo/Redo commands. It is well worth the effort to study this pull-down menu in detail, because you will be using these commands repeatedly. You should learn the keyboard shortcuts for the most important of these items: Undo, Cut, Copy, and Paste. Fortunately, many of these shortcuts are the standard ones you may know from other software packages. For example, Copy is Cmd-C (Ctrl + C in Windows).

Undo*

Keyboard Shortcut:
Mac (Cmd-Z), Windows (Ctrl + Z)

When you make a mistake, the Undo command is a convenient way of going back a step. Undo cancels your last command and, if used repeatedly, will cancel out a series of operations. It can take your project all the way back to the state it was in at the time of the last Save operation.

Note that the Undo menu prints the name of the command about to be undone. For example, if you have just used the Ghost Object command, the Undo menu will read "Undo Ghost Object." The Undo command affects only the modeling or drafting operations you have performed; option settings and graphic environment changes are not affected.

➥ **NOTE:** *The Undo command is a blessing for the beginner and the expert alike. The beginner can experiment without fear of making a fatal move. The expert can use it to repeat and tweak a procedure until it is just right. There are a lot of very expensive high-end 3D packages that lack unlimited Undos.*

✓ **TIP:** *Get used to using the Cmd-Z shortcut (Macintosh). This is one of the most useful commands in form•Z.*

The asterisk after the Undo command indicates that a dialog box is available. Hold down the Option key (Ctrl + Shift in Windows) to dis-

Edit Menu

play the Undo List dialog box (figure 3-18). The Undo List dialog box displays the commands you have executed, with the most recent one at the top. You can drag the mouse to select a range of commands to be undone all at once.

Fig. 3-18. Undo List dialog box.

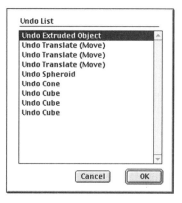

Keyboard Shortcut:
Mac (Cmd-B) , Windows (Ctrl + B)

Redo

The Redo command cancels out the Undo command. It is grayed out in the menu until an Undo command has been executed. If you have repeatedly used the Undo command, you can use Redo an equal number of times.

Redo All

The Redo All command will replay the operations you have canceled with a series of Undo commands. In other words, if you use the Undo command ten times in a row, the Redo All command will redo that sequence of ten operations. The Redo All command is available only after the Undo command has been used.

Replay

Think of the Replay command as a ready-to-go slide show; it will clear the screen and replay every move you have made since the last time the Undo record was reset. The operations are played back in the order you executed them, up to the most recent command. You can hold down the mouse button to stop the slide show if it plays back too fast. Another trick to slow down the Replay feature is to select a higher-quality rendering mode, such as Surface Render or RenderZone in the Display menu.

Reset Undo/Redo

Reset Undo/Redo will clear the records of the commands you have executed up to that point, thereby freeing up space on the hard drive. Use this command if you know you have to free up some space on the disk, but only if you are certain you will not need to use the Undo command.

Undo Options...

The Undo Options dialog box (figure 3-19) allows you to control and customize many aspects of the Undo and Redo commands. The Use Undos option lets you turn off the Undo feature altogether, which is not recommended because it will remove one of the handiest features of form•Z. An example of legitimately turning off the Undo feature might be when you are running out of space on the hard drive and do not want to keep writing large Undo files to the disk.

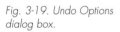
Fig. 3-19. Undo Options dialog box.

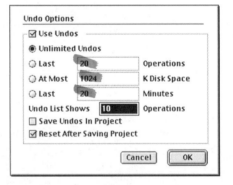

The Unlimited Undos option does what it says; it removes any limit on the number of operations that can be undone. This option is on by default. Last n Operations sets a numerical limit on the number of operations that can be undone. If your system is running low on memory, a reasonable value of n is 20. At Most n K Disk Space limits the Undo operation by kilobytes of space on the hard drive. A reasonable value of n is 1,024.

Last *n* Minutes remembers only the commands you have executed in the specified elapsed time. Anything you did before that time cannot be undone. Undo List Shows n Operations controls the number of lines shown in the Undo List dialog box.

Edit Menu

Save Undo in Project will cause the Undo records to be saved with the project file. This means that your Undos records will be available when you reopen the project. This can be very useful if you have to stop working on a project and are therefore forced to save the file but want the ability to perform a step-by-step review later. Reset After Saving Project will cause your Undo records to be lost when you save your work. This is the default setting.

> **NOTE:** *The beginner can benefit from the Undo Options by deactivating Reset After Saving Project and selecting Save Undo In Project. These two settings will allow you to experiment more confidently, knowing that you can always back up to a previous point in the modeling process, even if you have saved your file and quit form•Z. However, there is a cost: your project file will grow much larger if it contains a long history of Undo information.*

Keyboard Shortcut: *Mac (Cmd-X), Windows (Ctrl + X)*

Cut

The Cut command is part of a familiar trio of commands (Cut, Copy, and Paste) seen in many other software applications. The Cut command removes the selected objects and stores them in the system Clipboard. Cut and Paste are commonly used in sequence to move an object from one project to another. Copy and Paste are often used to duplicate objects.

The Cut command works with objects only; you cannot use it to remove points or segments from an object. If you want to remove all objects from your project, it may be faster to use the Clear command, as discussed in the material that follows. If you want to remove points, you will have to select Points in the Topological Levels tool, select the points with the Pick tool, and then use the Delete Topology tool.

Keyboard Shortcut: *Mac (Cmd-C), Windows (Ctrl + C)*

Copy

The Copy command is similar to the Cut command, except that it saves in the Clipboard a copy of the selected object, rather than the object itself. You can use the Copy and Paste commands in sequence to duplicate an object, which is equivalent to the Duplicate command (described in the material that follows). Remember that every time you use the Cut or Copy commands, you delete and replace the current content of the system's Clipboard.

Chapter 3: Menus

> **NOTE:** See also the Self/Copy modifier in the Command Tools, as an alternative to Cut and Paste, or Duplicate operations, as discussed in Chapter 5.

Keyboard Shortcut: *Mac (Cmd-V), Windows (Ctrl + V)*

Paste

The Paste command places the content of the Clipboard in the active window. The Paste command is usually invoked immediately after using the Cut or Copy command to save an object to the Clipboard. You can use the Paste command over and over again; using it does not empty the Clipboard.

There are two types of Paste operations. When objects are selected in the active window, the Paste command will replace them with the content of the Clipboard. When no objects are picked, the Paste command will simply add the content of the Clipboard to the project without deleting any other objects.

Keyboard Shortcut: *Mac (Cmd-D), Windows (Ctrl + D)*

Duplicate*

The Duplicate* command has an asterisk (*) to indicate that it contains several options that can be accessed by holding down the Option key (Mac) or Ctrl + Shift keys (Windows) while selecting the command. Duplicate is equivalent to using the Copy and Paste commands in succession. Copies of the objects you have selected will appear at a certain distance from the original objects as soon as the Duplicate command is invoked. The default setting of the offset distance is two feet. The X, Y, and Z values of this offset distance can be customized in the Object Duplication Offset dialog box (figure 3-20). The same is true for duplicated elements in the 2D Drafting module, except that there is no data field for duplication in the Z direction.

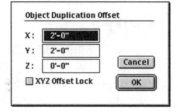

Fig. 3-20. Object Duplication Offset dialog box for the 3D Modeling module.

The Duplicate* command copies objects selected only at the Object or Group topological level. If you have picked Points, Segments, or Faces, the Duplicate* operation will ignore and deselect them. If you need to make and move multiple copies, you may prefer to use the Move command and the Copy or Multiple Copy modifier in the Command Tools.

☞ **NOTE:** *The Command Tools are discussed in Chapter 5.*

Duplication Offset...

The Duplication Offset command accesses the Object Duplication Offset dialog box, which can also be accessed by holding down the Option key (Mac) or Ctrl + Shift keys (Windows) while selecting the Duplicate command (figure 3-20).

Paste From Modeling*/Paste From Drafting

The precise wording of this command depends on whether you are in the modeling module, where it will read Paste From Drafting, or in the drafting module, where it will read Paste From Modeling*. This dual approach addresses the fact that there are really two Clipboards in form•Z: one for 3D modeling and one for 2D drafting. You can cut and paste objects within either module and you can cut and paste, in either direction, between the 3D and 2D environments.

Both the Cut and Copy commands save objects in the form•Z Clipboard. Objects cut or copied from the modeling module are remembered as 3D objects, whereas elements cut or copied from the drafting module are saved in the 2D Clipboard.

When you paste a 3D object into a 2D drafting window, using the Paste From Modeling* command, the object will appear as a flat (2D) image of the original 3D object. However, 2D elements pasted into a 3D modeling window using the Paste from Drafting command will not show up as 3D objects; they will remain flat (2D) objects on the xy reference plane.

Of course, you can always use the simple Paste command to place a 3D object within a modeling window, and a 2D element within a drafting window. Remember that the Paste command will replace any selected object with the content of the Clipboard. This does not occur with the cross-pasting between 2D and 3D modules previously described.

The asterisk after the Paste From Modeling* command indicates that there are options contained in a dialog box that can be accessed by holding down the Option key (Mac) or Ctrl + Shift keys (Windows). The Paste From Modeling dialog box (figure 3-21) lets you choose between Each Face/Outline As A Polyline and Each 3D Segment As A Single Line.

Fig. 3-21. Paste From Modeling dialog box.

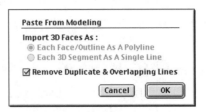

With the first option, the sides of closed polygons are described in single, closed polylines; with the second option, each line segment within a closed polygon becomes a separate drafting element. The first method is the default setting, and is more economical of points; the second may be useful if you want to edit line segments individually, or apply different line weights to each segment.

The Paste From Modeling dialog box also contains a Remove Duplicate & Overlapping Lines option, which eliminates redundant line segments when you place a 3D Clipboard object into a 2D modeling window. This is a useful option because a wireframe image actually includes two lines for each element, given that each edge of an object belongs to two faces. This option provides an easy method of eliminating the useless extra lines, which only slow down your machine.

Grab Image

Keyboard Shortcut: *Mac (Cmd-G), Windows (Ctrl + G)*

The Grab Image command provides a very convenient method of capturing a "snapshot" of anything on your screen. When the Grab Image command is invoked, you can move the cursor across the screen and define, with two mouse clicks, the two opposite corners of a rectangle, which will be saved as a color image file in the system Clipboard. Once an image is saved in the Clipboard, it can then be pasted into many other applications, from word processors to photo-manipulation software.

The Grab Image command is particularly useful for Web designers who want to quickly create buttons, backgrounds, and banners for Web sites. By creatively cropping the screen with the Grab Image command, it is relatively easy to extract interesting 2D graphical elements from even a relatively simple 3D view of a form•Z model. Images grabbed from the screen will naturally be at a screen resolution (usually 72 dots per inch), which is the same resolution at which they will appear on the Web. For Web publishing, you will have to paste the images into an image processing software such as Adobe Photoshop in order to edit and save them in JPEG or GIF format.

Edit Menu

Paste Image

The Paste Image command will take the image most recently saved in the Clipboard and place it within the form•Z drafting window. The image may have been saved with the Grab Image command, or it may have been saved into the Clipboard from some other program, perhaps after being digitized on a scanner. Whatever the source, the image will show up in your drafting window as an image element, which can be manipulated there with operations such as Move, Rotate, and Scale. The Paste Image command is active in the drafting environment only. The command will appear in gray and be inactive in the modeling module.

form•Z Clipboard...

The Clipboard command accesses the form•Z Clipboard dialog box, which shows you the current content of the Clipboard. That is, the dialog box shows you the portion of the system's memory that can be used to temporarily store text and graphics, which are obtained by means of the Cut or Copy commands. Upon selecting form•Z Clipboard while in the modeling module, you will see the 3D Clipboard. Selecting the same in the drafting module results in the 2D Clipboard. You would typically use this command to check the content of the Clipboard before beginning a series of Paste operations. (See figures 3-22 and 3-23.)

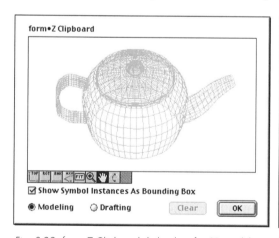
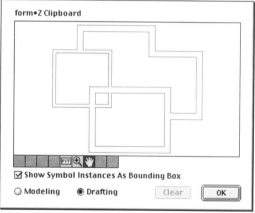

Fig. 3-22. form•Z Clipboard dialog box for 3D modeling.   Fig. 3-23. form•Z Clipboard dialog box for 2D drafting.

The form•Z Clipboard dialog box contains two radio buttons that allow you to switch between the Modeling and Drafting clipboards. Another option, Show Symbol Instances As Bounding Box, will simplify the drawing of objects taken from the Symbol Library, representing them as simple boxes. This feature can significantly speed up the Clipboard viewer if you are using a lot of items from a symbol library. The Clear button will erase the content of the clipboard.

The row of small square icons on the form•Z Clipboard dialog box provides viewing tools common to most form•Z preview windows. Included from the left are the TOP, RGT, and BAK settings for the Top, Right, and Back views. By holding down the Option key (Mac) or Ctrl + Shift keys (Windows), these icons are switched to BOT, LFT, and FRT, representing the Bottom, Left, and Front views.

The AXO icon will bring up a 30- to 60-degree axonometric view. The FIT option will cause the image of the object to be resized to fit the window. The magnifying glass is the Zoom In control. By holding down the Option key (Mac) or Ctrl + Shift keys (Windows), the plus sign becomes negative and the icon represents Zoom Out. The hand icon allows you to pan the image left, right, and up and down. The last icon has two curved arrows to signify view rotation. The first click in the preview area sets a view rotation point. The view is then changed by dragging the mouse. Note that in the 2D drafting module the icons related to 3D views are gray and inactive (figure 3-23).

Select Previous

Keyboard Shortcut:
Mac (Cmd-`), Windows (Ctrl + `)

The Select Previous command deselects whatever is currently selected and picks the previously selected items. You would probably use the Select Previous command only to recover from a mistake during a long process of selecting multiple points, segments, or objects. For a single selection, it is usually just as fast to reselect the object with the Pick tool.

Select All UnGhosted

Keyboard Shortcut:
Mac (Cmd-A), Windows (Ctrl + A)

The term *ghosted* refers to the inactive state in which some objects rest after certain operations, such as the Boolean Difference command. UnGhosted refers to normal, active objects. The Select All UnGhosted command picks and highlights all active objects. In this sense, you can think of Select All UnGhosted as equivalent to the familiar Select All command seen in many other programs. The shortcut, Cmd-A (Windows, Ctrl + A), may also be familiar to you from many other software packages.

Select All Ghosted

A ghosted object cannot be selected with the Pick tool; it must first be reactivated with the UnGhost tool. This can be a slow process if you wish to select several ghosted objects for a particular operation. Fortunately, the Select All Ghosted command picks and highlights all inactive objects in the model at once. You can use this command to reactivate ghosted entities. That is, after picking them with the Select All Ghosted command, you can make them all active again with a single UnGhost command.

Select By...

The Select By command accesses a large dialog box that allows you to specify the criteria for any selection process in form•Z (figure 3-24). These criteria are divided into categories, which include object types, topology, colors, layers, and names. This is a powerful function that will become important to you as your models become larger and your work more advanced.

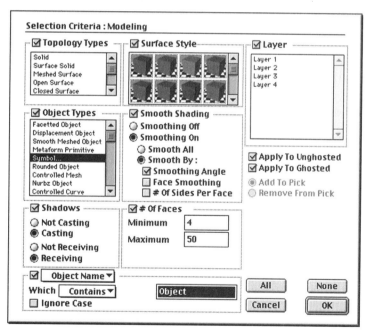

Fig. 3-24. Selection Criteria/Modeling dialog box with all categories selected.

In the Selection Criteria/Modeling dialog box, each category of selection criteria is presented in a separate section and begins with a check

Chapter 3: Menus

box to control whether or not it will be used in the selection process. By default, the boxes are not checked; all categories are deselected. The first selection category is Topology Types. This allows you to select objects that are Solid, Surface Solids, as well as Meshed, Open, and Closed surfaces.

➻ **NOTE:** *These terms are explained in the introduction to Chapter 5 and under the command tools that generate these topology types, such as Sweep, Revolve, C-Mesh, and Skin.*

The second selection category is Object Types. The list of 24 types varies from general geometrical forms such as Faceted, Displacement, Smooth Meshed, and Rounded Objects to specific architectural components such as Spiral Staircases and Bolts. By clicking on these object types, you can identify all entities of that type in a model.

One item in the Object Types list, Symbol, is different from the others, as indicated by the ellipsis that follows it. Upon selection with the Option key (Mac) or Ctrl + Shift keys (Windows) held down, the Symbol Instance Selection Criteria dialog box (figure 3-25) is accessed. This dialog box lets you make selections based on the attributes of symbols, such as the name of the symbol library in which the symbol is stored, symbol definition, color display attribute of the symbol, layer display attribute of the symbol, and detail level of the symbol.

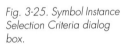
Fig. 3-25. Symbol Instance Selection Criteria dialog box.

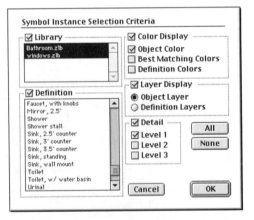

The third selection category is Shadows. This will allow you to select objects that do or do not cast shadows, as well as objects that receive (show shadows that are cast on them) and do not receive shadows. This will be mainly of interest to the advanced user creating complex renderings.

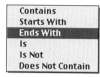

Fig. 3-26. The "Which" pop-up menu

The fourth selection category in the Selection Criteria/Modeling dialog box is Object Name/Group Name. This category allows you to specify objects or groups by name. The pop-up menu beside the word *Which* (figure 3-26) will help you to select by object or group name without necessarily providing the entire name. Six options are provided to facilitate your search: Contains, Starts With, Ends With, Is, Is Not, and Does Not Contain.

In the text box, enter the name, or partial name, of the objects or groups you want to select. The Ignore Case option is recommended if you do not remember whether you used upper or lower case when naming the object or group.

The fifth selection category is Surface Style. This category allows you to select the color, texture, or transparency/opacity the selected object must match.

Smooth Shading is the sixth selection category. This category allows you to select objects by their smooth shading attributes previously set with the Smooth Shade tool in the Attributes tool palette.

The seventh selection category, below the center of the large Selection Criteria/Modeling dialog box, is # Of Faces. With the Min and Max fields, you can set a range for the selection process. For example, if you wanted to select all tetrahedrons (four-sided solids) and cubes (six-sided solids) in your model, you would set Min to 4 and Max to 6.

The eighth and final selection category in the Selection Criteria/Modeling dialog box is Layer. This feature allows you to restrict your selection to objects that occur on the selected layers. Note that the Layers function is itself a mechanism for sorting and selecting objects in your model; you would use the Layer category of the Select By command only if you were selecting by other criteria at the same time. If you simply wanted to select everything on a particular layer, it would be faster to use the Layers palette to isolate that layer and select all objects on it.

The Apply To Unghosted and Apply To Ghosted options control whether the active or inactive objects are selected with the Select By command. The Add To Pick and Remove From Pick options determine whether the objects to be selected with the latest criteria settings are to be added to, or removed from, the current group of selected objects. This means that you could actually work "backward" and select all objects in the model first, and then deselect objects with the Remove From Pick option.

Chapter 3: Menus

The All and None options can be used to reset the dialog box. All will select all categories and attributes, whereas None deselects them.

The drafting module of form•Z incorporates a related Select By procedure that helps to demonstrate some of the inherent differences between the 2D and 3D environments. The Selection Criteria/Drafting dialog box (figure 3-27) is modified to accommodate the items specific to 2D work. Element Types includes geometrical entities such as Rectangles, Circles, and Arcs, as well as drafting conventions such as Leader Lines. The Layer category is similar to the one in the modeling module, as previously discussed. Selections can also be made by Color, Line Weight, and Line Style.

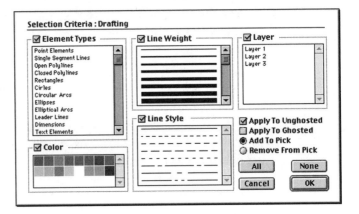

Fig. 3-27. The Selection Criteria/Drafting dialog box allows selection by line weight and style.

Deselect

Keyboard Shortcut:
Mac (Cmd-Tab), Windows (Ctrl + Tab)

The Deselect command deselects all objects, faces, segments, and/or points selected with the Pick tool. Use the Deselect command upon making a mistake in a series of Pick operations if you think it would be easier to start the selection process over. If you simply want to go back one step in a long selection process, you would use the Select Previous command, as previously described.

Clear

The Clear command deletes all unghosted objects. It is similar to Cut, except that it does not store the deleted objects in the Clipboard. This command should be used with caution. Because you could delete an entire model this way, the system will prompt you for confirmation before proceeding.

Clear All Ghosted

The Clear All Ghosted command deletes all inactive (ghosted) objects in your project. Ghosted objects are not usually as valuable as active objects; nevertheless, the system will prompt for a confirmation each time you use this command. If a particular ghosted object is valuable to you, unghost it and place in on a separate, inactive layer. Figure 3-28 shows a cylinder that has been ghosted after being subtracted from a box in a Boolean difference operation. Figure 3-29 shows the effect of using the Clear All Ghosted command.

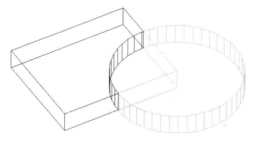 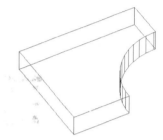

Fig. 3-28. Cylinder subtracted from a box, leaving both original objects ghosted.

Fig. 3-29. Ghosted objects removed.

Hide Ghosted

The Hide Ghosted command will make all ghosted objects invisible. These objects can be made visible again by revisiting this command to remove the check mark beside it. If the gray outlines of ghosted objects are causing a visibility problem on your screen, but you want to keep the objects for later operations, it may be beneficial to use this command to temporarily hide the ghosted objects. The visual effect of this command is the same as the one seen in figure 3-29.

Key Shortcuts...

The Key Shortcuts command allows you to create a customized list of shortcuts to represent commands you frequently use. Keyboard shortcuts can be extremely helpful as you attempt to increase your facility and speed in modeling. Because the keyboard can be a far more efficient input device than the mouse for oft-repeated commands, you should consider defining your own keyboard equivalent for any command that requires excessive mouse movement and clicking. At the very least, you should learn the most important shortcuts; you will find that you can immediately work more rapidly and with greater ease.

➻ **NOTE:** *The Key Shortcuts Manager does double duty as a means of customizing form•Z and as a helpful guide to the menus and commands. It is explained in detail in Chapter 2.*

Preferences...

Keyboard Shortcut:
Mac (Cmd-,), Windows (Ctrl + ,)

The last item in the Edit menu is Preferences, which allows you to customize many aspects of the form•Z modeling environment. You can choose to ignore this function as a beginner, but as you master form•Z you will increasingly find yourself wanting to adjust certain aspects of the form•Z system to better suit your working style.

The Preferences dialog box (figure 3-30) will allow you to save many of these custom settings in a file on a floppy disk. In this way, you could carry your favorite settings with you even if you have to work on a different workstation. Preference files permit you to save and recall as many sets of custom-defined system parameters as you want; you could have many collections of different settings for different types of projects.

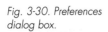

Fig. 3-30. Preferences dialog box.

The Preferences dialog box is large and detailed, and includes a column of eight options on the right that, in turn, invoke dialogs with additional options. The first item at the left of the dialog box is the Current Preferences File field, which indicates the preferences file, if any, is currently being used. The Save Preferences button allows you to save your current settings. The Load Preferences button lets you load and apply a preference file previously saved.

There are two types of preferences: the first affect the system and the second affect only the form•Z project. For this reason, you can separately control system and project preferences with the System Option From and Project Option From selections by specifying Defaults, which means your preference file settings will be ignored, or by specifying the Preference File, which will cause your preferences to be used. The system options can also be set to remember the previous setting. If you find yourself repeatedly adjusting system parameters each time you start up form•Z, you should invoke the Previous Session option.

The next section of settings in the Preferences dialog box controls the windows presented when form•Z starts up. You can specify that a New Project (Model), New Project (Draft), or "File Open" dialog be displayed at start-up.

The Project File Options allow you to specify that a backup file is always kept, and to compress your files. Although the automatic backup is recommended, the compression of your form•Z files is usually unnecessary unless you are running out of space on your hard drive. The compression process will make your files slower to save and open.

There are six additional settings at the bottom of the Preferences dialog box. The first is the Continuous Window Tool Control (Zoom By Frame/Hand/View Tools). This option is on by default. This means that after you complete a Zoom, Hand (panning), or View command, the Command Tools and the other Window Tools are grayed out and inactive and the system waits for you to do another Zoom, Pan, or View command. You may instinctively click in the modeling window after completing a Zoom In, only to find that you have then zoomed in even closer. You are required to click on the grayed-out menus to restore the interface to its normal condition. If you find this type of continuous execution of zoom operations counterintuitive, deselect this option.

The Save Prompts in TEXT File option does what it says: every word you see in the Prompts window is recorded in a file if this option is invoked. The file is named *form•Z Prompts* and can be found in the form•Z application folder. The Store New Image Elements in Project option affects only the drafting module, causing every image you Paste into a drafting project to be saved at its full resolution as a part of that project file.

The Always Open File Format Options Dialogs option controls whether or not you will need to click on the Options button in the Save Project As dialog box. When Always Open File Format Options is selected, a file format options dialog will always pop up when you use the Save As menu.

The Show Window Zoom Percentage option can be used to show or hide the small rectangular read-out at the bottom of the modeling window that displays the current magnification of the image. The Memory Display option can be used to display or hide the numerical or graphical display of available memory.

There are eight additional buttons on the right side of the Preferences dialog box; each of them invokes yet another dialog box. The Auto Save button accesses the Auto Save Options dialog box (figure 3-31). When the Auto Save option is checked, you can specify the frequency of saves in minutes or number of operations. Remember that you have to save a file at least once and give it a name in order to activate the Auto Save feature. The Save To Project setting causes the project file itself to be saved, whereas the Save As Copy option leaves the original file untouched.

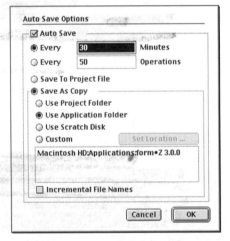

Fig. 3-31. Auto Save Options dialog box.

You have several choices as to where the files are saved: Use Project Folder, Use Application Folder, Use Scratch Disk, and Custom. If you select Custom, you can use the Set Location button to specify exactly where the files will be saved. The Incremental File Names option is useful if you want to save your project every step of the way; each time it is saved, a new file is written with a name that includes the date and time.

The second button on the right side of the Preferences dialog box invokes the Dialog Preferences dialog box (figure 3-32), which controls the appearance and screen location of form•Z dialog boxes. The slider bar allows you to control the size of preview boxes, which are used by commands such as Sweep and Smooth Mesh. The Use Dialog Buffer option speeds up the display of dialog boxes, but requires more memory. Only advanced users will want to concern themselves with these details.

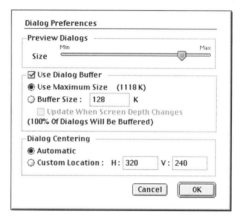

Fig. 3-32. Dialog Preferences dialog box.

The Fonts button on the right side of the Preferences dialog box accesses the Font Options dialog box (figure 3-33), which controls default fonts. You can separately control the default font of each type: TrueType, Postscript, and Bitmap. Select the Load Fonts At Launch option if you are frequently using the Place Text command or the drafting module.

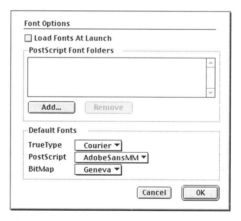

Fig. 3-33. Font Options dialog box.

The fourth button on the right side of the Preferences dialog box is labeled Scratch Disk, which accesses the Scratch Disk Preferences dialog box (figure 3-34), used to specify the storage location on your system. This can be useful when you have multiple drives or partitions and are running low on hard drive space.

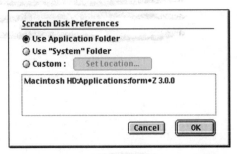

Fig. 3-34. Scratch Disk Preferences dialog box.

The fifth button is labeled Textures and invokes the Texture Options dialog box, which will be of interest only to advanced users who want to control the amount of memory allocated to texture maps. If the Cache Textures setting is turned on, texture maps will be transferred from memory to a file on the hard drive, thereby freeing memory when the texture file is not needed.

NOTE: *Texture maps are discussed in Chapter 10.*

The sixth button on the right side of the Preferences dialog box is labeled Warnings. This button accesses the Warnings Preferences dialog box (figure 3-35), which controls the manner in which warnings are presented on screen. The default settings are ideal if you are a beginner; they will warn you before clearing ghosted and unghosted objects, and before clearing rendering memory. The advanced form•Z user may want to disable some or all of the warnings.

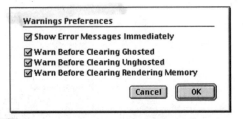

Fig. 3-35. Warnings Preferences dialog box.

Radiosity, the seventh button on the right side of the Preferences dialog box, affects the way the Radiosity renderer functions. This is applicable to the RenderZone RadioZity version of form•Z only.

NOTE: *Radiosity is discussed in Chapter 10.*

Windows Menu

The eighth and final button on the right side of the Preferences dialog box is Recent Files. This controls the number, naming style, and sorting of the files displayed by the Open Recent command in the File menu.

Below the Recent Files button, at the right of the Preferences dialog box, the Cursor option gives you a choice between the hourglass cursor and the watch cursor.

The Windows menu is the third item in the menu row across the top of the form•Z screen. The items in the pull-down Windows menu (figure 3-36) control the various types of windows that let you view your modeling or drafting projects, as well as the behavior and appearance of the frames, cursor, axes, grid, rulers, grid snap, and underlays.

Fig. 3-36. The pull-down Windows menu.

New Model Window

The New Model Window command opens up a new modeling window into your active project. Window names are listed at the bottom of the Windows menu. Having more than one window open can be very helpful. You can, for example, draw in a window that shows a Top view and instantly see the results in perspective in another window. Figure 3-37 shows a form•Z screen manually arranged, with three windows viewing the same model from different angles. The Window Frames command, new in form•Z version 3.0, discussed in material that follows, can also be used to set up multiple windows.

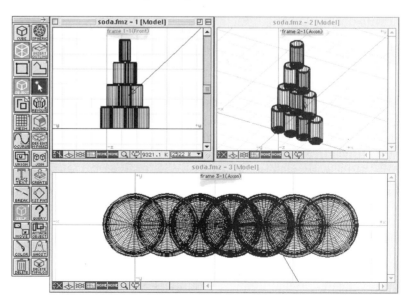

Fig. 3-37. Three modeling windows open to the same project.

New Draft Window

The New Draft Window command opens a new drafting window into the active project, providing an additional 2D viewing window. The Drafting Tools replace the Modeling Tools at the left side of the screen.

Tile Windows

This is the third command in the pull-down Windows menu. The black triangle to the right of Tile Windows indicates that this menu expands sideways. In fact, four additional items are revealed when the mouse is held down over Tile Windows: Open, Close, Arrange, and Align & Scale Views (figure 3-38).

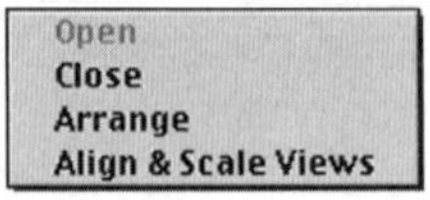

Fig. 3-38. Tile Windows pop-up menu within the Windows menu.

Open

In the Tile Windows menu item, the Open command opens up four new windows that are superimposed over the existing window and divide the screen into four quadrants. Three of the four windows show Top, Front, and Side views. The upper right window displays a 3D view.

Each of these window tiles can be moved and resized. Each comes with its own set of Window Tools, so that reference planes, snaps, geometrical constraints, rendering styles, and view angles can be customized in each window. Each window can be expanded to fill the entire screen by clicking on the small expand button at the right end of the window's gray header bar (figure 3-39).

Close

The Close command will shut all four windows of the Tile Windows arrangement. If you want instead to shut just one or two of the windows, you can use the Close box at the left end of the window's gray header bar.

Arrange

The Arrange command will restore the Tile Windows to their original positions in the four quadrants of the modeling window.

Fig. 3-39. Tile Windows with different display settings in each window.

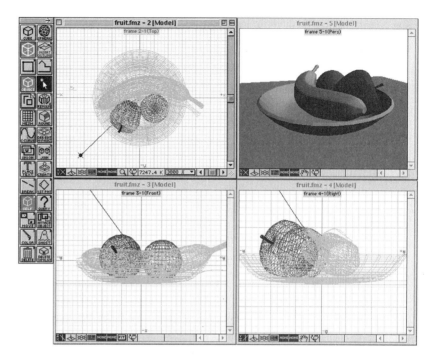

Align & Scale Views

The Align & Scale command will adjust the zoom setting so that all objects will fit within the boundaries of the three 2D windows. It also aligns the axes in each of the 2D views so that the parts of the model line up.

Close

The Close command will close the active window. You may find it easier to simply click on the close box at the left side of the window's header bar. If you attempt to close the last window without saving your project, you will be prompted to save the file.

Close All

Close All will shut all windows. If you have not saved your changes, you will be prompted to do so before all windows are closed.

Window Frames

This is a new feature in form•Z version 3.0, which is often more convenient than the Tile Windows command. Window Frames are adjustable windows that provide multiple views of the modeling or drafting project (figure 3-40). The boundaries between Window Frames are easily moved and rearranged with a click-and-drag operation (figure 3-41). This can facilitate the modeling of long or tall objects.

Fig. 3-40. Window Frames, default setting.

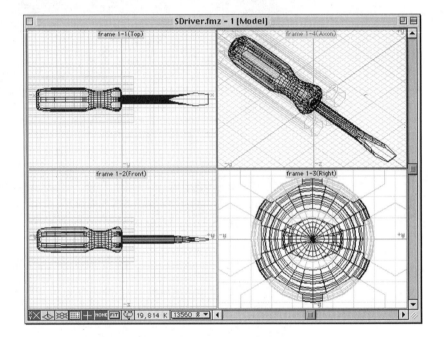

Fig. 3-41. Window Frames, with windows adjusted to a custom layout.

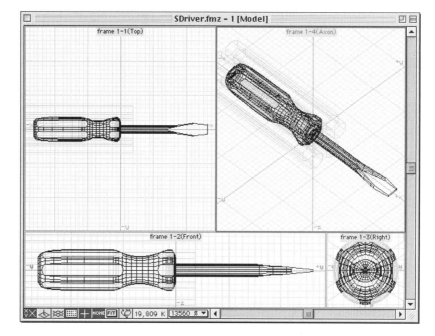

Unlike the Tile Windows, the four window frames always remain within the confines of the larger project window, thereby ensuring a more organized work space. The Window Tools are displayed only once, at the bottom of the entire group, rather than being repeated four times, as they are with Tile Windows. The Window Frames option is turned off by reselecting that item in the Windows menu, which removes the check mark beside the item and restores the previous project window.

Extended Cursor*

Extended Cursor activates the extended cursor, which has guidelines that extend from the regular cross-shaped cursor all the way to the edge of the screen, or as far as specified by the current setting of the Custom Cursor Size option in Extended Cursor Options. Extended Cursor facilitates alignment of points and objects, but many people find it to be visually distracting. Try both Normal and Extended Cursor and decide for yourself. Figures 3-42, 3-43, and 3-44 compare three cursor styles: Normal, Extended, and Extended To A Custom Cursor Size.

Fig. 3-42. Normal cursor in 3D view.

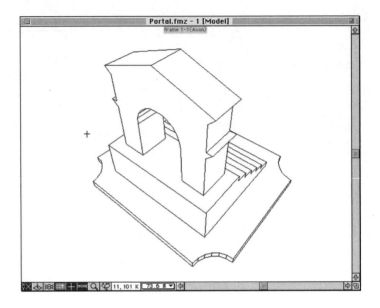

Fig. 3-43 Extended cursor.

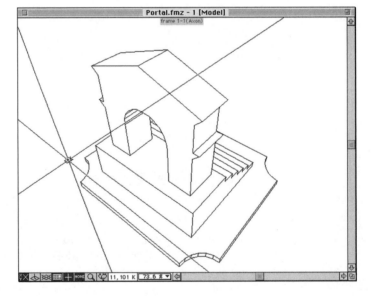

Windows Menu

Fig. 3-44. Cursor extended to custom size of 200 pixels.

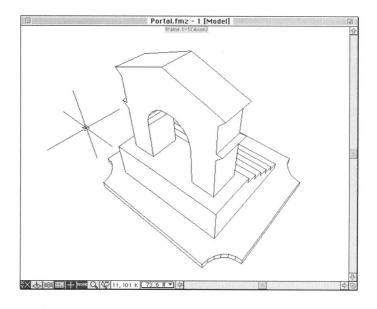

Extended Cursor Options...

The Extended Cursor Options dialog box (figure 3-45) allows you to customize the appearance and function of the cursor. Checking the Show Extended Cursor option box is identical to selecting the Extended Cursor* menu item.

Fig. 3-45. Extended Cursor Options dialog box.

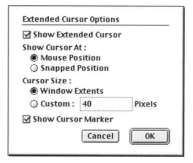

The Show Cursor At option offers the choice of Mouse Position and Snapped Position. The former dynamically displays the cursor at its actual position as you move the mouse; the latter shows the cursor lines at the closest Grid Snap location. If Grid Snap is not selected, it has no effect.

The Cursor Size option allows you to select Window Extents, or to set a custom dimension measured in pixels. The former means that the cursor lines run to the edge of the active window, whereas the latter sets the length of the cursor cross hairs.

Auto Scroll

Auto Scroll turns on or off the automatic scrolling function that causes the view in the active window to follow the motion of the cursor. Auto Scroll is a great convenience when you need just a little more room to fit in the lines or objects you are drawing. As the cursor goes beyond the edge of the active window, the view is adjusted in that direction. One alternative to Auto Scroll is the rather clumsy technique of interrupting the act of drawing to manually adjust the horizontal and vertical scroll bars at the edges of the active window. Another is to use the zoom and pan tools in Window Tools.

Show Plane Axes*

Show Plane Axes controls the visibility of the reference plane axes in the active window. By default, these three lines (labeled x, y, and z) appear as blue lines slightly darker and heavier than the blue lines of the reference plane grid. In figure 3-46, the cone sits on the current, arbitrary reference plane, whose angle conicides with the sloping side of the pyramid. The cube sits on the World coordinate system plane.)

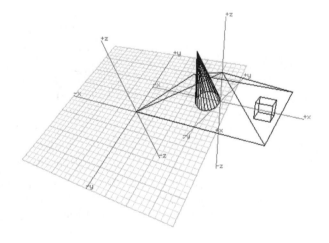

Fig. 3-46. View showing the reference plane Grid, and the Plane and World axes.

Show World Axes*

The Show World Axes command turns the World axes on or off. These axes appear as three red lines, labeled x, y, and z (figure 3-47).

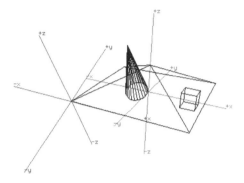

Fig. 3-47. View with reference plane Grid hidden.

Show Grid*

Show Grid turns the reference plane grid on or off in the current window (figure 3-48).

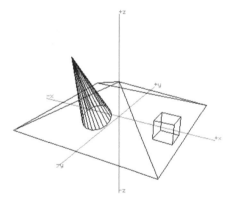

Fig. 3-48. View with Grid and Plane axes hidden.

Window Setup...

The Window Setup command accesses the Window Setup dialog box (figure 3-49). The first four options allow you to control the visibility of the World Axis, the Reference Plane Axis, the Grid, and Axis Marks. Axis marks are small triangles that slide along the two axes of the current reference plane and specify the position of the cursor.

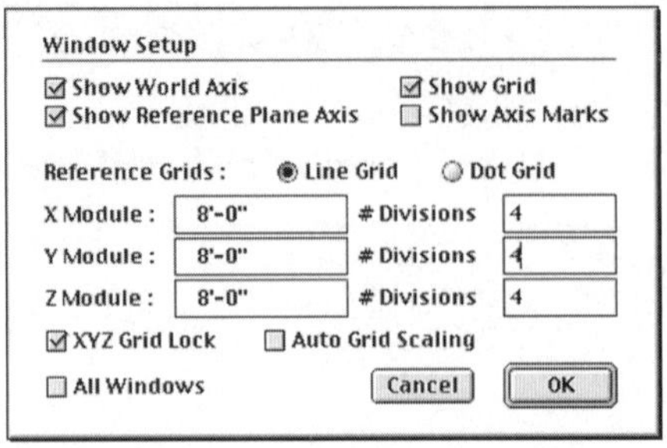

Fig. 3-49. Window Setup dialog box.

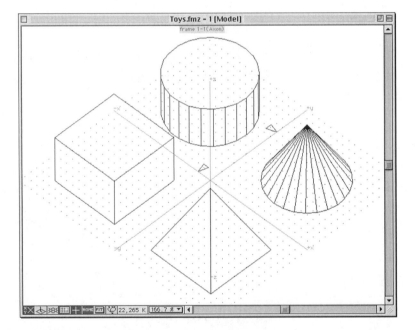

The Reference Grids option lets you specify Line Grid or Dot Grid. Dot Grid is more difficult to see, but may be desirable if you find that Line Grid is obscuring elements in your drawing (figure 3-50). For modeling convenience, Grid Snap in Window Tools can be made to coincide with the parameters of the Grid module in Window Setup. To do this, double click on the Grid Snap icon and select the Match Grid Module option.

Fig. 3-50. Modeling window with Dot Grid and Axis Marks applied.

➥ **NOTE:** *The Grid Snap tool is described in Chapter 6.*

The third section in the Window Setup dialog box controls the spacing of the major and minor grid lines. The grid module can be separately controlled for the *x, y,* and *z* directions. The # Divisions field sets the spacing of the minor lines. To avoid the thinner minor grid lines, set this option at 1.

There are three additional controls in the Window Setup dialog box. XYZ Grid Lock forces the Y and Z modules to use the X Module values, thus ensuring a square grid.

Show Rulers*

Show Rulers controls the display of rulers at the perimeter of the active window, a particularly useful feature in 2D views wherein you want a constant reminder of scale and dimension (figure 3-51). In 3D views such as Isometric or Axonometric, however, it can be confusing to see and measure from a 2D ruler along the edge of a 3D window. In a perspective window, it makes no sense to show a 2D ruler because the scale varies with the foreshortening of the perspective view.

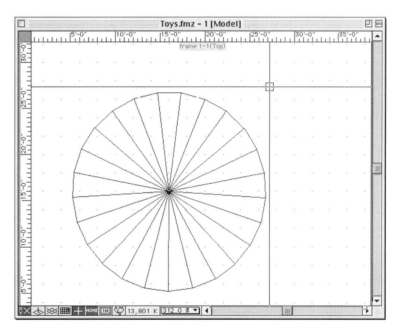

Fig. 3-51. Top view with Rulers and Extended Cursor turned on.

The asterisk in the command name indicates that you can hold down the Option key (Mac) or Ctrl + Shift keys (Windows) while selecting the command to open the options box. This is equivalent to selecting the next command, Ruler Options.

Ruler Options...

The Ruler Options dialog box allows you to customize the rulers appearing around the active window. The first option is Show Rulers, equivalent to the Show Rulers* item in the Windows menu. Ruler Display lets you specify rulers on the Top, Left, Bottom, and Right edges of the window. The default settings Top and Left are adequate in most cases.

The Coordinate Mode setting allows Absolute and Relative coordinates. With Absolute, the zero point on the ruler always aligns with the origin of the reference plane. With Relative, the zero point of the ruler is always set to the location of the previous mouse click. The Reference option allows you to specify that the ruler is coordinated with either the World coordinate system or the Reference Plane. The Show Text and Show Unit indicators determine the style of the ruler text and tick marks.

The Ruler Increment section of the Ruler Options dialog box (figure 3-52) allows you to further customize the alignment and increments of the ruler. You can even separately control the horizontal and vertical modules, divisions, and subdivisions of the rulers. The final options control the placement of the ruler Origin, the use of Plot Scale, and the application of these settings to All Windows.

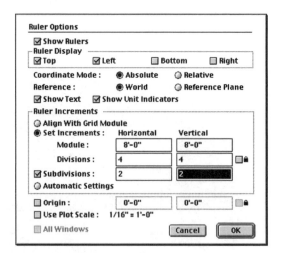

Fig. 3-52. Ruler Options dialog box.

Snap Options...

Snap Options accesses the Snap Options dialog box (figure 3-53), which controls grid, directional, and object snapping. Note that the panel accessed via the Grid Snap icon in Window Tools is called the

Windows Menu

Grid Snap Options dialog box and is less extensive, but contains a duplicate of the most important element of the Snap Options box, the Grid Snap Module.

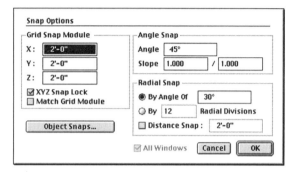

Fig. 3-53. Snap Options dialog box.

The Grid Snap Module lets you separately control the increment of the snap in the *x, y,* and *z* directions. The XYZ Snap Lock enforces the same module in the three directions. The Match Grid Module option ensures that the visible reference plane grid, controlled by Window Setup, coincides with the geometrical grid of the Grid Snap.

To the right of the Snap Options dialog box (figure 3-53) are the Angle and Radial Snap controls. Angle Snap specifies in degrees the angle at which lines will be constrained if the Snap to Angle/Slope tool is selected in Window Tools. Radial Snap does the same for the Radial Snap in Window Tools, with the additional alternative of using Radial Divisions and a Distance Snap.

The Object Snaps button brings up another dialog box containing advanced snap options, which include tolerance settings, segment snap controls, and adjustments for Projection and 3D views.

✓ **TIP:** *The Snap Option controls can be evoked most quickly by double clicking on the Grid Snap icon in the Window Tools, as described in Chapter 6.*

Keyboard Shortcut:
Mac (Cmd-U), Windows (Ctrl + U)

Underlay...

The Underlay command allows you to use as a drawing guide any photograph or drawing that can be scanned and saved in a standard image file format. By providing a visual reference that can literally be traced over with a variety of form•Z drawing tools, the Underlay command is often the most efficient method of modeling a complex shape. It can just as easily be used to build 3D forms on the basis of gestures in a freehand sketch, scanned from the paper original. The Underlay command is particularly useful in the generation of 3D site models from aerial photographs or maps (figure 3-54).

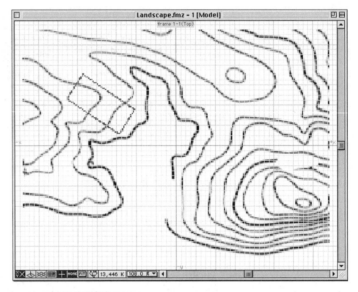

Fig. 3-54. Image scanned from a map used as an underlay in creating a site model.

Whatever the visual information to be used as an underlay, it must be saved in one of the standard file formats, including PICT, JPG, TIFF, or Targa for Macintosh and BMP, JPG, Metafile, TIFF, or Targa for Windows. The underlay file is simply a reference picture; it is not part of your 3D model.

The Underlay dialog box (figure 3-55) appears as soon as the Underlay command is invoked. The Show Underlay option is on by default. If you turn it off, the items below it are made inactive and the underlay image will not be visible. Using this feature regularly is useful as you build a model on top of an underlay image. This gives you a clear view of the model, allowing you to spot any problems that may be hidden by the visual clutter of the underlay image.

The Select Underlay File button opens the Open File dialog box and allows you to browse the hard drive for the underlay file you want. Although you normally use an underlay in Top, Front, or Side view, in some cases you may want to use an underlay in a 3D view. If the Match Image Size To Underlay Size option is not selected, the underlay image will be stretched to fit the size and aspect ratio of the 3D view window. If the option is selected, the window will be adjusted to match the underlay.

Fig. 3-55. Underlay dialog box.

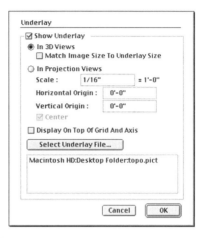

Normally you will import underlay images into a Projection View such as Top, Front, or Side view. Because these are 2D views, a different set of controls is available to control the scale and placement of the underlay image. The Scale field controls the size of the underlay image. The default setting is 1/16" = 1'-0" (1:200 if metric units are used).

✓ **TIP:** *For maps, site plans, and architectural plans, it is recommended to include a graphical scale on the scanned drawing so that the scale of the underlay can be easily double checked against the units implied by the form•Z reference plane grid. If the scales do not match, adjusting the value in the Scale field is much easier and faster than rescanning the image or saving it at a different resolution.*

The Horizontal Origin and Vertical Origin fields allow you to move the underlay image left, right, or up and down on the screen in a 2D view. This adjustment can be very useful when modeling buildings on a large site map. For convenience, you should model the building at the center of the reference plane near the origin (0,0), even if the building is placed +500 units to the right along the x axis according to the site map. Rather than move the building, you can shift the underlay image –500 units horizontally, or to the left.

The Center option will cause the center of the underlay image to be placed at the point defined by the Horizontal Origin and Vertical Origin fields, previously described. When deselected, the upper left corner of the underlay image is placed at that point.

The Display On Top Of Grid And Axis option affects the visibility of the reference plane grid and axes. If the grid is interfering with the visibility of the underlay image, you can select this option and hide the grid and axes behind the underlay image.

Below the Underlay command at the bottom of the Windows menu, a list appears with the names of the form•Z files currently open. The list will grow as you open more form•Z files.

Heights Menu

The Heights menu, a permanent fixture at the top of the form•Z screen, is a constant reminder that you are dealing with a 3D modeling program designed for architectural modeling. The Heights menu (figures 3-56 and 3-57) makes it easy to assign different heights units to building components that lie on the same plane.

Fig. 3-56. Pull-down Heights menu with English units.

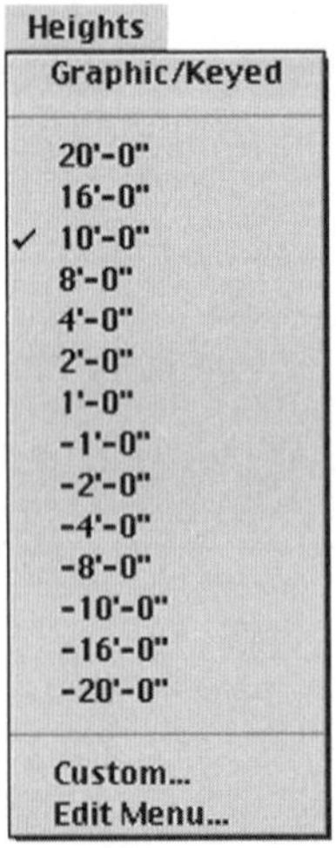

Heights
Graphic/Keyed
20'-0"
16'-0"
✓ 10'-0"
8'-0"
4'-0"
2'-0"
1'-0"
-1'-0"
-2'-0"
-4'-0"
-8'-0"
-10'-0"
-16'-0"
-20'-0"
Custom...
Edit Menu...

Fig. 3-57. Pull-down Heights menu with metric units.

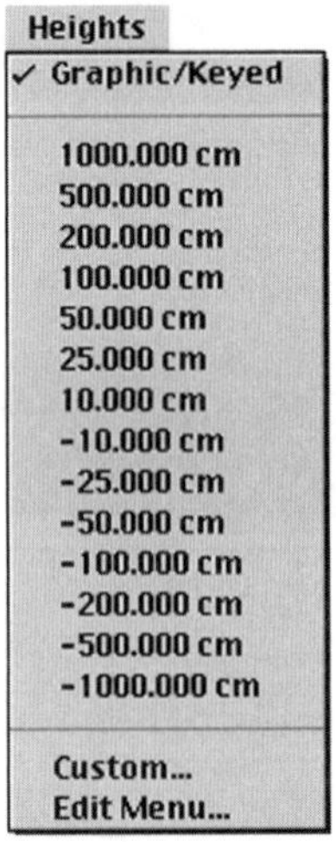

Heights
✓ Graphic/Keyed
1000.000 cm
500.000 cm
200.000 cm
100.000 cm
50.000 cm
25.000 cm
10.000 cm
-10.000 cm
-25.000 cm
-50.000 cm
-100.000 cm
-200.000 cm
-500.000 cm
-1000.000 cm
Custom...
Edit Menu...

Height settings in the Heights menu are numerous, and you can add custom settings to the menu. The numbers can be positive or negative, and they represent the distance shapes will be extruded upward or downward along the z axis. Remember that you can switch to metric units with the Working Units command in the Options menu. Note that in the Drafting Module the Heights and View menu items are gray and inactive.

Graphic/Keyed

You will use the Graphic/Keyed setting frequently because it allows you to interactively set object height. The alternative is to select a preset height from the menu. Furthermore, in combination with the Grid Snap tool, the Graphic/Keyed setting can give you dynamic and exact height readouts in the Prompts palette, which makes it easy to use this interactive method with precision.

The settings shown in figure 3-56 are the default values for heights, if you are using English measurements. The Metric menu is shown in figure 3-57.

Custom...

The Custom command accesses the Custom Height dialog box (figure 3-58). With this dialog box you can define a new Heights setting and add it to the menu. Note that as soon as you click on the Add button, the new setting is added to the existing Heights menu.

Fig. 3-58. Custom Height dialog box.

Edit Menu...

The Edit Menu item within the Heights menu accesses the Heights Menu dialog box (figure 3-59) and provides all of the tools you need to customize the Heights menu. You can specify a new value and Add it to the list, or Remove an existing value. You can also use the Save and Load buttons to store your settings in a file, or to retrieve previously saved settings. Storing the Heights settings can be extremely useful when switching modeling tasks. For example, if you are modeling kitchen cabinets you would save a Heights menu that included all of the standard kitchen dimensions for counter heights, thickness, and so on. If you wanted to switch to an urban-scale project, you would load a Heights menu that listed useful building heights.

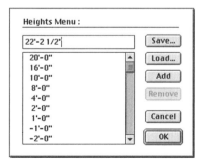

Fig. 3-59. Heights Menu dialog box.

View Menu

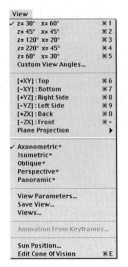

Fig. 3-60. The pull-down View menu.

Keyboard Shortcut:
Mac (Cmd-1), Windows (Ctrl + 1)

The pull-down View menu (figure 3-60) provides a choice of preselected viewing angles and projections, as well as access to dialog boxes that allow you to customize the 3D views of your model. It also allows you to adjust the shadows in your model by specifying the location of an imaginary sun. One of the most useful items in the View menu is Edit Cone Of Vision, an interactive, visual interface that facilitates setting up and adjusting 3D views. The View menu applies to the 3D modeling module only, and is inactive in the drafting module.

An important new addition to the View menu in form•Z version 3.0 is the Animation From Keyframes item, which allows you to generate an animated sequence based on previously saved views that have been selected in the Views palette.

NOTE: *See Chapter 12 for a complete discussion of animation.*

z=30° x=60°

The first of a series of five preset viewing angles, the z=30° x=60° menu item will vary, depending on the active projection selected in the third section of the View menu. See the menu sections labeled Axonometric, Isometric, Oblique, Perspective, and Panorama for discussion and examples of these specific projection types.

If either Axonometric or Perspective has been selected, this menu choice will read z=30° x=60°. If Isometric is the active projection, the menu changes to x=30° y=60°. If Oblique is selected, the choice becomes Inclination=30°. If Panoramic is selected, the first two sections of the View menu are grayed out and inactive.

The view represented by the label z=30° x=60° can be imagined as being constructed with a fixed camera looking down from above in Top view. The object is first rotated 30 degrees around the z axis, then 60 degrees around the x axis. This results in a 3D view, as shown in figures 3-61, 3-62, and 3-63.

View Menu

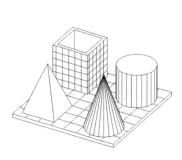

Fig. 3-61. A preset Axonometric view (z=30° x=60°).

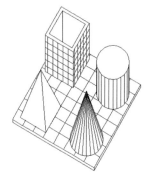

Fig. 3-62. A preset Isometric view (x=30° y=60°).

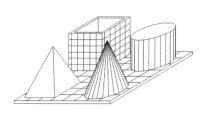

Fig. 3-63. A preset Oblique view (Inclination=30°).

Keyboard Shortcut:
Mac (Cmd-2), Windows (Ctrl + 2)

z=45° x=45°

z=45° x=45° is the second in the series of preset viewing angles. If either Axonometric or Perspective is selected, this menu item will read z=45° x=45°. If Isometric is the active projection, the menu changes to x=15° y=75°. If Oblique is selected, the choice becomes Inclination=45°. (See figures 3-64, 3-65, and 3-66.)

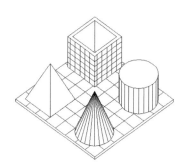

Fig. 3-64. A preset Axonometric view (z=45° x=45°).

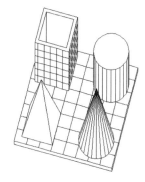

Fig. 3-65. A preset Isometric view (x=15° y=75°).

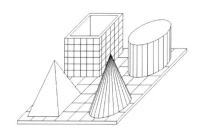

Fig. 3-66. A preset Oblique view (Inclination=45°).

Keyboard Shortcut: *Mac (Cmd-3), Windows (Ctrl + 3)*

z=120° x=20°

z=120° x=20° is the third of the preset viewing angles. If either Axonometric or Perspective is selected, this menu item will read z=120° x=20°. If Isometric is the active projection, the menu changes to x=60° y=30°. If Oblique is selected, the choice becomes Inclination=60°. (See figures 3-67, 3-68, and 3-69.)

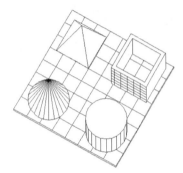

Fig. 3-67. A preset Axonometric view (z=12° x=20°).

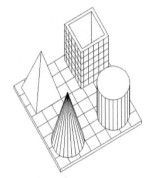

Fig. 3-68. A preset Isometric view (x=60° y=30°).

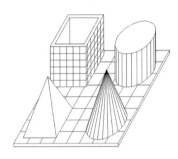

Fig. 3-69. A preset Oblique view (Inclination=60°).

Keyboard Shortcut: *Mac (Cmd-4), Windows (Ctrl + 4)*

z=220° x=45°

220° x=45° is the fourth of the preset viewing angles. If either Axonometric or Perspective is selected, this menu item will read 220° x=45°. If Isometric is the active projection, the menu changes to x=15° y=15°. If Oblique is selected, the choice becomes Inclination=120°. (See figures 3-70, 3-71, and 3-72.)

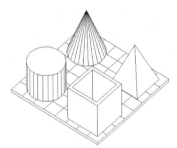

Fig. 3-70. A preset Axonometric view (z=220° x=45°).

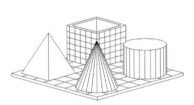

Fig. 3-71. A preset Isometric view (x=15° y=15°).

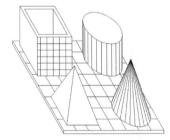

Fig. 3-72. A preset Oblique view (Inclination = 120°).

View Menu 85

Keyboard Shortcut: *Mac (Cmd-5), Windows (Ctrl + 5)*

z=60° x=30°

The fifth menu item of the preset viewing angles is z=60° x=30°. If either Axonometric or Perspective is selected, this menu item will read z=60° x=30°. If Isometric is the active projection, the menu changes to x=30° y=30°. If Oblique is selected, the choice becomes Inclination=135°. (See figures 3-73, 3-74, and 3-75.)

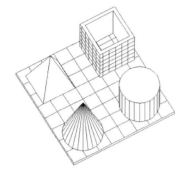

Fig. 3-73. A preset Axonometric view (z=60° x=30°).

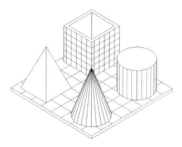

Fig. 3-74. A preset Isometric view (x=30° y=30°).

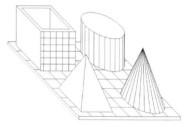

Fig. 3-75. A preset Oblique view (Inclination=135°).

Custom View Angles...

Custom View Angles brings up the Custom View Angles dialog box (figure 3-76), which provides fields in which you can enter values (in degrees) to control the rotation of the view. If you enter zero in each field, the result is a Top view. If you enter the values in the five preset views previously discussed, you will replicate these views.

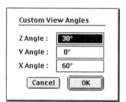

Fig. 3-76. Custom View Angles dialog box.

Custom View Angles is available for Axonometric views and Perspective views only; it is not available in the drafting module, or for Panoramic projections. If you select either Isometric or Oblique modes, this menu item will be replaced by an additional preset view angle setting: x=45° y=45° or Inclination=150°.

This command is recommended to advanced users only, who are interested in the minute adjustments of various 3D projections. The beginner will have no need to create custom view angles because the menu of standard views is already quite extensive. In any case, you will probably find that the Edit Cone Of Vision command is a more useful and intuitive method of customizing views.

Chapter 3: Menus

Keyboard Shortcut:
Mac (Cmd-6), Windows
(Ctrl + 6)

[+XY] : Top

This command Heights menu causes the view in the active window to be a plan (Top) view. This is a view of the *xy* plane from above (positive *z* axis). This is one of the most useful view settings because it gives you a clear overview of your model. Note that the view type appears in the animation-frame label at the top center of the modeling window; that is, frame 1-1(Top).

✓ **TIP:** *Memorize the keyboard shortcut for the Top View (+XY) command: cmd + 6. You will use it often.*

The visual result of this command will depend on the active projection selected in the third section of the View menu. If Perspective is selected, the Top view will display the foreshortened lines in an aerial perspective. The Axonometric setting gives the expected results: an undistorted, 2D Top view. If either Isometric or Oblique modes are selected for any of the six preset 2D views (Top, Bottom, Right, Left, Front, or Back), the result will be a 3D isometric or oblique view. (See figures 3-77, 3-78, 3-79, and 3-80.)

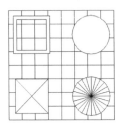

Fig. 3-77. A preset Top (+XY) view in Axonometric mode.

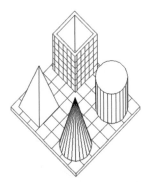

Fig. 3-78. A preset Top (+XY) view in Isometric mode.

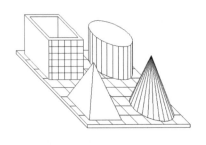

Fig. 3-79. A preset Top (+XY) view in Oblique mode.

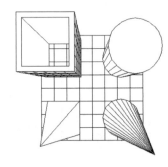

Fig. 3-80. A preset Top (+XY) view in Perspective mode

Keyboard Shortcut:
Mac (Cmd-7), Windows
(Ctrl + 7)

[–XY] : Bottom

This command changes the view in the active window to be a view from underneath the *xy* plane (negative *z* axis). This is one of the least useful projections. It generates an image that may be mistaken for a Top view, except that the image is flipped horizontally; that is, the positive *x* axis appears on the left. (See figures 3-81 and 3-82.)

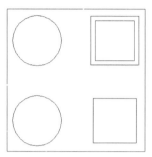 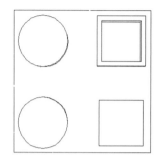

Fig. 3-81. A preset Bottom (–XY) view in Axonometric mode.

Fig. 3-82. A preset Bottom (–XY) view in perspective mode.

[+YZ] : Right Side

Keyboard Shortcut:
Mac (Cmd-8), Windows (Ctrl + 8)

This command generates an Elevation (side) view. This is a view of the *yz* plane from the right (positive *x* axis) side. This is a very useful view in cases where the object is defined by its side profile. The traditional drafting arrangement of Top, Front, and Side view usually favors the choice of Right Side instead of Left Side.

If your project is one that involves orientation to the cardinal points, orientation toward cardinal points, and you follow the convention of aligning North with the positive *y* axis, the Right Side view will display the east side of your model. (See figures 3-83 and 3-84.)

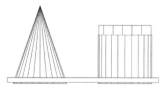 

Fig. 3-83. A preset Right Side (+YZ) view in Axonometric mode.

Fig. 3-84. A preset Right Side (+YZ) view in Perspective mode.

Keyboard Shortcut:
Mac (Cmd-9), Windows (Ctrl + 9)

[-YZ] : Left Side

The [–YZ] : Left Side command generates a side view that shows the yz plane from the left (negative x axis) side. If you follow the convention of aligning North in your project with the positive y axis, the Left Side view will display the west side of your model. (See figures 3-85 and 3-86.)

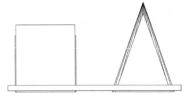
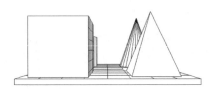

Fig. 3-85. A preset Left Side (+YZ) view in Axonometric mode.

Fig. 3-86. A preset Left Side (+YZ) view in Perspective mode.

Keyboard Shortcut: Mac (Cmd-0), Windows (Ctrl + 0)

[+ZX] : Back

The [+ZX] : Back command generates an Elevation, a view that shows the zx plane from the "back" (positive y axis) side. If you follow the convention of aligning North in your project with the positive y axis, the Back view will display the north side of your model. (See figures 3-87 and 3-88.)

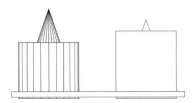
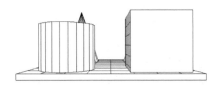

Fig. 3-87. A Back (+ZX) view Axonometric mode.

Fig. 3-88. A preset Back (+ZX) view in Perspective mode.

Keyboard Shortcut (using minus sign):
Mac (Cmd-"–"), Windows (Ctrl + "–")

[-ZX] : Front

The [–ZX] : Front command generates an Elevation (front) view. This is a view of the zx plane from the front (negative y axis). This is a very useful view in cases where the object is defined by its front profile. The Front view is always one of the three standard views in the traditional drafting arrangement of Top, Front, and Side.

If your project is one that involves orientation to the cardinal points, and you follow the convention of aligning North in your project with the positive *y* axis, the Front view will display the south side of your model. (See figures 3-89 and 3-90.)

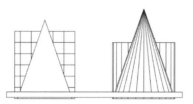

Fig. 3-89. A preset Front (–ZX) view in Axonometric mode.

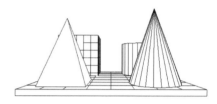

Fig. 3-90. A preset Front (–ZX) view in Perspective mode.

Plane Projection

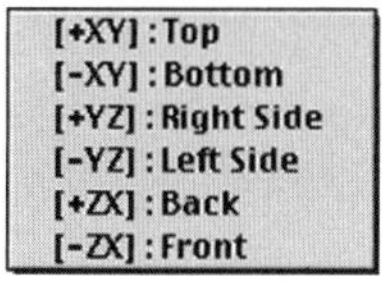

The Plane Projection command adds functionality to the Define Arbitrary Plane tool in the Window Tools. The six Plane Projection choices are accessed in the View menu through a slide-out submenu (figure 3-91).

Fig. 3-91. Plane Projection submenu, located within the View menu.

◆◆ **NOTE:** *The Reference Planes tools are located within the Window Tools, which are discussed in Chapter 6.*

If Top is selected in the slide-out Plane Projection submenu, the active arbitrary reference plane is shown in Top view. This can be useful if you need to define shapes in relation to an arbitrary plane. For example, if you have modeled a house with a sloping roof, you may want to project the roof plane as a Top view so that it is easier to measure and position the skylights. If you have not defined an arbitrary reference plane, the six menu items will be grayed out.

Axonometric*

This Axonometric command switches the current view to an axonometric projection (figure 3-92). The asterisk (*) immediately after the word *Axonometric* indicates that there are options available in the Axonometric View Parameters dialog box (figure 3-93) that can be accessed by holding down the Option key (Mac) or Ctrl + Shift keys (Windows) while selecting the Axonometric* command.

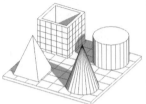

Fig. 3-92. An Axonometric projection.

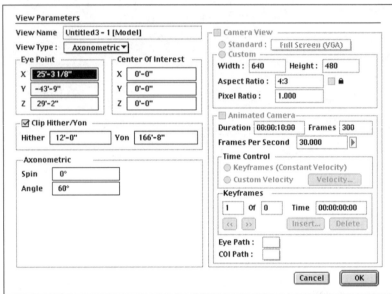

Fig. 3-93. View Parameters dialog box showing Axonometric settings.

Axonometric is a type of 3D projection popular among engineers and architects because it avoids the foreshortening effects of perspective. Axonometric also tends to present a 3D view that is less distorted than those created by the Oblique and Isometric projections.

The Axonometric View Parameters dialog box allows you to specify every parameter that affects the view, including the view type itself. View Eye Point is the position of the viewer as defined by the X, Y, and Z values of a point in space. Center Of Interest is the point the viewer is looking at. As you change a value in one of these six fields, the other five will update. There may be occasional situations when you will want to use the View Parameters dialog box to precisely specify the view, but in general you will find that the Edit Cone Of Vision command provides a much more intuitive interface for adjusting the view.

The lower left half of the View Parameters dialog box controls Clipping, Spin, and Angle of the cone of vision. The Clip Hither/Yon option is a very useful feature that allows you to eliminate from the view anything that lies in front of or beyond the clipping planes of the cone of vision.

The right half of the View Parameters dialog box contains settings related to animation, which are discussed in Chapter 12.

➥ **NOTE:** *The Hither and Yon "clipping" planes are shown in the Cone of Vision diagram in figure 3-113.*

The Spin option lets you tilt the view as if you were leaning to the side as you snapped a picture with a camera. This can be used to create unusual effects for final renderings, but it is not particularly useful during the modeling stage of a project. Center of Interest is the point at which the imaginary camera is pointing. This point will always show up at the center of your Axonometric command window. Angle controls the width of the cone of vision and applies only to Perspective views.

Isometric*

The Isometric command switches the current view to an isometric projection (figure 3-94). The View Parameters dialog box (figure 3-93) can be accessed by holding down the Option key (Mac) or Ctrl + Shift keys (Windows) while selecting the Isometric command. Isometric is traditionally the favorite 3D projection of mechanical engineering because it preserves the dimensions of the object as measured along the three axes. This means that you can measure directly from an isometric drawing as long as you keep your ruler parallel to one of the three axes. An "exploded" isometric projection is often used in patent document and parts-assembly diagrams.

Fig. 3-94. An Isometric view.

In the View Parameters dialog box, the Inclination Angles option controls the angles by which the *x* and *y* axes are inclined in the final view. Usually it is not necessary to generate custom inclination angles for isometric projection because there are six preset choices offered at the top of the View menu.

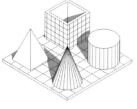

Fig. 3-95. A "traditional" isometric view with X and Y view angles of 30 degrees.

The Preserve Angles option ensures that the X and Y inclinations (Figure 3-95) total 90 degrees, which may be useful if you want to correctly display the proportions of shapes that were generated on the reference plane. However, the traditional isometric projection distorts the squares of the reference plane grid, turning them into diamond shapes.

Oblique*

The Oblique command switches the current view to an oblique projection. The Oblique View Parameters dialog box (figure 3-96) can be accessed by holding down the Option key (Mac) or Ctrl + Shift keys (Windows) while selecting the Oblique command. Elevation Oblique View is often used by architects because it preserves the shape of the facade (or elevation) of a building. This can be achieved by setting the Inclination to 60 degrees, and entering X, Y, and Z Eye Points of 0, –100, and 100 units, respectively (figure 3-97).

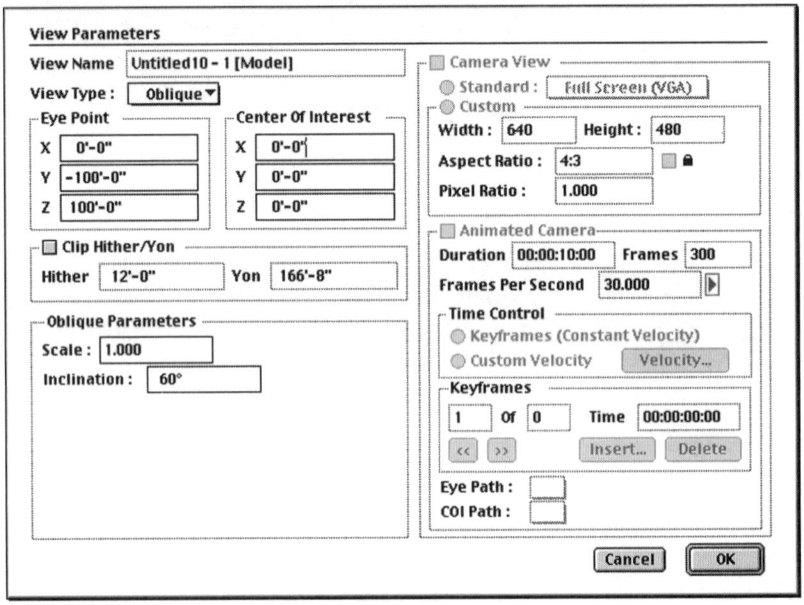

Fig. 3-96. View Parameters dialog box showing Oblique settings.

Fig. 3-97. An Elevation Oblique view with an inclination of 60 degrees.

Fig. 3-98. Plan Oblique view with an inclination of 90 degrees.

Plan oblique is another projection suited to architectural use because it does not distort the floor plan. This is accomplished in form•Z by setting Inclination to 90 degrees (figure 3-98).

The original rationale for oblique views was to produce a fast 3D view while minimizing time spent redrawing difficult facade or plan details. Oblique views were quickly drawn by hand, but they tended to produce shapes that looked too long if the oblique lines were projected back at the same scale as the front plane of the drawing. For this reason, draftsmen would often reduce the depth projection to 75%, and even 50%, of the full depth of the object. The Scale option in the Oblique View Parameters dialog box allows you to do the same. Note that if

View Menu

Oblique is selected, the View menu displays six different preset Inclination choices, all of which generate Elevation Oblique views.

Perspective*

The Perspective command switches the current view to a perspective view (figure 3-99). The Perspective View Parameters dialog box (figure 3-100) is accessed by pressing the Option key (Mac) or Ctrl + Shift keys (Windows) while selecting the Perspective command.

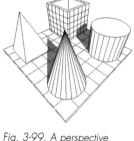

Fig. 3-99. A perspective view.

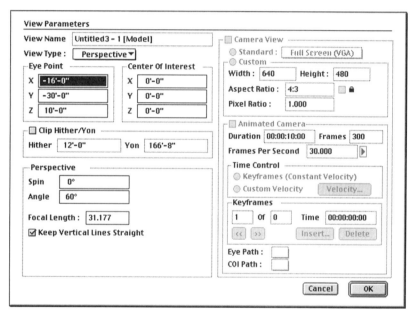

Fig. 3-100. View Parameters dialog box showing Perspective settings.

Perspective is the most realistic of the 3D projections because it most closely matches what the human eye sees: objects appear larger when they are closer and smaller when they are further away. Furthermore, we are well practiced in interpreting perspective documents; that is, we routinely obtain complex 3D information from photographs, video, and films, all of which use perspectival projection. The great failing of perspective, however, is that unlike isometric projection you cannot lay a ruler on top of such a drawing and confidently read off accurate dimensions.

The Perspective View Parameters dialog box allows you to specify every parameter that affects the view. The Keep Vertical Lines Straight option corrects the distortion problem that occurs in perspective when vertical lines appear to converge at a vanishing point below or above the image.

Architects especially like this corrected perspective because architectural renderings of buildings have traditionally been drawn in a two-point perspective, which maintains a "straight-up" perspective. Computers produce "distorted" three-point perspectives, but it is just as easy to generate straight-up two-point perspectives. There is something more stable and reassuring about a perspective that keeps its vertical lines straight (figure 3-101).

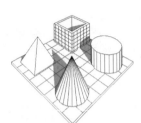

Fig. 3-101. A Perspective view with the Keep Vertical Lines option.

Eye Point View Point is the position of the viewer as defined by the X, Y, and Z values of a point in space. The Clipping, Spin, and Center Of Interest items are all identical to the options previously discussed in the Axonometric section. Angle, which controls the width of the cone of vision, is measured in degrees. The default angle is 60 degrees.

Try a low Angle setting such as 20 degrees to get a telephoto lens effect, and a high value such as 100 to get a wide-angle effect. In practice, it is usually easier to adjust the Eye Point, Clipping, Spin, Angle, and Center Of Interest options interactively by means of the Edit Cone Of Vision command, as described in the material that follows.

Panoramic*

Panoramic switches the current view to a panoramic, or 360-degree, display mode (figure 3-102). This mode is for viewing only; you cannot use the modeling tools, which are in fact grayed out and inactive when Panoramic is selected. The main reason to use this option is to generate panoramic QuickTime Virtual Reality (QTVR) movies, which are popular in various multimedia and Internet applications.

View Menu

Fig. 3-102. A panoramic view.

→ **NOTE:** *See the heading "Save QuickTime VR...," at the beginning of this chapter for more on QTVR.*

The Panoramic View Parameters dialog box (figure 3-103) is accessed by pressing the Option key (Mac) or Ctrl + Shift keys (Windows) while selecting the Panoramic command. The Perspective View Parameters dialog box allows you to specify the Eye Point by XYZ location.

Fig. 3-103. View Parameters dialog box showing Panoramic settings.

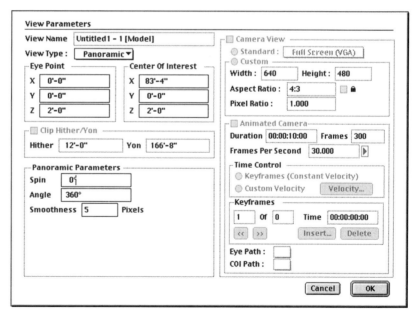

The Spin, Angle, and Center Of Interest items were previously discussed in the Perspective section. In this case, Angle is the extent of the panorama; a 270-degree setting produces a three-quarter panorama. The default Angle is 360°, a complete panorama.

The Smoothness option controls the intervals, in pixels, at which the sections of the panoramic image will be created. The higher the setting, the smoother the curved lines will appear. As usual, it will cost you; rendering times will be slower if the smoothness setting is higher. Start out with the default setting of 5, and increase it only if the resulting curves look too rough.

Note that the Edit Cone Of Vision command, which is found at the bottom of the View menu, will display a "cylinder of vision" if you are in Panoramic mode (figure 3-104). This interactive Panorama Cylinder is controlled in a similar fashion; you can click on the viewer position, line of sight, and center of interest and move them by clicking on a new position. You can adjust the Spin by tilting the cylinder in the Front or Side window.

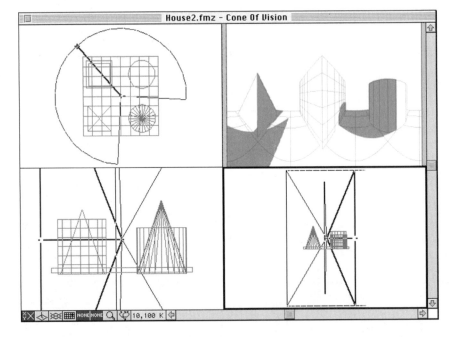

Fig. 3-104. Panorama Cylinder in the Edit Cone Of Vision window adjusted for a 270-degree field of view.

View Parameters...

The View Parameters command brings up the same dialog box previously discussed in the Axonometric, Isometric, Oblique, Perspective, and Panoramic sections. The View Type setting and the visible options

View Menu

you see will depend on which of the five projection settings you have selected. You can also call up the View Parameters dialog box from within the Edit Cone Of Vision window.

Save View...

The Save View command brings up the Name View dialog box (figure 3-105), which allows you to name and save the current view. The saved view will be listed in the Views palette.

Fig. 3-105. Name View dialog box.

➥ **NOTE:** *The Views palette is discussed in Chapter 4.*

Views...

The Views command brings up the Views dialog box (figure 3-106), which is also accessed through the Views palette. This palette will let you Save, Delete, Rename, Edit, Copy, Sort, Clear, and Load views.

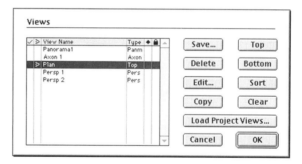

Fig. 3-106. Views dialog box.

➥ **NOTE:** *The various options of the Views palette are discussed in detail in Chapter 4.*

Animation From Keyframes...

This command is new to version 3.0. It brings up the Edit View dialog box, in which the animation parameters can be set.

➥ **NOTE:** *Animation is discussed in detail in Chapter 12.*

Sun Position...

The Sun Position command brings up the Sun dialog box (figure 3-107), which allows you to control light and shadows in two ways: by specifying the position of the "sun" in the "sky," or by specifying a location on earth and a date and time.

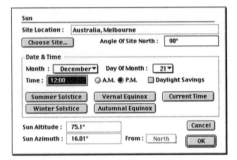

Fig. 3-107. Sun dialog box used to select the winter solstice lighting conditions in Melbourne, Australia.

The first method, defining the sun's position by altitude and azimuth, is not intuitive, to say the least, unless you happen to be an astronomer or perhaps a cartographer. The second method is the recommended one: Use the Choose Site button to bring up the Geographic Position dialog box (wherein you select an appropriate location from a list of more than 500 international cities); then specify the Month, Day Of Month, and Time for which the lighting and shadow-casting conditions should be calculated. (See figure 3-108.)

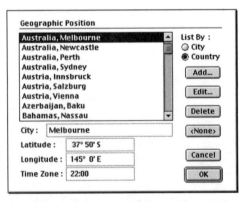

Fig. 3-108. Geographic Position dialog box.

If you are interested in showing extremes of actual lighting conditions, select the Summer Solstice or Winter Solstice buttons. Figures 3-109, 3-110, and 3-111 show the shadows a 16-foot cube would generate at noon at three different locations on the planet, as generated in form•Z's Surface Render display mode.

View Menu

Fig. 3-109. Shadows generated by a 16-foot cube at noon on the summer solstice in Acapulco.

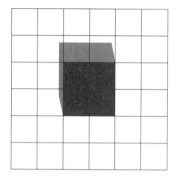

Fig. 3-110. Shadows at noon on the summer solstice in Toronto.

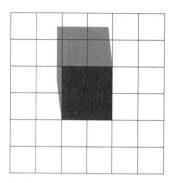

Fig. 3-111. Shadows at noon on the summer solstice in Oslo.

Edit Cone Of Vision...

The Edit Cone Edit Cone Of Vision command displays a graphical representation (figure 3-112) of the "virtual camera" that controls the view in the active modeling window. This is one of the most powerful features of form•Z. This command window lets you control the position of the viewer, the center of interest, and the "sun," as well as the hither and yon planes, the spin, and the perspective viewing angle. It also allows you to pick up and move both the line of sight and the light beam of the sun.

Fig. 3-112. Edit Cone Of Vision screen.

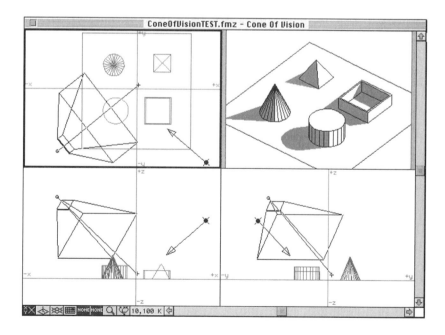

Keyboard Shortcut:
Mac (Cmd-E), Windows (Ctrl + E)

Edit Cone Of Vision Window

The Edit Cone Of Vision window comprises four equal areas that display 2D Top, Front, and Side views, as well as a 3D view in the upper right quadrant. The active quadrant is the one with a heavy black line as a border. To make another quadrant active, click in it. The cone of vision, of course, is actually a truncated pyramid; it appears in the 2D windows, whereas the 3D window shows the view visible through the "cone." As you move or adjust the cone of vision, the 3D view is instantly updated.

When the Edit Cone Of Vision window is open, the Window Tools appear at the bottom left corner of the screen. These tools are similar to the standard Window Tools, except that they affect only one of the three quarter-screen windows. In this case, the Window Tools are not used to directly adjust the view; they are used to adjust the view of the cone of vision tools that control the view.

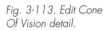 **NOTE:** *Window Tools are discussed in Chapter 6.*

It is sometimes difficult to differentiate between the symbols that represent the various Edit Cone Of Vision controls. You may find that the window is easier to read and adjust if you turn off the Show Grid option in the Window Setup dialog box under the Windows menu. The following list refers to the ten interactive controls of the cone of vision, which are labeled 1 to 10 in figure 3-113. All of these controls can be adjusted with two clicks of the mouse: one at the indicated point to activate the control and a final click to establish the new position.

Fig. 3-113. Edit Cone Of Vision detail.

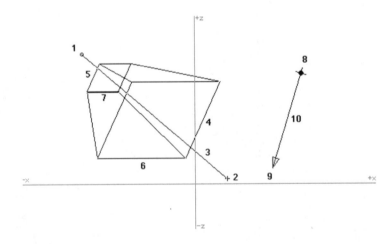

1. The Viewer Position, or Eye Point, is represented by the small circle with a dot at its center. Move this point in the Top window first to get the general direction, then in the Front or Side window to adjust the angle of view.

2. The Center of Interest is the point you are looking at. This point always ends up in the center of your screen. The Eye Point and the Center of Interest are the two fundamental points that determine your Line of Sight and set up your perspective. In general, this point should be dragged to the center of your model.

3. Moving the Line of Sight is a convenient way to adjust both the Viewer Position and the Center of Interest to a new, parallel position without changing the angle of view.

4. The View Angle is adjusted by dragging the sides of the cone of vision, creating effects that can range from telephoto to wide-angle-lens views.

5. The Hither Plane is the narrow end of the cone of vision. If the Clip Hither/Yon option is active, elements of the model positioned closer to the viewer than this plane will not show up in renderings.

6. The Yon Plane is the wide end of the cone of vision. If the Clip Hither/Yon option is active, elements of the model positioned further from the viewer than this plane will not show up in renderings.

7. The View Spin command controls the tilt of the image and is represented by the heavy line segment at the base of the Hither plane. If you want to undo a tilt effect, you can use the Reset View Spin command in the pull-down Edit Cone Of Vision menu.

8. The Light Source is represented by the large black dot. The position of this "sun" determines the angle at which shadows are cast.

9. The Light Target is the point at which the "sun" is aimed.

10. The Light Beam is the line between the Light Source and the Light Target, and is a convenient means of moving those two points at the same time. This allows you to move the light source symbols away from the cone of vision symbols to a more convenient location on the screen without affecting the angle of the sun.

Edit Cone Of Vision Pull-down Menu

When you enter Edit Cone Of Vision mode, the screen is filled with four viewing windows, and the normal form•Z menus, command tools, and palettes are not available. However, the Window Tools are available. In addition, a special selection of commands is made available via the Edit Cone Of Vision pull-down menu (figure 3-114), activated by clicking on the gray header bar at the top of the screen.

Fig. 3-114. Edit Cone Of Vision pull-down menu.

The Edit Cone Of Vision pull-down menu commands are all available through the normal command structure of the program. The first two items in the menu are Undo and Redo, which are related to the similar commands in the Edit menu, except that they affect manipulations of the Edit Cone Of Vision parameters only.

The next group of commands in the Edit Cone Of Vision pull-down menu is the same as those in the Display menu: Wire Frame*, Quick Paint*, Hidden Line*, Surface Render*, Shaded Render*, QuickDraw 3D, and OpenGL*. The selected display type will affect only the image in the Edit Cone Of Vision's preview window. The third group of items includes Perspective and Clip Hither/Yon, which are identical to the View menu commands previously discussed.

The fourth group of commands in the Edit Cone Of Vision pull-down menu are reset buttons that restore the default settings of parts of the cone of vision. Reset View Angle restores the cone of vision to a 60-degree viewing angle; Reset View Spin removes any camera tilt effects by restoring the Spin setting to 0 degrees. Reset Hither/Yon adjusts the front and back planes of the cone of vision so that all objects in front of the viewer are visible.

Align & Scale Views resets the pan and zoom settings so that the model fits entirely within the viewing area, and the axes are aligned between the three viewing windows. The "sun" is not included in this fitting operation, unless so specified in the Zoom Options dialog box. For this reason, it may be helpful to drag the Light Beam line closer to the model so that the sun is not left "off camera" when the Align & Scale Views command is used. The Window Setup item brings up the same Window Setup dialog box discussed in the "Windows" section of this chapter.

The next group of commands in the Edit Cone Of Vision pull-down menu provides access to the Save View and Views commands, which appear in the View menu previously discussed. The sixth group of commands in the Edit Cone Of Vision pull-down menu includes Display

Display Menu

Detail, View Parameters, and Sun Position. The last two are identical to their counterparts in the View menu, as previously discussed.

The Display Detail command brings up the Cone Of Vision Display Detail dialog box, which allows you to speed up the display of your model by simplifying the representation of objects within the Cone Of Vision window. Objects and entire layers can be replaced by simple boxes. This can be a valuable time saver if you are developing a large model on a slow computer and need to make frequent use of the Edit Cone Of Vision command.

The final command in the Edit Cone Of Vision pull-down menu is Close Cone Of Vision. You can close the cone of vision by clicking on this option, or by clicking on the close box at the upper left corner of the window. Note that as soon as you close the Edit Cone Of Vision window, your most recent adjustment of the view will be in effect in the active graphics window.

Display Menu

The Display menu provides control and options that affect the appearance of the model, such as the display scale, the rendering technique used, and the size of the rendering. The pull-down Display menu is shown in figure 3-115. Note the convenient keyboard shortcuts for the various rendering styles, such as Wire Frame and RenderZone.

Fig. 3-115. The pull-down Display menu.

Chapter 3: Menus

1/8" = 1' - 0" Scale, 1/16" = 1' - 0" Scale, and 1/32" = 1' - 0" Scale

These three menu items are grouped together because they actually represent one scaling concept. The second line displays the scale of the active window (1/16"=1'-0" by default). The first line always shows a scale that is double that of the middle line (1/8" = 1'-0" by default). The third line always shows a scale that is half that of the second line (1/32" = 1'-0" by default). Selecting the first or third lines changes the display scale of the active window. If you have selected metric units in the Working Units section of the Options menu, the Display menu will show the following three metric scales.

- 1 : 100 Scale
- 1 : 200 Scale
- 1 : 400 Scale

The functionality of these display scale commands is the same regardless of the unit system. Note that adjusting the display scale of the window does not affect the dimensions of the model. That is, for example, a 10-foot cube viewed at 1/32" = 1' - 0" scale will still be a 10-foot cube, even if it looms larger on the screen when seen at 1/8" = 1' - 0" scale. In practice, these display scale procedures are unnecessary because the zoom commands in Window Tools provide a more straightforward capability for adjusting the scale of the view of your model. Leave this command at its default setting and use the Zoom tools instead.

Custom Display Scale...

This command brings up the Custom Display Scale dialog box (figure 3-116), which lets you enter a custom display scale. It is recommended that the beginner ignore these display scale commands and concentrate instead on mastering the zoom controls in the Window Tools.

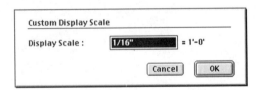

Fig. 3-116. Custom Display Scale dialog box.

➤ **NOTE:** *The zoom controls are discussed in Chapter 6.*

Wire Frame*

This command displays the model in wireframe (figure 3-117), as a line drawing with all lines visible. This is the default setting, and is the fastest display mode, but can sometimes be difficult to read because only the edges of surfaces are shown. The asterisk (*) indicates that the Wire Frame Options dialog box (figure 3-118) can be accessed by holding down the Option key (Mac) or Ctrl + Shift keys (Windows) while selecting the Wire Frame* command.

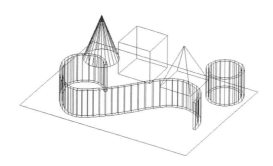

Fig. 3-117. A model displayed in Wire Frame.

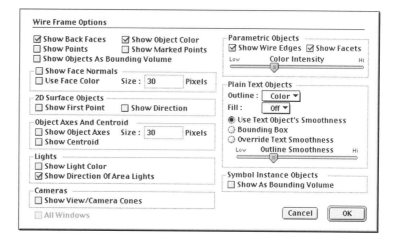

Fig. 3-118. The Wire Frame Options dialog box.

If you are using one of the other display modes and find the rendering is taking too long, press the Escape key to terminate the rendering, and then use the Cmd-W keyboard shortcut (Ctrl + W in Windows) to quickly bring back a wireframe view.

~ **NOTE:** *In most Macintosh programs, Cmd-W is the keyboard shortcut for closing the active window. Using the Key Shortcuts*

Manager in the Edit menu, you can define your own shortcut for Close, or any other form•Z command.

The Wire Frame Options dialog box (figure 3-118) provides display controls of particular interest to the advanced form•Z user because the controls reveal useful information about the fundamental geometry of points, lines, and surfaces.

The Show Back Faces option reveals the surfaces of solids that face away from the viewer; turning this option off can often make a wireframe drawing more understandable by eliminating unnecessary lines. The Show Object Color option is on by default. Turn it off if you want a purely black-and-white wireframe image.

If the Show Points option is selected, a small black diamond appears at the vertices of every line and object when displayed in wireframe (figure 3-119). This can be very useful for visualizing the density of points in curved lines. The Show Marked Points option makes marked points visible by surrounding them with a small diamond-shaped frame (figure 3-120). Points are often marked in preparation for the Skin operation, by means of the Topological Attributes tool. The Show Objects As Bounding Volume option will represent each object as a simple cubic form so that a large model can be manipulated.

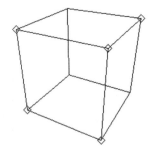

Fig. 3-119. Cube with points displayed.

Fig. 3-120. Cube with five marked points displayed.

◆◆ **NOTE:** *The Topology toolbar, which includes the Reverse Direction tool, is described in Chapter 5.*

The Show Face Normals option reveals the usually invisible vectors that are perpendicular to each face, indicating whether it is an inside or an outside surface (figure 3-121). This information is often useful in advanced surface modeling. The Reverse Direction command in the Topology toolbar can be used to reverse the direction of surfaces. The Use Face Color option can be turned on to control the color of the nor-

Display Menu

mal vectors. The length of the normals can be set in pixels in the Size n Pixels field. Turn off the Show Back Faces option to make the normals easier to see, and select individual faces to highlight the corresponding normals (figure 3-122).

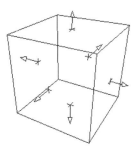
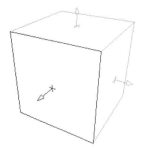

Fig. 3-121. Cube with Face Normals displayed.

Fig. 3-122. Cube with one Face Normal highlighted and back faces hidden.

The 2D Surfaces section of the Wire Frame Options dialog box provides two methods of visualizing the geometrical properties of 2D lines and surfaces. The Show First Point option surrounds the starting point of each line with a diamond shape. The Show Direction option adds directional arrows to each segment (figure 3-123). In the Object Axes And Centroid section you can make visible these geometrical points that form•Z calculates for each object.

Fig. 3-123. Line, rectangle, and circle with first point and direction displayed.

The Lights options determine whether lights are drawn in their own assigned colors instead of black, and whether area lights have a directional arrow. The Cameras section controls the visibility of the cones of vision; the corresponding views must be selected in the Views palette, as indicated by the small black diamond to the right of the view name. The All Windows option determines whether the rendering parameters you have selected will be applied to all viewing windows, not just the active one.

•◆ **NOTE:** *The Primitives modeling tools are discussed in Chapter 5.*

The Parametric Objects section of the Wire Frame Options dialog box controls the appearance of objects created with the Primitives modeling tools, which allow you to create shapes that can be repeatedly adjusted later. Select Show Wire Edges and Show Facets, and adjust the Color Intensity slider bar to about 50% to obtain the most useful representation of parametric objects.

The next section of the Wire Frame Options dialog box is labeled Plain Text Objects. Beside the words *Outline* and *Fill* are pop-up submenus that offer three choices: Off, Black/White, and Color. These control the manner in which the text will be drawn, and how they will be filled. The default setting produces color outlines without fill. Additional options allow you to Use Text Object's Smoothness, represent text with a simple Bounding Box, or Override Text Smoothness and set your own smoothness level by adjusting a slider bar.

The Symbol Instance Display option allows you to view symbols as bounding boxes. Eliminating the detail of the symbols can speed up your system if your model uses a large number of symbols.

Quick Paint*

Quick Paint is a fast preview mode that provides a low-quality color rendering of a project (figure 3-124). Because Quick Paint does not resolve intersections between objects, it is not appropriate for final images. However, remember the Cmd-T (Ctrl + T in Windows) keyboard shortcut because, even with a large model, Quick Paint is incredibly fast and can sometimes give you useful information a wireframe view cannot provide.

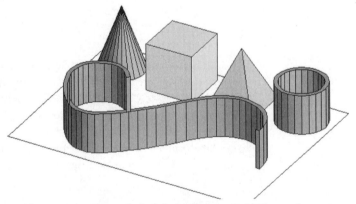

Fig. 3-124. Model displayed in Quick Paint.

The Quick Paint Options dialog box (figure 3-125) can be accessed by holding down the Option key (Mac) or Ctrl + Shift keys (Windows)

Display Menu

while selecting the command. The Show Color option is on by default, and there is little reason to turn it off because fast color viewing is really the only reason to use Quick Paint.

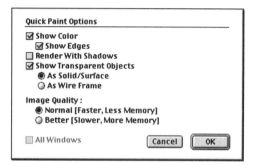

Fig. 3-125. Quick Paint Options dialog box.

The Show Edges option draws black lines around each surface. The Render With Shadows option is better left in its default (off) position with this low-quality renderer. Show Transparent Objects allows you to treat transparent objects as opaque solids or wireframes. The Image Quality option is set at Normal by default. Select Better for a slower but slightly improved view. The All Windows option applies these settings to all windows of a project.

Hidden Line*

Hidden Line view (figure 3-126) displays a line drawing that shows only those lines visible from your current viewpoint. This display mode is excellent for plotter or laser-printer output. The keyboard shortcut for Hidden Line in Mac is Cmd-H (Ctrl + H in Windows).

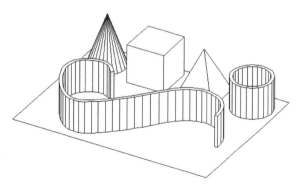

Fig. 3-126. Model displayed in Hidden Line.

The Hidden Line Options dialog box (figure 3-127) is accessed by holding down the Option key (Mac) or Ctrl + Shift keys (Windows) while selecting the command. The Show Color option is useful if you want to avoid the usual black-and-white look in favor of a colored line drawing. The Include Open Shapes option determines whether or not lines will be visible if they do not form closed shapes.

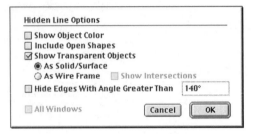

Fig. 3-127. Hidden Line Options dialog box.

Show Transparent Objects allows you to treat transparent objects as opaque solids or wireframes. The Hide Edges With Angle Greater Than option will simplify the rendering of curved areas, hiding the edges between faces that meet at an angle greater than that specified. The All Windows option applies these settings to all windows of a project.

Surface Render*

The Surface Render view is a relatively fast color-shaded rendering that also provides a shadow option (figure 3-128). The Mac keyboard equivalent is Cmd-R (Ctrl + R in Windows). The Surface Rendering Options dialog box (figure 3-129) can be accessed by holding down the Option key (Mac) or Ctrl + Shift keys (Windows) while selecting the command.

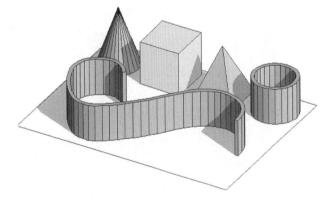

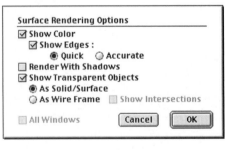

Fig. 3-128. A model displayed in Surface Render.

Fig. 3-129. Surface Rendering Options dialog box.

Display Menu

The Show Color option can be used to turn off the color. The Show Edges option provides a choice of Quick or Accurate black lines, or none at all, at the edges of all surfaces. Render With Shadows will cast shadows based on the position of the "sun" light source. Show Transparent Objects allows you to treat transparent objects as opaque solids or wireframes. The All Windows option applies these settings to all windows of a project.

Shaded Render*

The Shaded Render view displays a high-quality color-shaded rendering that can be used for final presentation images (figure 3-130). It features antialiasing, transparency, and shadow options. The Mac keyboard shortcut is Cmd-L (Ctrl + L in Windows).

Fig. 3-130. Model displayed in Shaded Render.

The Shaded Rendering Options dialog box (figure 3-131) can be accessed by pressing the Option key (Mac) or Ctrl + Shift keys (Windows) while selecting the command. Set Image Size is a very valuable option that allows you to render small rectangles within the active window, as defined by two mouse clicks, rather than waiting for the entire view to be redrawn (figure 3-132). If you are working on a fairly substantial model that redraws slowly, you should always select this option.

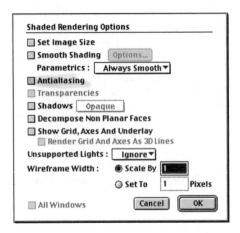

Fig. 3-131. Shaded Rendering Options dialog box.

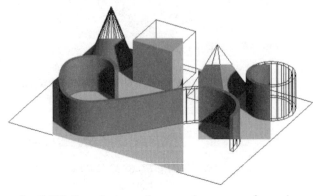

Fig. 3-132. Using Set Image Size to render a portion of a window.

The Smooth Shading option offers different settings for parametric objects, as well as a button that brings up the Smooth Shading Options dialog box (figure 3-133). This applies, and lets you adjust, the smooth shading feature, which can significantly enhance the appearance of a model containing curved surfaces. You can use these options to enhance, or to override, the smoothing settings previously applied with the Set Smooth Shading tool.

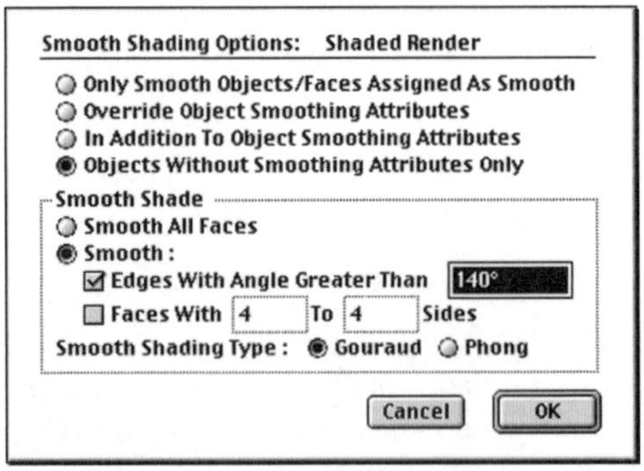

Fig. 3-133. Smooth Shading Options dialog box for Shaded Render mode.

➥ **NOTE:** *The Smooth Shade tool, found in the Attributes tool palette, is discussed in detail in Chapter 5.*

The Shaded Render Options dialog box contains an Antialiasing option that will eliminate the "jaggies" that appear at edges that are not perfectly vertical or horizontal. The Transparencies option is available if Antialiasing has been selected.

Display Menu

The Shaded Render Options dialog box has a Wireframe Width setting that allows you to define the line weight of lines that have been selected to remain as wireframes in the shaded view. The line width can be adjusted with a scale factor or as an absolute number of pixels. The Unsupported Lights option lets you Ignore or Approximate lights other than those of the distant type. The Render With Shadows option turns shadow casting on and off, and a related option will create Colored Shadows From Transparent Objects.

The Decompose Non Planar Faces option is recommended if your model contains many nonplanar faces, which typically occurs when shapes have been deformed. The Show Grid, Axes And Underlay option is useful if you want the final rendering to show the reference plane and axes, and even the scanned image you have used as an underlay. A related option allows you to Render Grid And Axes As 3D Lines, which gives them greater weight. The All Windows option applies these settings to all windows of a project.

RenderZone*

RenderZone is a powerful and full-featured rendering environment that will allow you to create photorealistic images with effects such as reflections, refraction, environmental backgrounds, bump mapping, and texture mapping (figure 3-134). The Mac keyboard shortcut for RenderZone is Cmd-K (Ctrl + K in Windows).

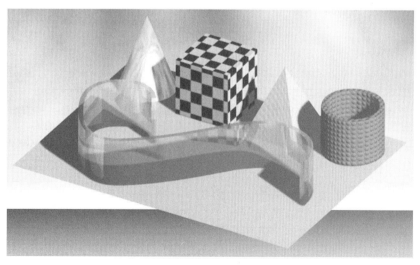

Fig. 3-134. A model displayed with textures in RenderZone.

The RenderZone Options dialog box is accessed by pressing the Option key (Mac) or Ctrl + Shift keys (Windows) while selecting the Render-

Zone* command in the Display menu. The most important option to select when first experimenting with RenderZone is Set Image Size, described in the foregoing "Surface Render" section.

Beginners should not hesitate to use RenderZone right away; even with its default settings, it will give you results that are superior to the other rendering styles. When you learn how to work with texture maps, you will get spectacular results.

◦● **NOTE:** *RenderZone is discussed in detail in Chapter 10.*

QuickDraw 3D*

QuickDraw 3D (QD3D) is a real-time-shading display technology for the Macintosh that lets you view your model with color shading even as you move, scale, and rotate objects, or rotate the view of the model (figure 3-135). The keyboard shortcut for QuickDraw 3D is Cmd-J. OpenGL, the Windows equivalent of QD3D, is discussed in the next section and is now also available for the Mac.

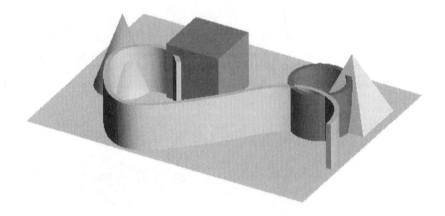

Fig. 3-135. A model shown in QuickDraw 3D. Color shading and intersections are dynamically updated as objects are moved.

The QuickDraw 3D Rendering Options dialog box (figure 3-136) is accessed by pressing the Option key (Mac) or Ctrl + Shift keys (Windows) while selecting the QD3D command in the Display menu. The Rendering Type pop-up submenu offers you the choice of viewing the model in Wire Frame or Interactive Renderer mode. The latter provides the spectacular real-time surface shading that can enhance your design process by letting you see the intersections of forms as you apply various transformations, such as Move, Scale, and Rotate.

Display Menu

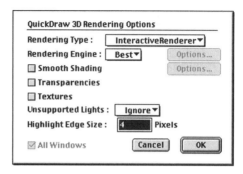

Fig. 3-136. QuickDraw 3D Rendering Options dialog box.

In the Rendering Engine pop-up submenu, Best will choose the most advantageous mode for your system, and the Software option or the Hardware option will force the application of software or hardware acceleration card modes. The Smooth Shading option is identical to its counterpart in the Surface Render command previously discussed. The Transparencies option is applicable only to systems with hardware acceleration graphics cards that support transparency. The QuickDraw 3D software developed for Macintosh computers will not support transparency. The Unsupported Lights option lets you Ignore or Approximate projector, custom, and area lights, which are not applicable in QuickDraw 3D but are supported by RenderZone.

The Highlight Edge Size n Pixels option lets you specify the width of the line used to highlight selections. The Textures option will display image-based textures in real time.

OpenGL*

OpenGL is a display technology originally supported on Windows 95/98 and Windows NT, and now available for the Mac. OpenGL is most effective if you have installed a hardware accelerator card on your PC. The functionality of OpenGL is similar to QD3D, as previously described, and includes a Render With Fog option.

Radiosity Options..., Initialize Radiosity*, Generate Radiosity Solution*, and Exit Radiosity

These four Display menu items concern the Radiosity renderer, discussed in Chapter 10.

Display Options...

The Display Options command allows you to choose your mode of rendering. It brings up the Modeling Display Options dialog box (figure 3-137), the seven main buttons of which provide quick access to the options dialog boxes for all eight Display modes. The modes, previously discussed, are Wire Frame, Quick Paint, Hidden Line, Surface Render, Shaded Render, RenderZone, QuickDraw 3D, and OpenGL. Additional options allow you to control the management of rendering memory in Hidden Line and Surface Render modes.

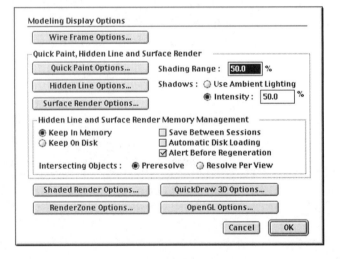

Fig. 3-137. Modeling Display Options dialog box.

Of the other items in the Modeling Display Options dialog box (figure 3-137), the most immediately useful option is Shading Range, which sets the intensity of the color shading on rendered surfaces. A value higher than the default 50% will give you more differentiation between shading of the lightest and darkest surfaces. The Shadows option lets you Use Ambient Lighting or set an Intensity. A value higher than the default 50% setting will create lighter shadow.

Generate Animation...

This command renders the frames generated by the Animation From Keyframes command in the View menu. The Display type and other settings are defined in the dialog box that is invoked by the Generate Animation command. The result of this command is a form•Z animation file that uses the file extension .fan.

➥ **NOTE:** *See Chapter 12 for a complete discussion of animation.*

Play Animation...

This command takes the rendered animation sequence created by the Generate Animation command and plays it back on the form•Z screen.

Draft Layout Mode

This command turns the drafting window into a 2D layout window in which various "panes" or rectangular areas can be created for displaying different parts of a project. This command is grayed out when in 3D mode.

➥ **NOTE:** *The form•Z drafting module is discussed in Chapter 11.*

Clear Rendering Memory...

Clear Rendering Memory clears the memory of information concerning previous rendering operations and gives you a wireframe view. This may improve the performance of your system.

Always Clear Rendering Memory

If Always Clear Rendering Memory is selected, the system will automatically clear the rendering memory whenever a new display mode is chosen. If your system is low on RAM, you should select this menu item.

Show Surfaces As Double Sided

The Show Surfaces As Double Sided command ensures that both sides of one-sided surface objects are visible in all rendering modes. Leave this option in its default (on) position so that you are never faced with the problem of forms that appear to be inside out, or to be missing faces.

Redraw Buffers...

Redraw Buffers brings up the Window Redraw Buffers dialog box (figure 3-138), which provides options for adjusting the way the screen redraws. This involves keeping an image of the form•Z window in memory, or in the buffer. Select Window Redraw Buffer Active and Match Buffer Depth To Screen Depth, but not the Update Screen During Redraw option, if you want to get the best possible screen refresh performance.

Fig. 3-138. Window Redraw Buffers dialog box.

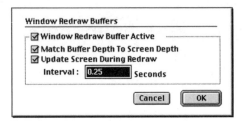

Image Options...

Image Options brings up the Image Options dialog box (figure 3-139), which controls the image size, color depth, and exposure. The size of a rendered image does not have to match the viewing window on the form•Z screen, although that is the default setting. Select the Use Custom Size option, and specify a different format By Size And Resolution, or use the pop-up submenu to define a new image size by number of pixels (figure 3-140).

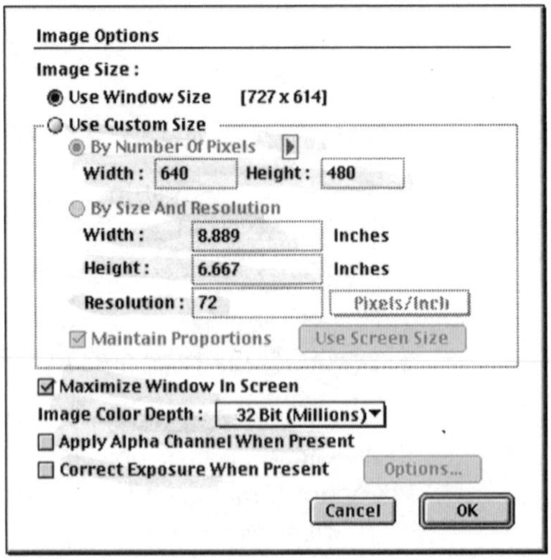

Fig. 3-139. Image Options dialog box.

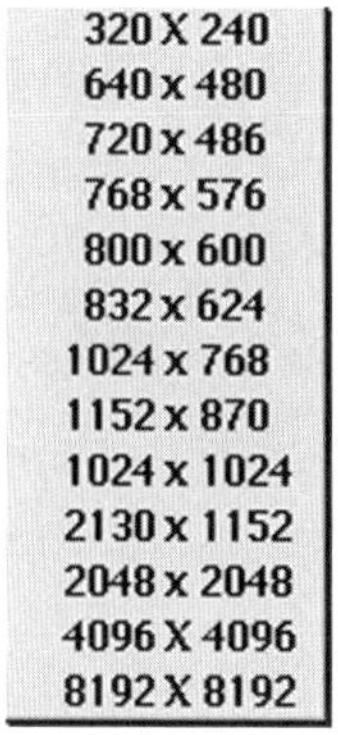

If you have selected the Maintain Proportions option, the aspect ratio of the project window will be preserved no matter what values you type into the Width or Height fields. If you have specified a custom image size larger than the screen, select the Maximize Window In Screen option so that the viewing window will automatically jump to the largest size that fits on the screen. The Image Color Depth option lets you specify 8-, 16-, or 32-bit color. For most renderings, you will want the highest setting, which provides millions of colors.

The last two items in the Image Options dialog box are Apply Alpha Channel When Present and Correct Image Exposure. These deal with advanced rendering issues, which are discussed in Chapter 10.

Fig. 3-140. By Number

Fig. 3-141. The pull-down Options menu.

Options Menu

The Options menu (figure 3-141) provides access to dialog boxes that control a variety of form•Z operations. Because of the built-in redundancy of the form•Z interface, the same dialog box can usually be opened from several different menus, tools, or palettes. In such cases, the text provides a brief discussion and refers you to the appropriate chapter in this book. The Options menu provides a handy means of quickly reviewing the many dialog boxes of form•Z, and can be especially helpful if you cannot remember exactly which Command Tool or Palette contains the dialog box you are looking for.

Input Options...

The Input Options command brings up the Input Options dialog box (figure 3-142), which affects both the modeling and drafting modules of form•Z. The Allow Intersecting Lines option lets you draw self-intersecting shapes. You will appreciate this option if you have noticed the program beeping and displaying an error message as you attempted to draw a line that crossed itself. The Allow Colinear Points option lets you preserve points that fall on the same line. If this option is not selected, points that fall on the same line will be deleted, except for the end points that define the line.

Fig. 3-142. Input Options dialog box.

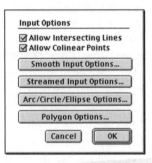

The four major buttons of the Input Options dialog box control the faceting and curve-smoothness settings for the Smooth Line, Stream Line, Arc/Circle/Ellipse, and N-Sided Polygon tools found in the Command Tools. Double clicking on the corresponding tool icon is equivalent to clicking on one of these four buttons, and will bring up the same dialog boxes.

➥ **NOTE:** *These tools and their corresponding dialog boxes are discussed in Chapter 5.*

Pick Options...

The Pick Options command is equivalent to double clicking on the Pick tool in the Command Tools. Either action will bring up the Pick Options dialog box (figure 3-143), which controls selection methods, "nudge" keys, "click and drag" mode, as well as the Frame and Lasso selection options.

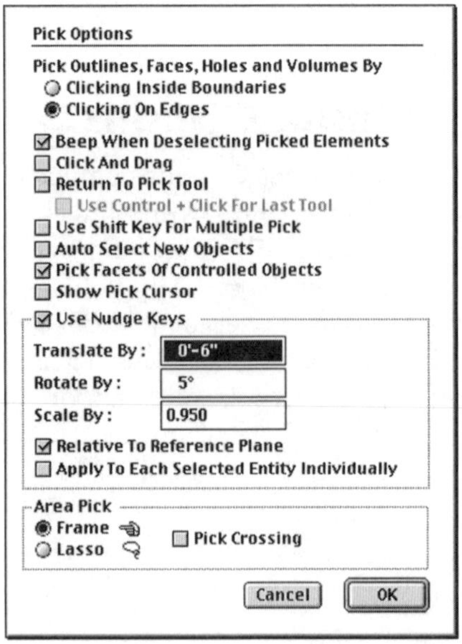

Fig. 3-143. Pick Options dialog box.

Options Menu

⇝ **NOTE:** *Selection methods, nudge keys, "click and drag" mode, and Frame and Lasso selection options are discussed in Chapter 5.*

Zoom Options...

This command brings up the Zoom Options dialog box (figure 3-144), which can also be accessed by double clicking on any of the zoom tools in Window Tools. These options control the geometry and behavior of the zoom and fit operations.

Fig. 3-144. Zoom Options dialog box.

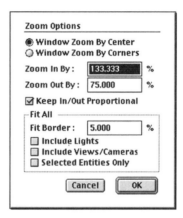

⇝ **NOTE:** *The zoom and fit operations are discussed in Chapter 6.*

Layers...

The Layers command brings up the Layers dialog box (figure 3-145), which allows you to add, load, edit, copy, group, purge, rename, delete, and sort layers (subsets of the model), as well as change their attributes, such as visibility, snapability, and selectability. This dialog box can also be reached by clicking on the word *Name* in the header bar of the Layers palette.

Fig. 3-145. Layers dialog box.

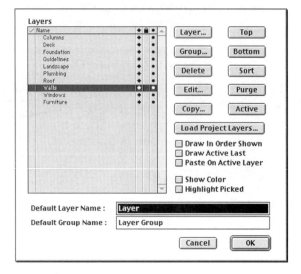

→ **NOTE:** *The Layers palette is discussed in Chapter 4.*

Project Colors...

The Project Colors command brings up the Project Colors dialog box (figure 3-146), which displays the colors assigned to each aspect of the form•Z environment. These include colors for grids, axes, ghosted objects, and the Highlight (selection) color. To change a color, double click on the color rectangle, which will open the Color Picker dialog box (figure 3-147). Use the color circle, the slider bars, or the Hue Angle, Lightness, and Saturation fields to define a custom color, depending on which of the color-definition modes you have selected in the column at the left side of the dialog box. The choices include CMYK, Crayon, RGB, and HTML colors.

Fig. 3-146. Project Colors dialog box.

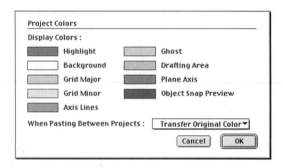

Options Menu

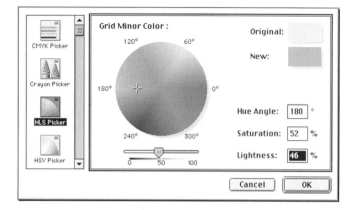

Fig. 3-147. Color Picker dialog box used to change the color of the minor grid lines.

Symbol Libraries...

The Symbol Libraries command brings up the Symbol Libraries dialog box (figure 3-148), which can also be accessed by clicking on the word *Library* in the header bar of the Symbols palette. The dialog box provides methods of viewing, editing, sorting, saving, deleting, duplicating, loading, and unloading symbols.

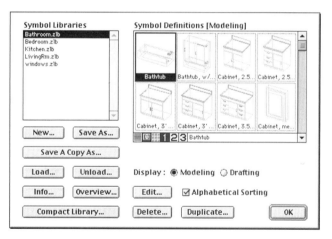

Fig. 3-148. Symbol Libraries dialog box displaying bathroom symbols.

✏ **NOTE:** *The Symbols palette is discussed in Chapter 4.*

Symbols, which are stored in 2D or 3D symbol libraries, are entities drafted or modeled only once, although identical copies of them can be distributed hundreds or even thousands of times throughout your project. Symbols allow you to copy "instances" of an object, rather than reproduce the entire geometry of an object, thus facilitating computation.

The related command tools that allow you to work with symbols are found in the Symbols tool palette. These four tools are Symbol Create, Symbol Place, Symbol Edit, and Symbol Explode.

∞ **NOTE:** *Create Symbol, Place Symbol, Edit Symbol Instance, and Explode Symbol are discussed in Chapter 5.*

Working Units...

The Working Units command brings up the Project Working Units dialog box (figure 3-149), which determines whether English or Metric units are used, and sets numeric accuracy. You can choose English units, with either Inches or Feet as the Base Unit, or Metric units, with Millimeters, Decimeters, Centimeters, or Meters as the fundamental unit.

The value in the Numeric Accuracy field sets the accuracy module used for numeric values entered with the mouse or keyboard. If you are modeling a very small object at scale, enter a small value here to ensure accuracy. For example, the default setting of 1/16" (or 1 mm for metric units) should be replaced by a smaller value, such as 1/64", if you are modeling a tiny piece of jewelry. The Numeric Options button opens up the Numeric Display Options dialog box (figure 3-150), which allows you to choose a style for expressing fractions and to adjust the trailing zeros and number of decimal places.

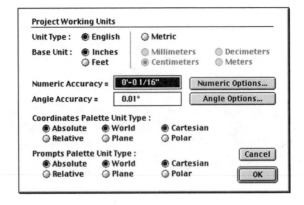

Fig. 3-149. Project Working Units dialog box.

Options Menu

Fig. 3-150. Numeric Display Options dialog box.

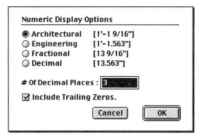

The Angle Accuracy field sets the accuracy of angles in degrees. The Angle Options button displays the Angle Display Options dialog box (figure 3-151), which controls the Angle Display style, Positive Angle Direction, Location of 0° Angle, Measurement Method, and # Of Decimal Places. The remaining items in the Project Working Units dialog box control the display of units in the Coordinates and Prompts palettes.

Fig. 3-151. Angle Display Options dialog box.

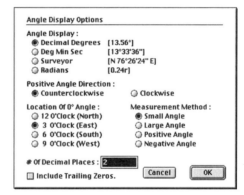

☞ **NOTE:** *The related topic of entering units directly into the Prompts and Coordinates palettes is discussed in Chapter 4.*

Color Palette...

The Color Palette command is inactive and grayed out in all versions of form•Z except the basic, regular version, which does not include RenderZone. Note that in the more advanced versions of form•Z colors are replaced by Surface Styles, which control color and texture. In versions of form•Z where it is active, this command brings up the Modeling Color Palette dialog box and offers a choice of restricted and extended color selections, as well as a Load button for importing colors from a saved project.

Lights...

The Lights command opens up the Lights dialog box (figure 3-152), which can also be accessed by clicking on the words *Light Name* in the header bar of the Lights palette. This dialog box allows you to create, delete, edit, copy, sort, clear, and load various types of lights, as well as to adjust the color and intensity of the ambient light.

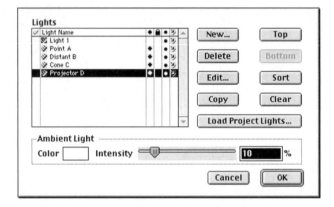

Fig. 3-152. Lights dialog box.

➛ **NOTE:** *The Lights palette is discussed in Chapter 4. The use of lights in rendering is described in Chapter 10.*

Macro Transformations...

This item in the Options menu brings up the Macro Transformations dialog box (figure 3-153), which is also accessed by double clicking on any of the three Macro Transformation icons in the Geometric Transformations tool palette. A macro is a recording of a series of transformations to an object that can be applied to any other object.

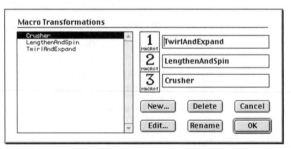

Fig. 3-153. Macro Transformations dialog box.

Macros are ideal for creating shapes (such as spirals) that require repetitive move, rotate, and scale operations. To start recording a series of operations as a macro, select an object and then click on the Define Macro tool in the Self/Copy tool palette. Any subsequent Move, Rotate, Scale, and Mirror procedures will be recorded. Terminate the recording operation by selecting any other icon in the Self/Copy tool palette. The Macro Transformations dialog box provides tools for listing, creating, editing, deleting, and renaming macros; you can create and save many macros but only three macro shortcuts can be displayed in the Geometric Transformations box at any one time.

➥ **NOTE:** *Macros are discussed in Chapter 5.*

Objects...

The Objects command brings up the Objects dialog box (figure 3-154), which allows you to create, delete, edit, copy, sort, and purge objects and groups, as well as to adjust their attributes, such as visibility, snapability, selectability and default names. The default location of each object's automatically recorded "object axis" can be set here, as well as the way the object's centroid is calculated. This dialog box can also be opened by clicking on the word *Name* in the header bar of the Objects palette.

Fig. 3-154. Objects dialog box.

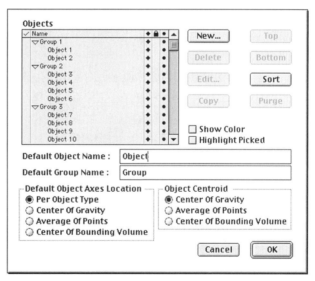

➥ **NOTE:** *The Objects palette is discussed in Chapter 4.*

Chapter 3: Menus

Profiles...

The Profiles command brings up the Profiles dialog box (figure 3-155), which allows you to name, resize, and delete profiles. These are 2D shapes you can use with the Deform tool to reshape meshed surfaces. Profiles are created with the Define Profile tool in the Meshes tool palette. The Profiles dialog box can also be opened from the Profiles palette by holding down the Option key (Mac) or Ctrl + Shift keys (Windows) while clicking on the icon of a profile.

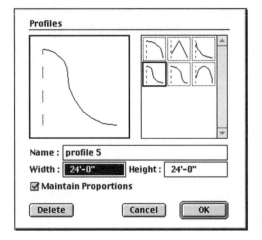

Fig. 3-155. Profiles dialog box.

> **NOTE:** *The Meshes tool palette is discussed in Chapter 5. The Profiles palette is discussed in Chapter 4.*

Reference Planes...

The Reference Planes command opens the Reference Planes dialog box (figure 3-156), which lists the reference planes, or active drawing surfaces, you have saved. Command buttons are provided to save, delete, rename, load, sort, and clear reference planes. This dialog box can also be opened by clicking on the words *Plane Name* in the header bar of the Planes palette.

Options Menu

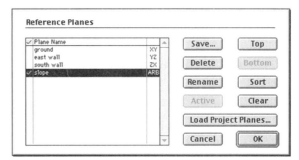

Fig. 3-156. Reference Planes dialog box.

Reference Planes tools allow you to define an arbitrary reference plane other than the standard Cartesian planes. The reference plane commands in Window Tools let you quickly switch between standard planes.

➥ **NOTE:** *The Planes palette is discussed in Chapter 4. The Reference Planes command tools are discussed in Chapter 5. The reference plane commands in Window Tools are discussed in Chapter 6.*

Status Of Objects...

Status Of Object brings up the Status Of Objects dialog box (figure 3-157), which controls the way objects behave during certain form•Z operations, such as Boolean Union and Difference. The options Keep, Ghost, and Delete determine whether objects used in these operations are preserved, ghosted, or erased, and whether new objects can be composed of multiple volumes. These settings can be applied globally to all commands, or locally to particular commands, by means of the pop-up menu (figure 3-158) at the top of the Status Of Objects dialog box.

Fig. 3-157. Status Of Objects dialog box.

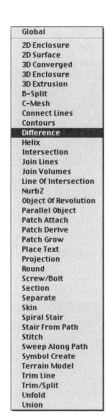

Fig. 3-158. Status Of Objects pop-up menu.

Surface Styles...

The Surface Styles command opens the Surface Styles dialog box (figure 3-159), which can also be accessed from the Surface Styles palette by holding down the Option key (Mac) or Ctrl + Shift keys (Windows) while clicking on the icon of one of the color or texture surface styles.

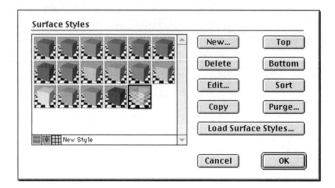

Fig. 3-159. Surface Styles dialog box.

This dialog box lets you add, delete, edit, copy, sort, purge, and load surface styles. Note that the New and Edit buttons bring up the Surface Style Parameters dialog box, which contains settings for color, reflections, transparency, and bump maps. This is the same dialog box accessed by double clicking on one of the icons in the Surface Styles palette (without holding down the Option key).

✒ **NOTE:** *Surface Styles are available in RenderZone only. The Surface Styles palette is discussed in Chapter 4. The Color tool, used to apply surface styles to an object, is found in the Attributes tool palette, which is discussed in Chapter 5. The use of surface styles in rendering is discussed in Chapter 10, and is demonstrated in the Coin, Paper Clip, and Ice Cream Cone exercises in Chapter 7.*

Color Palette..., Line Styles..., and Line Weights...

The Color Palette, Line Styles, and Line Weights menu items are inactive and grayed out in the 3D modeling module of form•Z. They control drafting colors, line weights, and line types, and apply only to the 2D drafting module.

✒ **NOTE:** *The 2D drafting module is discussed in Chapter 11.*

2D and 3D Digitizer Options

Six menu items deal with the use of 2D and 3D digitizers for graphical input to form•Z: Digitizer Setup, Screen Digitizer Calibration, Screen Digitizer Input, World Digitizer Calibration, World Digitizer Input, and Capture Digitizer Mesh. Digitizer Setup opens the Digitizer Setup dialog box (figure 3-160), which lets you control the parameters of the digitizer interface. Screen Digitizer Calibration adjusts the digitizer's workspace to the form•Z screen with the help of the Screen Digitizer calibration dialog box.

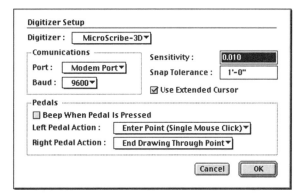

Fig. 3-160. Digitizer Setup dialog box.

Screen Digitizer Input activates the digitizer for graphical input, allowing it to control the position of the cursor on the form•Z screen. World Digitizer Calibration adjusts the digitizer's workspace to the project's modeling environment. This calibration can be configured in three ways: reference point and scale, reference plane and scale, or three points. World Digitizer Input activates World Digitizer mode so that the digitizer controls all graphic input in the 3D modeling space. Capture Digitizer Mesh creates a mesh from a sequence of objects drawn with the digitizer.

Palettes Menu

Palettes are the small, adjustable control panels that appear to float on the form•Z screen. They always remain visible unless they are purposely hidden, because they cannot be obscured by modeling windows. Palettes provide quick access to a variety of essential controls. The form•Z menu structure reserves an entire pull-down menu for palettes and provides a quick interface for making any or all of them disappear or reappear. Figure 3-161 shows the Palettes menu in its default state, with visibility check marks beside 9 of the 13 palettes. Note the

Chapter 3: Menus

Fig. 3-161. The pull-down Palettes menu.

Customize Tools item at the bottom; this is a convenient new feature of form•Z version 3.0.

NOTE: *Because of their importance in the form•Z modeling environment, Chapter 4 is devoted to the palettes.*

The first 14 items in the Palettes menu are toggle switches for those commands: Animation, Coordinates, Layers, Lights, Objects, Planes, Profiles, Prompts, Surface Styles, Symbols, Tool Options, Views, Window Tools, and Hide Palettes. By selecting them, you change their position to on or off, depending on their current state. A check mark indicates that a palette is active, and visible on the screen. Selecting the Hide Palettes item makes all of the palettes temporarily disappear from the form•Z window; selecting it again brings them back.

Customize Tools...

This menu item invokes a large dialog box that allows to you customize the modeling, drafting, and window tools of form•Z (figure 3-162). Once they have been modified, the new toolbar settings are retained for the next session and can be saved. Customization can include the following: removing and including tools, rearranging the tool palettes, changing the number of columns in the toolbar, and selecting different icons (figure 3-163).

Fig. 3-162. Customize Tools dialog box with standard gray icons.

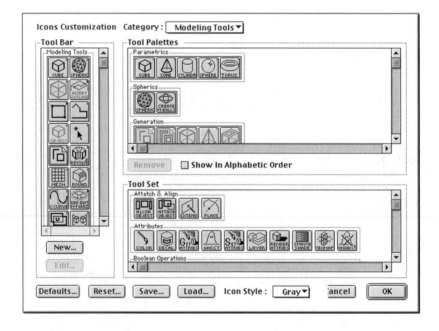

Help Menu

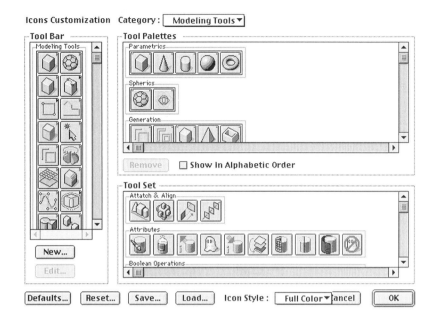

Fig. 3-163. Customize Tools dialog box showing the new optional color icons.

Help Menu

Fig. 3-164. The pull-down Help menu.

At the far right end of the form•Z menus is the Help menu (figure 3-164). This pull-down menu provides access to brief text descriptions of the program's features. The most important items for the beginner are the three main components of the form•Z interface: Modeling Tools, Menus, and Window Tools. The Help menu includes the items General, Introduction, Keyboard, Menus, Modeling Tools, Drafting Tools, and Window Tools, as well as two additional items discussed in the material that follows.

➥ **NOTE:** *The first seven items in the Help menu are discussed in Chapter 2.*

Error Messages...

The Error Messages command brings up the Error Messages dialog box (figure 3-165), which lists the error statements form•Z has issued in the current session. A scroll bar appears at the right side of the dialog box when the text exceeds the length of the box. The Save button lets you write the list to your hard drive as a text file. The Clear button deletes the list.

Fig. 3-165. Error Messages dialog box.

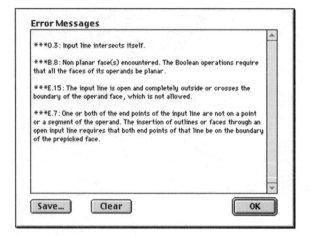

Project Info...

Project Info opens the Project Info dialog box (figure 3-166), which provides various statistics about the current project. The amount of free memory is indicated at the bottom left corner. The Topology Count in the Modeling section is especially useful for keeping track of the complexity of your model.

Fig. 3-166. Project Info dialog box.

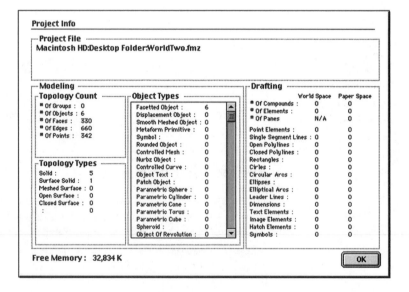

chapter 4

Palettes

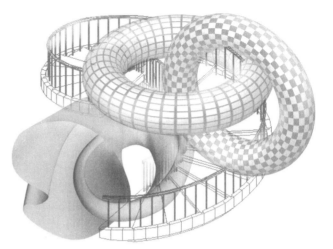

Introduction

When you open the form•Z program, you will notice nine small windows, or palettes, distributed inside and around the main modeling window. These palettes, which can be dragged to any location on your form•Z window, are convenient control panels for particularly important aspects of the form•Z working environment, such as visibility and control of groups and layers. They are sometimes called "floating" palettes because they sit above the modeling window and always remain visible. The Palettes are such a vital component of the form•Z interface that this entire chapter is devoted to them. Note that the Customize Tools item was discussed in Chapter 3 as part of a brief review of the pull-down Palettes menu.

The form•Z palettes vary in usefulness depending on the task at hand and your personal modeling style; it is quite possible to do significant form•Z work without any of the palettes being visible. Explanations of the palettes are presented in this chapter in the order of their importance. Naturally, every expert will have a different opinion about any such ranking. The order here is designed from the point of view of the beginner, based on classroom-teaching form•Z experience.

Many of the palettes are redundant in the sense that the information or controls they provide are available through other means. For example, the Layers palette is a shortcut to information provided by the Layers command in the pull-down Options menu. Furthermore, there are two Set Layer tools, one for modeling and one for drafting. For this reason, certain discussions in this chapter are brief, referencing more detailed information elsewhere in the book.

Note that the form•Z drafting module includes palettes for hatching, line styles, and line weights that do not apply to the 3D modeling environment. Drafting is discussed in Chapter 11. Also note that the 2D module does not include the following palettes: Lights, Objects, Planes, Profiles, and Surface Styles. In addition, the Surface Styles palette applies only to form•Z RenderZone; in the basic version of form•Z it is replaced by the Colors palette.

Palettes and Screen Size

Figures 4-1 and 4-2 show two different arrangements of palettes, which are the default settings for computer screens of 800 x 600 (or 832 x 624) and 1024 x 768 pixels, respectively. The older, smaller 640- x 480-pixel screen setting is not recommended for form•Z. In each illustration, a bulky model of a house is shown in the same perspective view to demonstrate the fact that the best configuration of the palettes may depend on your screen size.

With the smaller screen size, such as 800 x 600 pixels (figure 4-1), parts of the model may be hidden by the floating palettes. This is because palettes are designed at a fixed size measured in pixels. If the screen is smaller, the palettes do not shrink accordingly; they take a larger proportion of the screen area and hide relatively larger parts of your model. If your modeling window is obscured by palettes, as shown in figure 4-1, you should reduce its size, as shown in figure 4-3, or, if possible, adjust the monitor to a higher resolution, such as 1024 x 768 pixels, as shown in figure 4-2. Note the Macintosh system's Help menu at the top of the screen, beside the form•Z Help menu.

Palettes and Screen Size

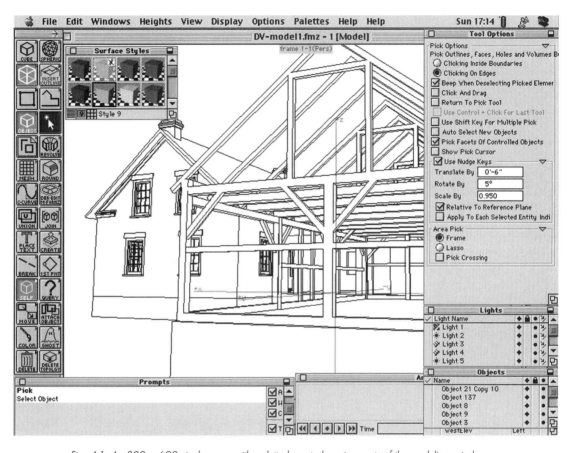

Fig. 4-1. An 800- x 600-pixel screen with palette layout obscuring parts of the modeling window.

Chapter 4: Palettes

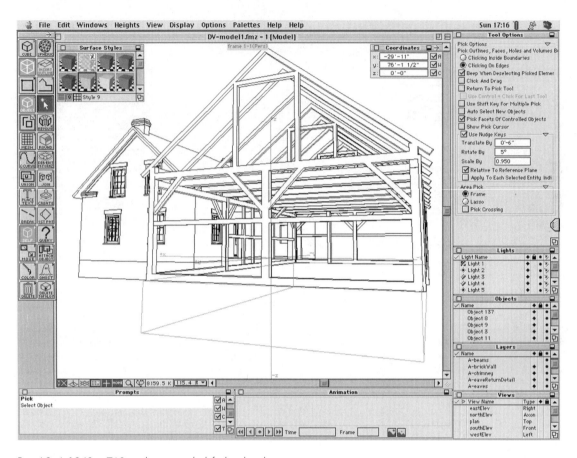

Fig. 4-2. A 1040- x 768-pixel screen with default palette layout.

When form•Z senses a large-size screen, such as 1024 x 768 pixels, it will arrange the palettes around the modeling window, as shown in figure 4-2. A higher-resolution monitor will be even more convenient.

Nevertheless, you may find that the Surface Styles or Coordinates palettes obscure important areas of the modeling window. In this case, you will probably want to remove or reposition certain palettes, depending on your work style. To give you a hint as to which palettes you may or may not need, they are discussed in this chapter in order of importance. You should know which palettes are essential, and you should be familiar with ways of rearranging and minimizing the palettes you decide to keep on your screen.

Manipulating Palettes

The palettes in form•Z can be a wonderful convenience, especially if you have a large monitor. The secret of using the palettes to increase your productivity is to learn how to manipulate them quickly and easily. At times, palettes will block the view of objects you are modeling. You need to be able to collapse, close, and hide palettes.

Similarly, you must know how to bring palettes back, expand them, and rearrange them to suit your needs. Just as important is the knowledge of which palettes to ignore completely in the early stages of your form•Z training. These issues are discussed in the material that follows, which ranks the palettes according to their usefulness to the beginner.

Rearranging Palettes

All palettes can be moved to a different location by dragging the gray bar at the top of the palette. If necessary, the modeling window can be resized and moved to accommodate the palettes. Figure 4-3 shows an 800- x 600-pixel screen; the modeling window was made smaller and all the palettes, except for surface styles, were positioned outside it. This arrangement is better than that shown in figure 4-1 because of the improved visibility of the modeling window.

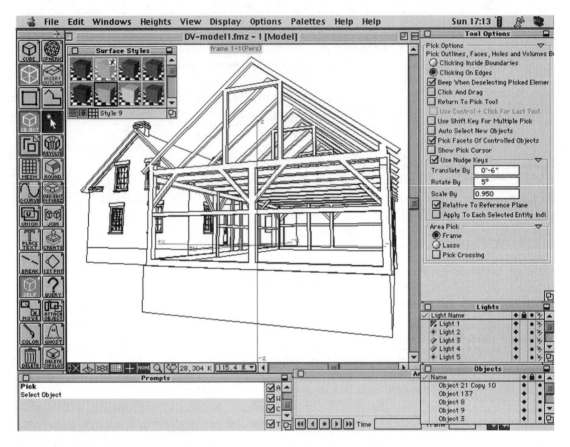

Fig. 4-3. An 800- x 600-pixel screen with palettes arranged around a reduced modeling window.

Collapsing Palettes

The small black-and-white rectangle in the upper right-hand corner of each palette allows you to minimize or collapse the palette, leaving just the gray header bar visible on the screen. This is convenient when you are running out of space on screen but do not want to go to the trouble of removing and later retrieving each palette. Figure 4-4 shows a custom arrangement that provides the largest possible 3D work area: a full-screen modeling window with nine visible palettes minimized.

✓ **TIP:** *It is best to position the minimized palettes near the top of the screen; if you place them at the bottom, they may extend beyond the edge of the screen when expanded.*

Manipulating Palettes

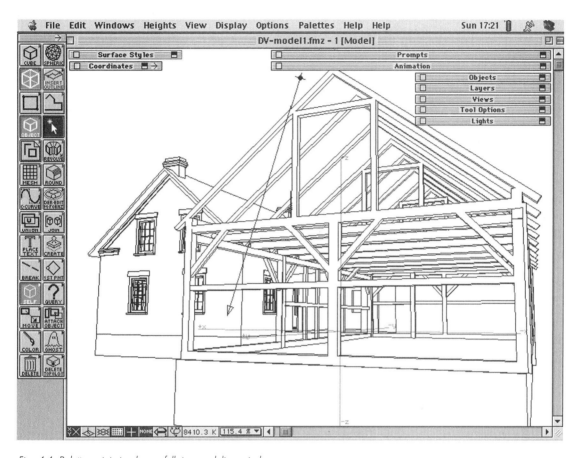

Fig. 4-4. Palettes minimized on a full-size modeling window.

Closing and Opening Palettes

Each or all of the palettes can be easily closed by clicking on the Close box at the top left-hand corner of the palette. If you hold down the Option key (Ctrl + Shift in Windows) while you do this, all of the open palettes will close. You can also close palettes by deselecting them in the pull-down Palettes menu. Note that closing palettes is not the same as hiding them, which is discussed in the next section.

To recover one the palettes, select the palette name in the pull-down Palette menu at the top of the screen. There are thirteen palettes listed in the menu, but only nine show on the screen by default. The four that do not appear automatically are Planes, Profiles, Symbols, and Window Tools.

Hiding Palettes

The Hide Palettes option is found in the pull-down Palettes menu at the top of the screen. Hiding the visible palettes is not the same as closing them. This option hides, but does not shut down, the palettes that are open. The fact that they are hidden is then indicated by a check mark beside Hide Palettes on the pull-down Palettes menu (figure 4-5). Select the same option to make the hidden palettes become visible again. This is a convenient way to temporarily clear the modeling environment of palettes when you want to improve the visibility of your model or of a full-screen rendered image.

Fig. 4-5. Pull-down Palettes menu.

Expanding Palettes

The Resize box at the bottom right-hand corner of some palettes allows you to expand or contract the palette. This is especially useful when working with palettes that contain expandable lists of information, such as the Layers and Views palettes. For long lists, use the vertical scroll bar at the right side of the palettes. Note, however, that the Window Tools, Animation, and Coordinates palettes do not expand in this manner.

Transferring Palettes

During any significant form•Z project, you will probably store a lot of valuable information in the various palettes: your favorite camera angles, surface styles, lights, and so on. You can transfer this accumulated wisdom to another project. To do this, you could duplicate your entire form•Z file, erase the model, and start a new model, thereby retaining the content of your carefully named Layers palette (or well-chosen set of drawing surfaces in a Planes palette). However, there is a more elegant way to transfer information, transferring between projects: using the Load From Project button, found in the dialog boxes of most of the palettes. This method enables you to grab the content of a palette saved in another file and apply it to an active project, by means of the standard Open File dialog box.

Ranking Palettes

You will find that some of the palettes are essential to almost every operation you perform as a beginner, and others seldom used. The palettes are rated in the following list from most to least commonly used, from the beginner's perspective. Experts will have their own ideas about the frequency of use of each palette, based on their working style and the type of project they are doing. For example, for certain projects involving repetitive predesigned components, the Symbols palette would be on the screen at all times, but for other projects it might never be needed. Note that the essential Animation palette is placed last, only because it is assumed that you will learn modeling before attempting animation.

1. Prompts
2. Tool Options
3. Views
4. Colors/Surface Styles
5. Layers
6. Lights
7. Coordinates
8. Objects
9. Planes
10. Profiles
11. Symbols
12. Window Tools
13. Animation

In contrast to this ranking by importance to the beginner, figure 4-5 shows the actual order of the palettes in the pull-down Palettes menu.

The Prompts Palette

The Prompts palette is by far the most significant for the beginner. This floating window is your guide and teacher as you attempt to master form•Z. The Prompts palette does what it says: it prompts you through numerous tasks. Get in the habit of reading each line of text as it appears. These lines explain which command you are using and what step comes next. For example, if you were attempting to draw a cube with the Heights command set at Graphic/Keyed, you would be guided by the following prompts.

```
First Corner Point (x,y,z)
Second Corner Point (x,y,z)
```

As you click the mouse to locate the two opposite corners of the cube, the Prompts window will tell you exactly what points you have chosen. Figure 4-6 shows what the Prompts palette would look like if you drew a single 8-foot cube with one corner at 0,0. Figure 4-7 shows an expanded Prompts palette, which reveals a longer trail of recent commands. This may be useful when you want to review exactly which commands you have executed.

Fig. 4-6. Generation of an 8-foot cube as documented in the Prompts palette.

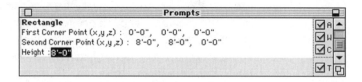

Fig. 4-7. An expanded Prompts palette showing recently executed commands.

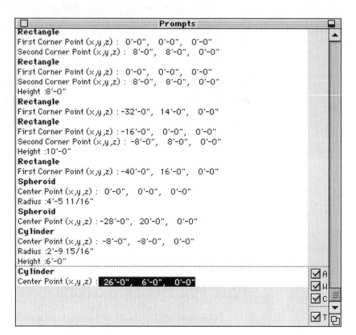

Note that as soon as any form•Z command is selected, its name appears in bold in the Prompts palette. In the previous illustration, for example, the command Extrude Rectangle is displayed. This is an excellent way to learn the full names of the commands, and to become familiar with their functions. Get into the habit of watching the Prompts window carefully; it is a real-time transcript of every operation you perform.

- **NOTE:** *The Prompts window also performs an unusual archival function. In the Preferences item in the Edit menu, you will find an option labeled Save Prompts in TEXT. This option allows you to save the entire history of your commands. See Chapter 3 for more information on this option.*

The four small boxes on the right side of the Prompts palette can be very useful. The A beside the top box indicates absolute units, which are locations expressed in relation to the origin of the coordinate system. If this is not selected, the Prompts window will display relative coordinates, which are distances relative to the previously entered point. This means that, for example, you can see the relative length of a rectangle as you draw it, rather than the absolute value of the coordinates at the corner of the rectangle.

Note that the Coordinates palette duplicates some of the features of the Prompts palette. Therefore, by specifying relative coordinates in one palette, and absolute in the other, you can get the best of both worlds. The Coordinates palette is further discussed in material that follows.

The W beside the second small square box in the Prompts palette stands for World coordinates. This means that the palette will display coordinates in reference to the original reference plane, no matter what temporary or arbitrary plane you are drawing on at that time. However, there are many cases for which this coordinate option would not be appropriate; you would therefore deselect (uncheck) this box. For example, if you were positioning a skylight on a sloping roof, you might be more interested in the local coordinate system of the roof surface than the World coordinate system that indicates the ground plane.

The C beside the lower-most small box allows you to switch between Cartesian coordinates (*xyz* values) and polar coordinates (length and angle). The difference is immediately visible in the Prompts window as soon as you draw a line. With Cartesian coordinates, the prompt will display the three values that define the location of the cursor, such as the following.

```
Point #3 (x,y,z) : 8'-0", -40'-2", 0'-0"
```

However, if you specify polar coordinates, the prompt will list distance (dist) and angle (ang), such as the following.

```
Point #4 (dist,ang,z) : 74'-1 5/16", -24.26 , 0'-0"
```

The box labeled T controls mouse tracking. If you deselect this option, the dynamically changing mouse input will not be printed in the Prompts window as the cursor moves.

In fact, you can actually model in form•Z by manually entering all locations, radii, and lengths directly in the Prompts window, but you still have to use the mouse to select the desired commands. In practice, this approach is too slow to be used except in rare instances in which accuracy is critical and all locations and dimensions are already known, documented, and ready to be typed in.

When executing a command (such as Vector Lines) that usually requires a double clicking to end the line, or a triple click to close the shape, you can type an *e* or a *c* in the Prompts window. Moving from graphical to numeric input in form•Z is as easy as moving the cursor within the confines of the Prompts window. Over time, you will probably find it is most efficient to mix these methods. For example, you may define the center of a circle interactively with the mouse, then enter a precise radius value in the Prompts palette.

The Tool Options Palette

The Tool Options palette is one of the most important new interface elements in form•Z version 3.0. This palette provides information previously accessed through the dialog boxes invoked by double clicking on each tool icon. Compared to previous versions of form•Z, the Tool Options provide a much greater level of convenience during the modeling process. However, this additional panel duplicates information found elsewhere. In this sense, the Tool Options palette adds another level of redundancy to the form•Z interface.

You may prefer the legibility of the old option dialog boxes, and not having to scroll down the long, narrow Tool Options palette to find the right setting. If so, you can choose to ignore, or to close, the Tool Options palette, and continue to double click on tool icons in order to change their option dialog boxes. However, it is recommended that you use the Tool Options palette because it constantly and instantly displays the appropriate tool options without any further effort on your part.

The various sections of the Tool Option palette, indicated by small triangles, can be closed or opened, creating various layouts. Figures 4-8 through 4-10 illustrate the Tool Options palette displayed when the 3D Enclosure modifier is selected and the Polygon tool is selected. The equivalent option dialog box for the Polygon command is shown in figure 4-11. The settings accessed by the two methods are not always identical; the Tool Options palette includes the 3D enclosure options if that modifier has been selected (compare figures 4-10 and 4-11).

> **NOTE:** *It would require an entire book to show all combinations of Tool Options palettes; there is one for each item in the modeling and drafting toolbars, and each of these can be configured in a variety of ways. Also note that because there is no proper scroll bar in the Tool Options palette, you scroll by dragging the mouse up or down in the palette.*

Manipulating Palettes

Fig. 4-8. Tool Options palette for the Polygon tool with 3D Enclosure mode selected.

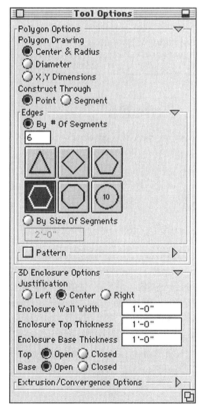

Fig. 4-9. Tool Options palette for the Polygon tool with additional sections visible.

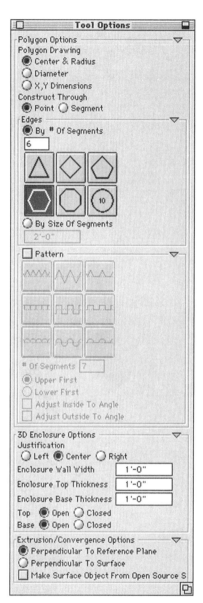

Fig. 4-10. Tool Options palette for the Polygon tool, extended with all sections visible.

Fig. 4-11. Alternative to the Tool Options palette: double click on each tool to invoke its options dialog box.

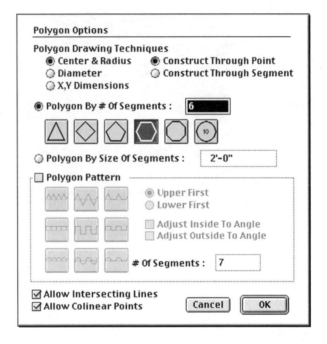

The Views Palette

The Views palette is essential to both the modeling and presentation stages of a project. By letting you quickly name, store, and retrieve favorite views, the Views palette gives you access to the full 3D power of form•Z. Furthermore, the Views palette is now an integral part of the animation capability that has been added to form•Z version 3.0.

↦ **NOTE:** *See Chapter 12 for a complete discussion of animation.*

The Views palette has six columns. The first column allows you to pick a view. A check mark indicates the selected view and highlights its graphic representation in the modeling window. The second column indicates the active view, the view that you are actually seeing in the modeling window. The third column displays the name of the view. A view-type code is displayed to the right of the view name (figure 4-12) so that you can distinguish, for example, axonometric (Axon) views—such as Top, Side, and Front from perspective (Pers) views.

The fourth column displays a small diamond shape that controls the visibility of the representation of the picture plane and virtual camera for each view. A click of the mouse allows you to rotate among visibility (black diamond), ghosted (white diamond), or invisible (no diamond)

Manipulating Palettes

settings. Seeing images of cameras that correspond to each saved view will be useful when setting up animations. The sixth column in the Views palette allows you to lock views so that they are not selectable.

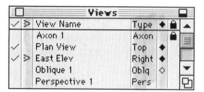

Fig. 4-12. Views palette with five user-defined views.

Double clicking on the View Name label will bring up the Views dialog box, shown in figure 4-13. This box redisplays the current saved views exactly as they appear in the Views palette, but also allows you to Save, Delete, Edit, Copy, Sort, Clear, and Load views. There are also Top and Bottom buttons for moving the selected view to the top or bottom of the list of sorted views.

⇒ **NOTE:** *For further discussion of the concept of Views in form•Z, see Chapter 3. For the role of the Views palette in animation, see Chapter 12.*

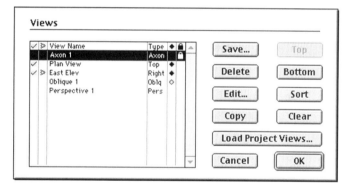

Fig. 4-13. Views dialog box.

The Colors/Surface Styles Palette

The regular version of form•Z and the drafting module contain a Colors palette. RenderZone, on the other hand, contains a more advanced Surface Styles palette, which controls color and texture. Both palettes are briefly described here.

⇒ **NOTE:** *The use of the Surface Styles palette in rendering is described in detail in Chapter 10.*

The Colors palette appears by default in the upper left-hand corner of the screen and displays 16 colors in small squares. It can be moved or hidden like any other palette. If your monitor can support more than 8-bit (256) color, you can resize the palette and display up to 240 colors, 16 of which are reserved for the system itself.

In both the Colors palette and the Surface Styles palette, the active color is indicated by the presence of a black outline around the color square (figure 4-14). All objects are drawn in the active color. You must click on any other color square if you want to select a new color. Colors of previously drawn objects can be changed with the Color tool in the Attributes menu.

•◦ **NOTE:** *The Attributes menu is discussed in Chapter 5.*

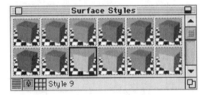

Fig. 4-14. Default Surface Styles palette.

From the Colors palette you can redefine a new color by double clicking on an existing color square. The Edit Color dialog box will appear, offering a variety of ways to define colors, such as CMYK, HLS, and Crayon (figure 4-15). The same dialog is available through the Surface Styles palette by clicking on a surface style, then on the Color Option button. You can then customize the Color or Surface Styles palette by clicking on the color wheel, choosing a color crayon by moving the horizontal slider, or by specifying Hue Angle, Saturation, and Lightness.

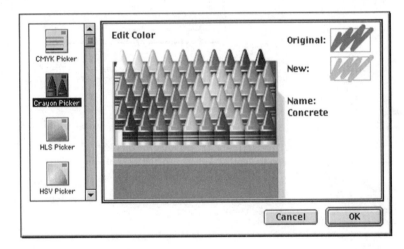

Fig. 4-15. Color Picker dialog box.

Manipulating Palettes

The Surface Styles palette allows you to control textures as well as colors, and is an essential tool when using RenderZone. The Surface Styles palette is, at first glance, similar to the Colors palette, but it also includes controls for texture maps, reflections, transparency, refraction, and bumps.

The Surface Styles palette displays the familiar title bar, close box, resize box, scroll bar, and hide box common to most palettes (figure 4-16). At the bottom of the Surface Styles palette are three small icons that, from left to right, allow you to view the styles by name, small icon, or large icon. The icons allow you to see a preview of the resulting style, and can show effects such as transparency, in addition to color.

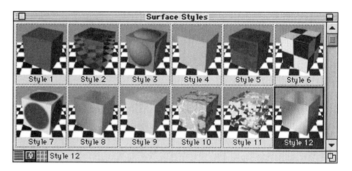

Fig. 4-16. A customized Surface Styles palette, with the large icon setting.

To edit a surface style, double click on the icon to bring up the Surface Style Parameters dialog box (figure 4-17). In this dialog box, you can click on Options in order to change the basic color or apply a texture map, Reflection for mirror effects, Transparency for see-through effects, and Bump for a variety of textured surfaces.

> **NOTE:** *The use of Surface Styles in rendering is explained in detail in Chapter 10.*

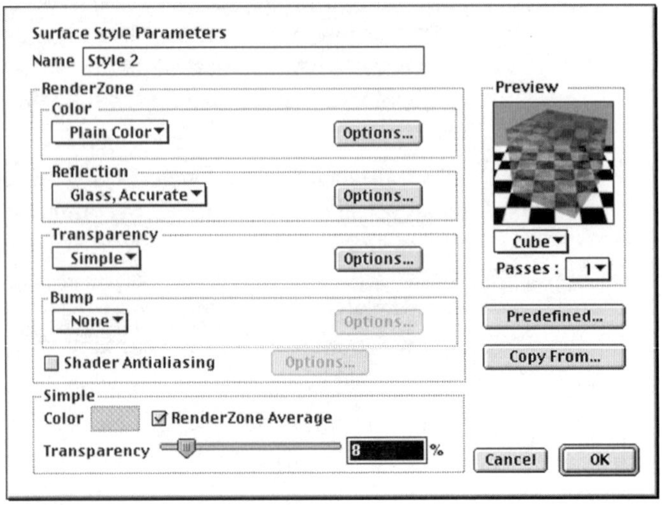

Fig. 4-17. Surface Style Parameters dialog box.

The Layers Palette

The Layers palette is essential to any serious modeling project. Layers are designated subsets of your entire model and are categorized to help organize a large project. For example, you might divide a model of a house into layers for Columns, Stairs, Walls, Windows, and Slabs. The Layers palette is a convenient interface for manipulating and keeping track of the various components and subsets of your model as you build it, and again later as you prepare for final renderings. form•Z version 3.0 has introduced Layer Groups, a convenient way of arranging layers in natural groupings, as shown in figure 4-18.

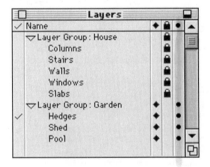

Fig. 4-18. Layers palette showing two Layer Groups.

The Layers palette displays the typical form•Z palette structure: gray header bar at the top, with a close box and a hide box, a scroll bar at the right, and a resize box at the bottom right. The Layers palette is divided

Manipulating Palettes

into five columns: (1) the status column, which shows which layer is active; (2) the layer and layer group names; (3) the visibility of the layer, which can be hidden, ghosted, or visible; (4) the selectability of the layer, which can be locked or unlocked; and (5) the "snapability," which can be on or off.

The Layers dialog box (figure 4-19) is invoked in one of two ways: by double clicking on the title Layer Name in the Layers palette, or via the Options menu. The Layers dialog box is similar in structure to that of the Views or the Lights dialog box: it repeats the content of the Layers palette on the left side while on the right side providing tools for creating and managing Layers.

✓ **TIP:** *When you turn on or turn off the visibility of multiple layers, the program always processes one layer at a time and then refreshes the screen in between. This refresh can be time consuming with complicated models. To turn off multiple layers without the screen refreshing, hold down the Shift key while toggling the Visibility option in the second column of the Layers palette. If you continue holding down the Shift key, you can turn off several layers without the screen redrawing.*

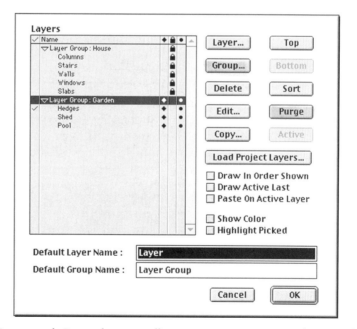

Fig. 4-19. Layers dialog box.

The Layer and Group buttons allow you to create a new layer or layer group. The Delete and Purge buttons allow you to remove a selected layer or all of the empty layers. The Edit button brings up the Layer Attributes dialog box, which allows you to change various settings for an entire layer or layer group. The Layer Attributes are off by default until you place a check mark in the box beside them. For example, if the option Surface Style is selected, you can assign a "layer" color to all objects in a layer, regardless of their previous "object" color (figure 4-20). Other attributes that can be globally changed in this way include Shadow Casting, Shadow Receiving, and Smooth Shading. This is a very powerful feature that will be of great utility in managing large and complex modeling projects.

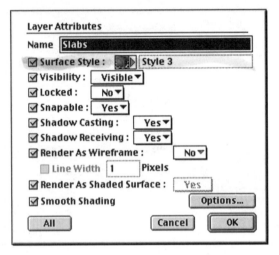

Fig. 4-20. The Layer Attributes dialog box lets you set attributes, such as color, for an entire layer.

Three more Layer dialog box buttons (Top, Bottom, and Sort) let you rearrange the list of layers. Copy will duplicate a layer without copying the actual objects on the original layer. Load Project Layers grabs the layer settings from a specified project. An important option is Paste On Active Layer, which will ensure that objects used in cut-and-paste operations are deposited on the current layer. The Show Color option will display the text of the Layers palette labels in the actual colors that have been globally set for them. This may be confusing if the colors match the gray of the palette background (figure 4-21).

Manipulating Palettes

Fig. 4-21. The text labels in the Layers palette can display the colors set for each layer.

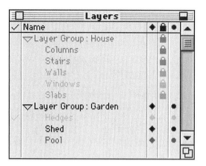

Layers and layer groups can be created directly from the Layers palette without opening the Layers dialog box. Click on the white space under the last layer name and type in the name of a new layer. Do the same while pressing the Ctrl key (Mac) or Ctrl + Alt keys (Windows) to generate a layer group. To move layers into a layer group, drag the label name on top of the layer group name.

The careful use of layers distinguishes expert modelers from beginners. A little extra time taken at the outset of a project to organize a model carefully, and to select and name layers logically, will translate into great time savings later. Expert form•Z users constantly refer to the Layers palette, switching visibility, selectability, and snapability on and off as the modeling activity progresses. The use of layers becomes even more important when adding lighting and texture maps and preparing for final renderings. By suppressing certain layers, the interactive rendering and testing cycle can be speeded up considerably.

NOTE: *See also the Layers tool, discussed in Chapter 5.*

The Lights Palette

The Lights palette is a valuable tool as you move into the rendering stage of a project in which you must repeatedly add and adjust lights to achieve a desired visual effect. The Lights palette gives you a handy seven-column list of all lights in your scene. These columns indicate (1) which light is selected, (2) the icon of the light type (that is, Distant, Point, Cone, Projector, Area, or Custom), (3) the light name, (4) the visibility of the light, (5) the selectability of the light, and (6) whether the light is shining.

✓ **TIP:** *Option click (Ctrl + Shift and click in Windows) on the title of a column to change the status of all lights in that column.*

Double click on the title Light Name at the top of the Lights palette (figure 4-22) to bring up the Lights dialog box (figure 4-23), which allows you to create your own custom collection of lights for a project. The New button brings up the Light Parameters dialog box for creating new lights. The Delete and Clear buttons let you delete unwanted lights or clear the entire list.

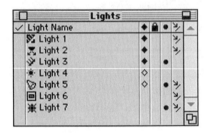

Fig. 4-22. Lights palette with seven user-defined lights.

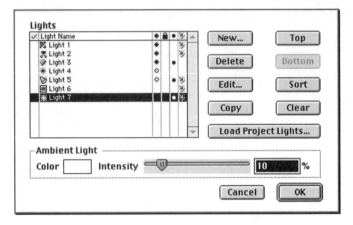

Fig. 4-23. Lights dialog box.

Options under the Copy button duplicate a light so that you can create a variant of an existing light at will. The Top, Bottom, and Sort buttons allow you to rearrange the list of lights. Load Project Lights lets you grab a set of lights from another project. The Intensity slider bar allows you to adjust the brightness of the ambient light.

◆◆ **NOTE:** *In the regular (non-RenderZone) version of form•Z, the only available light type is Distant. The use of lights in rendering is discussed in detail in Chapter 10.*

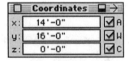

Fig. 4-24. Coordinates palette.

The Coordinates Palette

The Coordinates palette is low on the list of the most frequently used palettes because it duplicates functions of the Prompts window. It displays the *x*, *y*, and *z* values of the current position in real time as the mouse is moved (see figure 4-24). You can use this palette to specify precise values by moving the cursor into one of the Coordinate palette fields and typing in the desired value. Use the Tab key to move to the next field.

Press Return when you are finished filling in the fields, and the values will be applied. The same values will be displayed in the Prompts palette. All of this can be done equally well from the Prompts palette. If you are running out of space on the screen, it is recommended that you hide the Coordinates palette and use the Prompts window.

Small boxes labeled A, W, and C at the right side of the Coordinates palette allow you to select Absolute, World, and Cartesian or polar coordinates, as previously discussed for the Prompts palette. The Coordinates palette accepts and displays measurements in the unit system you define in the Project Working Units dialog box, found in the Options menu. These include English and metric systems. In the English system, Numeric Options can be adjusted to display Architectural (e.g., 2'-0 1/8"), Engineering (e.g., 2'-0.125"), Fractional (e.g., 24-1/8"), or Decimal (e.g., 24.125") values.

The Objects Palette

Every time you create an object in form•Z, it is recorded in a list that can be viewed and manipulated separately from the visual representation of the objects in the modeling window. The Objects palette gives you quick access to the names and attributes of each object in your project (figure 4-25). The five columns of the Objects palette allow you to control various aspects of each object.

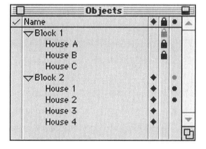

Fig. 4-25. Objects palette showing seven objects arranged in two groups.

The first column shows check marks beside selected groups or objects. Putting a check mark beside an object in the Object palette is equivalent to using the Pick tool to select it in the graphics window. The second column displays the name of the object or group. The triangular marks beside the names can be rotated in order to hide or reveal the content of a group. The third column has black diamond symbols to show that the objects are active, or white diamonds to show that they are ghosted. The fourth column lets you lock the object or group. The bullets in the final column control snapability.

You can assemble objects into sets (or groups) to facilitate handling them. Objects can be regrouped by clicking on the object names in the Objects palette and dragging them to the desired group. This functionality is similar to that previously described for the Layers palette. Objects can also be directly grouped within the modeling window by means of the Group command in the Grouping tool palette.

➥ **NOTE:** *The Group command is described in Chapter 5.*

The Objects dialog box is invoked by double clicking on the header bar of the Objects palette, or through the Objects item in the Options menu. This dialog box lets you edit and sort the list of groups and objects (figure 4-26). The New button lets you create a new group but not a new object, which requires the use of the Modeling Tools option. The Delete button removes the selected object or group. The Edit button (or double clicking on the object name in either the Objects dialog box or the Objects palette) brings up the Query Object dialog box, which allows you to modify the object.

Fig. 4-26. Objects dialog box with tools for handling objects.

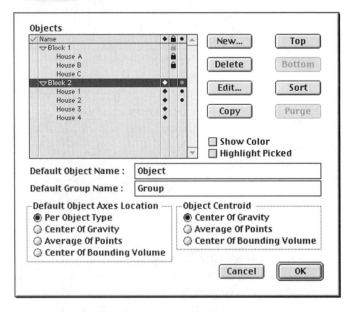

The Copy button makes a copy of the selected object or group in the same location. The Top, Bottom, and Sort buttons allow you to manipulate the list of objects. Purge clears away all empty groups. The Default

Object Name/Default Group Name field allows you to set a new standard naming system for groups and objects. The geometrical properties of object axes and centroids can be controlled by the selections at the bottom of the Objects Dialog box. The Show Color option will display the text labels within the Objects palette in the color assigned to the object in the modeling window. This can make it easier to locate specific objects via the palette.

The Planes Palette

The Planes palette allows you to move quickly between reference planes or drawing surfaces you have previously defined and saved. The Planes palette provides a list of the names of saved reference planes and their corresponding axes, and a check mark to indicate the active plane. Arbitrarily defined planes that do not lie on the *xy*, *yz*, or *zx* axes are labeled ARB. By clicking in the left column to move the check mark, you can instantly switch reference planes. Figure 4-27 shows the Planes palette with four user-defined planes.

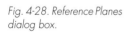

Fig. 4-27. Planes palette with four user-defined planes.

To bring up the Reference Planes dialog box, double click on the words *Plane Name* in the header of the Planes palette. The Reference Planes dialog box lists the names of the planes you have defined, if any, and gives you the tools to Save, Delete, Rename, Sort, Clear, and Load others (figure 4-28).

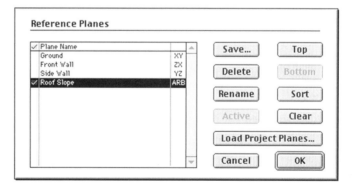

Fig. 4-28. Reference Planes dialog box.

Use the Save button to save and name your current reference plane. The Delete button removes the plane currently highlighted in the left-hand column and prompts you for a new name. The Active button makes the selected plane active, although this is an operation that would normally be done through the Planes palette, conveniently located on the interface. The Top, Bottom, and Sort buttons allow you to rearrange the list of planes. The Load Project Planes button brings up the Open File dialog box and lets you import planes you have saved in a different project.

☞ **WARNING:** *The Clear button can be dangerous; it will erase all arbitrary reference planes you have saved. Do not use this button unless you intend to define a new set of reference planes.*

☞ **NOTE:** *The creation of arbitrary planes is discussed in Chapter 5.*

The Profiles Palette

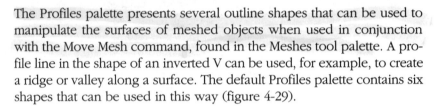

Fig. 4-29. Default Profiles palette showing six shapes.

The Profiles palette presents several outline shapes that can be used to manipulate the surfaces of meshed objects when used in conjunction with the Move Mesh command, found in the Meshes tool palette. A profile line in the shape of an inverted V can be used, for example, to create a ridge or valley along a surface. The default Profiles palette contains six shapes that can be used in this way (figure 4-29).

You can also add your own custom profiles to the palette using the Define Profile tool in the Meshes tool palette. With this tool, any 2D line can become a custom profile. The Profiles palette can be expanded or scrolled to reveal the additional profiles (figure 4-30).

☞ **NOTE:** *The Deformation and Define Profile tools are discussed in Chapter 5.*

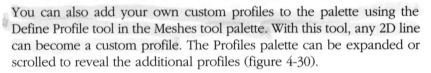

Fig. 4-30. Profiles palette expanded to show the addition of three user-defined profiles.

The Symbols Palette

Symbol Libraries constitute a form•Z system to handle multiple instances of an object in order to minimize computation time, as well as to permit greater freedom to edit an object. An object is described and saved in a symbol library and can then be repeated many times in the project. If the original symbol is modified, all instances of that symbol

Manipulating Palettes

will also be changed. If you have loaded a symbol library, or specified that an object is to be part of a library, the Symbol palette will appear, as shown in figure 4-31. If the Symbol palette is not visible, you can invoke it through the Palettes menu.

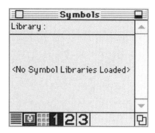

Fig. 4-31. Symbols palette with no libraries loaded.

Once you have loaded a preexisting symbol library (or have begun to create new symbols and have saved them in a library), the Symbol palette will display images of the objects (figure 4-32).

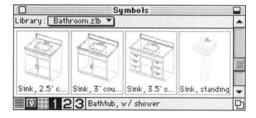

Fig. 4-32. Symbols palette expanded to show four symbol icons.

The small square icons at the bottom left-hand corner of the Symbol palette allow you to view the symbols as large icons, as small icons, or as a list of names (figure 4-33). The numbers 1, 2, and 3 represent levels of detail at which you can save each symbol. Level 1 should be used for the simplest representation of objects used for distant views or quick studies. Level 3 should be reserved for high-detail versions of these objects. Level 2 is used to achieve a mid-range level of detail. The Symbol palette also displays at its top the name of the library, and at its bottom the name of the currently active symbol.

Fig. 4-33. Symbols palette displaying a bathroom symbol library in list format.

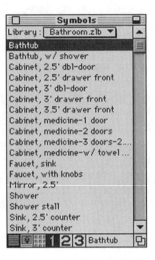

The Symbol Libraries dialog box is invoked by double clicking on the word *Library* in the header of the Symbol palette, or through the Symbol Libraries item in the Options menu. The dialog box repeats the names of the symbols and their icons, and provides buttons for creating, loading, deleting, and editing symbols (figure 4-34).

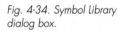 **NOTE:** *Symbols commands—including Create, Place, Edit, and Explode—are discussed in Chapter 5. The Symbol Library dialog box is discussed in more detail in Chapter 3. Chapter 9 contains a space-frame example that demonstrates the use of Libraries.*

Fig. 4-34. Symbol Library dialog box.

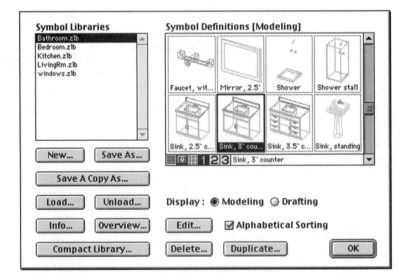

The Window Tools Palette

The Window palette, accessed through the Palettes menu, is another way of accessing the window tools located at the bottom left-hand corner of the form•Z window (figure 4-35). Window Tools includes visualization controls such as Set View and Zoom, Reference Plane manipulation tools, and geometrical constraints such as Grid Snap and Object Snap. These tools are of such importance that Chapter 6 is devoted entirely to them.

➥ **NOTE:** *The Window Tools are discussed in detail in Chapter 6.*

Fig. 4-35. Window Tools palette in horizontal mode.

You may find it convenient to access Window Tools through the Palettes menu and leave the palette floating above your modeling window. This will prevent having to repeatedly drag the cursor to the bottom left-hand corner of the form•Z screen in search of the original (and smaller and less legible) Window Tools. The Window Tools palette can also be adjusted to a vertical position by clicking on the small arrow in the upper right-hand corner of the palette. You can position this "floating" palette anywhere on your screen, which allows you to consolidate all of the Window Tools in a single location of your choosing. Additionally, individual menus can be extended to the right, as shown in figure 4-36.

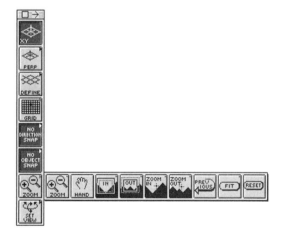

Fig. 4-36. Window Tools palette in vertical mode with the Zoom & Pan palette extended.

The Animation Palette

The Animation palette, new in form•Z version 3.0, is displayed by default at the bottom center of the screen. If you have no plans to use animation, you might consider closing this palette and replacing it with another, such as Surface Styles. It is recommended, however, that you read Chapter 12 to discover how easy it is to use the animation feature to create a short movie or animated sequence (figures 4-37 and 4-38).

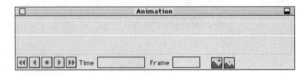

Fig. 4-37. An inactive Animation palette.

Fig. 4-38. Animation palette after views have been selected in the Views palette and the Animation From Keyframes command used.

➻ **NOTE:** *Animation is discussed in detail in Chapter 12.*

chapter 5

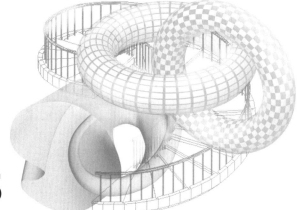

Modeling Tools

Introduction

This chapter covers all of the modeling tools of form•Z. Modeling tools are used for generating and manipulating objects and models. This chapter constitutes an overview of the modeling tools, explaining how, when, and why to use them. This chapter can be used as a quick reference guide or can be read from beginning to end to gain a basic understanding of the modeling potential and methods of form•Z.

The Interface

When you first open a model in form•Z, the modeling tools are located on the left of the screen in two vertical columns. This is referred to as the tool bar. Each icon in the tool bar shows one tool; by clicking on each icon, a range of related tools is projected to the right, across the screen. By sliding the cursor along this list, each tool can be selected and made active. These groups of similar tools are called the tool palettes (figure 5-1). Clicking and dragging a tool while keeping the mouse button held down will separate the selected tool palette from the tool bar, allowing it to be repositioned on the screen.

Tool icons with small red dots in their top right corners have options associated with them. These options are accessed by double clicking on the tool icon and opening the options dialog box. Alternatively, you can use the Tool Options palette. Dialog boxes and the tool options palette are used for selecting additional options and numerical inputs for the tool. Throughout this chapter, wherever an options dialog box is referred to, the Tool Options palette can usually be used instead.

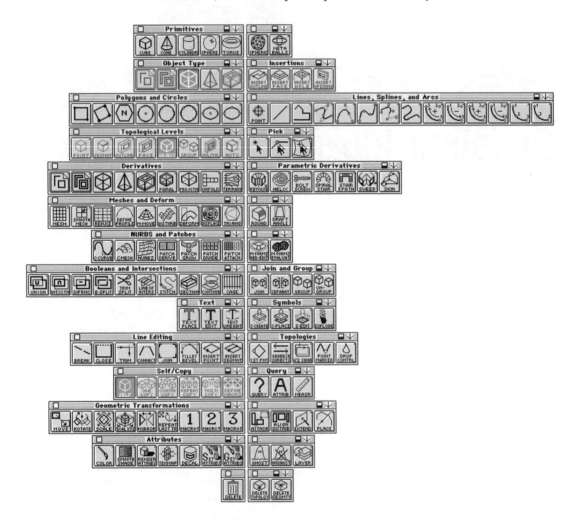

Fig. 5-1. form•Z Modeling Tools.

An overview of all the tool palettes (see figure 5-1) can be accessed through the Help pull-down menu, as described in Chapter 2. Through this Help overview you can find the name and a brief description of each tool.

Concepts and Terminology

Fundamental computer-aided design concepts and terminology are referred to throughout this chapter. These ideas are basic to form•Z modeling and are inherent to the coordinate geometry that form•Z uses to describe objects.

Polygons

form•Z is a polygon-based modeling program. In general, this means that every object you model in form•Z is described as a collection of edges and faces. Even rounded surfaces are actually defined by polygonal faces, meaning they are not actually curved but very finely faceted. This is important to keep in mind because the smoother you want a curved surface to appear, the more facets you need. However, there is a trade-off to also keep in mind: the more polygons you have in an object, the larger the model is and hence the longer it takes to render.

✓ **TIP:** *To mitigate against slow rendering times, you can add more RAM to your system or add a graphics card.*

Solids and Surfaces

There are two types of objects in form•Z: surfaces and solids. A solid object is one that is completely closed on all sides, forming an enclosed volume. When a volume of this type such as a cube or a sphere is cut, the interior will not be hollow.

A surface object is any object that is not closed on all sides, including planes and open shapes. These objects are ultimately 2D, and if they were cut in half you would see they have no thickness. In that they are not solid objects, simple 2D lines and curves are also considered surfaces.

Surfaces in form•Z can be defined as either a Surface or Surface Solid. A Surface is defined as a One-Sided object (titled Meshed Surface if inspected with the Query tool). One-Sided objects do not interact well with solid objects in form•Z. For example, One-Sided surfaces cannot be used with the Boolean Union and Boolean Difference tools. However if an object is a One-Sided surface it can be used with the Trim/Split tool.

A Surface Solid is a Two-Sided surface, behaving more like a solid and less like a surface. This means that a surface that is defined as a Surface Solid can be used in conjunction with other solid objects in operations such as the Boolean Difference or Boolean Union. An object can be converted between One-Sided and Two-Sided surface using the Make One/Two-sided surface tool (discussed later in this chapter).

Parametric Objects and Wires and Facets

Many types of objects generated with the form•Z tools (primitives and Nurbz surfaces, to name two) are parametric. Parametric objects are objects that retain information about the process that created them. The "parameters" of this information can be edited to change the shape of the object. The parameters of a parametric object can be edited using its controls, accessed using the Edit Controls tool. See the Pick tool palette later in the chapter for more information about the Edit Controls tool.

These parametric objects are fundamentally different than most other objects in form•Z, and because of this difference they are represented in a different manner, using "wires" along with polygons or "facets" (see the previous "Polygons" section). The wires are graphical representations of the parametric controls. They are usually displayed along with the "facets" or polygons of the object.

The wires are purely for wireframe display purposes and do not have an effect on the ability to control an object or its rendered qualities. The rendered quality, as with other objects in form•Z, is still determined by the number of facets. The density of and the number of facets for parametric objects are determined by the Wires and Facets settings in the Tool options palette for each tool.

Modifiers and Generators

The tools in form•Z can be roughly categorized into two types: those that perform transformative or generative actions and those that modify actions (referred to in this book as modifiers) to determine how an object is generated or transformed. These two types are represented differently in the tool palettes; the modifiers are shown in turquoise and magenta, whereas the generative and transformative tools are shown in white.

These two types of tools are always used in conjunction with each other. Modifiers determine how a tool behaves and then the action tool performs the function. Together, these tools determine whether an object is to be generated as a 2D object, a 3D object, or an insertion (Object Type palette or Insertion tool palette), or whether transformations are to be performed on points, lines, faces, or objects (the Topological Levels tools). The transformative and generative tools include tools such as Move, Rotate, and Scale, as well as drawing tools (e.g., Vector Line and N-sides Polygon).

Derivative Tools

Many modeling tools in form•Z can be considered derivative tools, and there is even a tool palette named Derivatives. Derivative tools are those that derive 2D or 3D forms from objects that have been previously created. These already existing objects are referred to as source objects. It is often difficult to design an object in a single step. Derivative tools allow you to derive complex objects from a collection of basic source objects that are more easily constructed.

These Derivatives, by default, keep the original source objects around after the new object has been derived, but in a "ghosted" form. Ghosted objects are colored gray and cannot be selected or used in any more modeling operations but can be "unghosted" when the need arises to use them again.

✏ **NOTE:** *For more information on ghosted objects, see discussion of the Ghost and Unghost tools in this chapter.*

Using the Tools

The sections that follow cover the various tool palettes. Each palette is in a different row and column of the main tool bar and can be dragged to any location on the screen. In figure 5-1, all palettes have been arranged in two columns. You can use the graphical key at the beginning of each section to help locate the tool palette. The title of the palette is displayed in the gray window bar at the top of the palette when it is pulled off the toolbar.

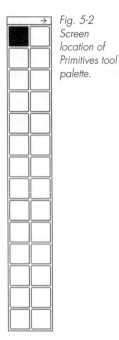

Fig. 5-2 Screen location of Primitives tool palette.

Primitives Tool Palette

The tools in the Primitives tool palette (screen location, figure 5-2; palette, figure 5-3) allow you to quickly model and edit five fundamental 3D forms, or primitive objects: cube, cone, cylinder, sphere, and torus. The palette's screen location is shown in figure 5-3.

Fig. 5-3. Primitives tool palette.

Tools 1: Cube

LEARNING OBJECTIVE 5-1: *Generating controlled cubic objects.*

TOOL COVERED:

- Cube

Fig. 5-4. Cube tool.

The Cube tool (figure 5-4) generates a cubic primitive that can be modified or edited through the use of the Edit Controls tool. The primitive Cube is defined by three parameters: Height, Width, and Depth. These can be edited after their generation using the Edit Controls tool. The object has editing vectors, activated with the Edit Controls tool, which emanate from its center of mass and affect the object as a whole, uniformly editing/scaling when selected (figure 5-5). Using the Edit Surface tool, the primitive can be edited by translating single faces. One of the benefits of using the Cube tool over the 3D Extrusion tool is that it can later be edited using these controls.

Fig. 5-5. A Primitive cube with its control vectors, after being selected with the Edit Controls tools.

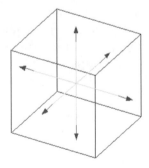

Using the Tools

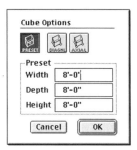

Fig. 5-6. Cube Options dialog box.

The Cube Options dialog box (figure 5-6) provides three methods for generating the object. The Preset option allows for input of predetermined width, depth, and height values, so with a few simple clicks, numerous identical shapes can be generated with ease. Other options include the Diagonal (3-point) and Axial method of generating an object. Diagonal is the default method, and it allows you to define a cube by its Length, Width, and Height with three clicks. Axial uses four points to determine the size of the cube.

Tools 2: Cone, Cylinder, Sphere, and Torus

LEARNING OBJECTIVE 5-2: Quickly generating and editing curved primitive objects.

TOOLS COVERED:

- Cone
- Cylinder
- Sphere
- Torus

Fig. 5-7. Tools covered.

The four other Primitive tools generate objects with curvature. These "curved" objects typically have more numerous generation options including Preset, Radius, Diameter, and Ellipse (figure 5-8). This provides for a wide variety of object generation possibilities (figure 5-9).

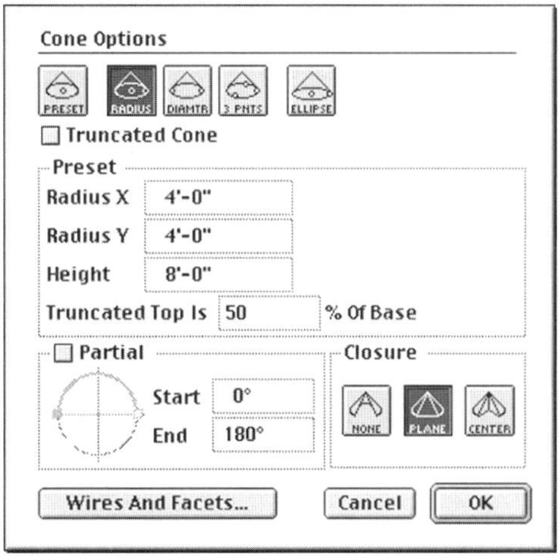

Fig. 5-8. Cone Options dialog box showing various generation choices.

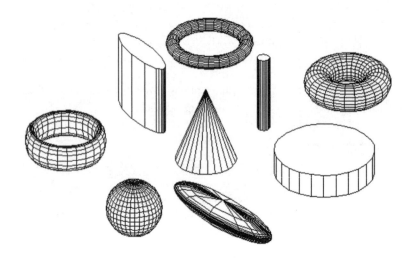

Fig. 5-9. A variety of primitives generated with various options.

All of the curved primitives (that is, all the tools in the Primitives tool palette excluding the Cube) have the option of partial closure. Partial closure allows you to adjust degree of completion of an object. The settings for Partial completion determine the amount of completion as degrees in two directions. Because each of the objects is 360 degrees around in one (Cone, Cylinder) or two (Sphere, Torus) directions, a partial section can be described as a portion of 360 degrees.

The settings for Partial in the options determine the section of the primitive as these degrees. Using the Partial option (along with various settings of Closure) can generate hollowed-out, segmented, or sliced/cutoff objects (figure 5-10). Partial objects generated with the Partial option can be edited with the Edit Controls tool, which lets you set the degree of the partial section.

The polygonal and control "wire" densities can be set in the Wires And Facets Options dialog box (figure 5-11).

The Wires options set how densely the Primitive objects will appear in the display window, but has no effect on the object's geometry or resolution. The Facets options increase or decrease the default settings for the number of polygons an object will have when it is rendered. The higher the number of facets, the higher the resolution of the object, making it smoother in appearance but slower to render.

Fig. 5-10. Various partially closed primitives.

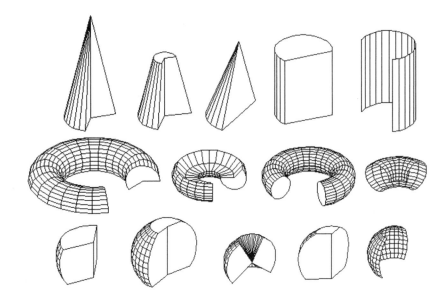

Fig. 5-11. Wire And Facets Options dialog box for the Parametric Cone primitive.

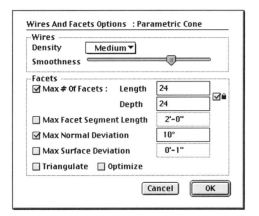

TIPS AND STUDY AND APPLICATION NOTES

1. The Cone tool includes an additional option for modifying the initial shape of the primitive, Truncated Cone. This option will allow you to truncate the cone as a function of a percentage of its base area.

2. The type of closure for curved objects is set with the Closure options. The Closure of None option will produce an open surface object when the Partial option is used. The choice of Plane

or Center will produce a slightly different solid object based on the primitive.

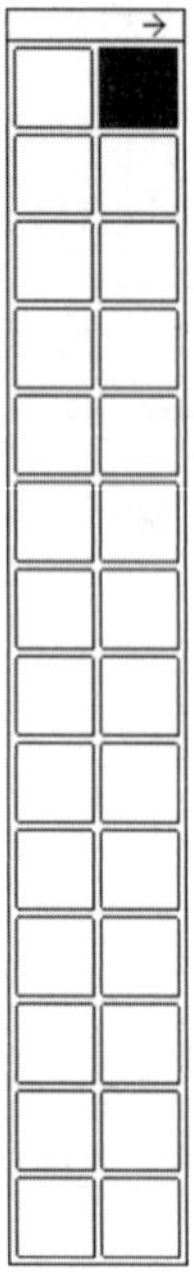

Balls Tool Palette

Fig. 5-12. Screen location of the Balls tool palette.

The tools of the Balls tool palette (screen location, figure 5-12; palette figure 5-13) are used to create spheres, or objects that consist of points equidistant from the objects' centers. The Balls tool palette screen location is shown in figure 5-13.

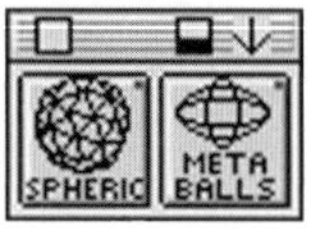

Fig. 5-13. Balls tool palette.

Tools 1: Spherical Objects

LEARNING OBJECTIVE 5-3: *Generating objects consisting of points equidistant from the objects' centers.*

TOOL COVERED:

- Spherical Objects

Fig. 5-14. Spherical Objects tool.

The Spherical Objects tool (figure 5-14) generates objects consisting of points equidistant from the objects' centers. This not only includes spheres but a collection of objects that by virtue of sharing a process of generation have been categorized under the same tool.

Figure 5-15 shows the Spherical Object Options dialog box. The six icons near the top of the dialog box determine the object generation method.

Preset allows you to enter values for x-, y-, and z-axis radii before selecting an origin point. Once the values are entered, a spherical object is created with a single mouse click defining the object's center. This option can be used to create elliptical objects or perfect spheres.

Using the Tools 175

Fig. 5-15. Spherical Object Options dialog box.

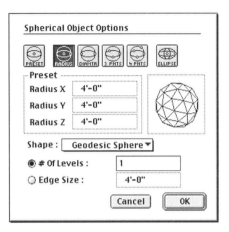

Radius creates a perfectly spherical object in two clicks, the first determining the center and the second determining the size of the object. Diameter creates a perfectly spherical object in two clicks, the first setting one endpoint on the surface of the sphere and the second defining the opposite end.

3 Points creates a perfectly spherical object by selecting three points that define the object's perimeter. These points can all lie on a single reference plane or occur anywhere within 3D space.

4 Points creates a perfectly spherical object in four clicks. If all four points lie on the same reference plane, the sphere is actually generated with three points, ignoring the point defined by your third click. However, this tool can be used to create a sphere that passes through four points in 3D space.

Ellipse allows you to create an elliptical object in four mouse clicks. The first click determines the object's center, whereas the second, third, and fourth defines the object's radii along the *x, y,* and *z* axes.

Under the Shape pull-down menu in the Spherical Objects Options dialog box, there are eight types of spherical objects that can be generated (figure 5-16). In a Revolved Sphere, the Length Resolution determines the number of segments in the longitudinal direction, and Depth Resolution determines the number of segments in the latitudinal direction.

Fig. 5-16. Types of spherical objects.

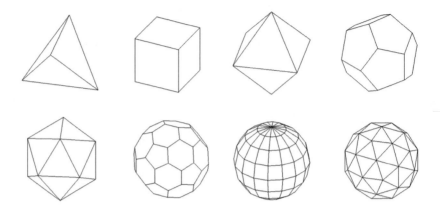

Geodesic Spheres are spheres consisting of equilateral triangles. The # Of Levels setting determines the level of resolution of the sphere, level 1 being the lowest, level 4 the highest. The value entered in Edge Size determines the length of the segments that constitute the sphere (figure 5-17).

Fig. 5-17. Geodesic spheres at level 1 (left) and level 3 (right).

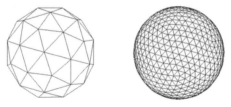

TIPS AND STUDY AND APPLICATION NOTES

1. The Sphere tool in the Primitives tool palette offers some additional methods for creating spherical objects, many of which allow for more control over creation parameters.

2. A geodesic sphere of # Of Levels 1 consists of 80 triangles, # Of Levels 2 of 320 triangles, # Of Levels 3 of 1,280 triangles, and # Of Levels 4 of 5,120 triangles. A higher-level sphere will render much more slowly than a level 1 sphere.

3. When a geodesic sphere is generated with the Edge Size option, the resolution of the sphere cannot exceed that of a level 4 geodesic sphere.

Tools 2: Metaballs

LEARNING OBJECTIVE 5-4: *Creating organic forms using Metaballs.*

TOOL COVERED:

- Metaballs

Fig. 5-18. Metaballs tool.

Metaballs are parametric objects with "mass" and a "region of influence" designed for creating organic forms. When these objects are combined, smooth, blended surfaces result, called Metaformz. The result does not have an exact size, but is a negotiation between the Metaball's mass and region of influence and its proximity to other Metaballs. Metaballs are combined using the Group tool in the Join and Group palette discussed later in this chapter.

Metaballs are created with the Metaballs tool. The type of Metaball, as well as its control parameters, is defined in the Metaball Type dialog box (figure 5-19). There are four types of Metaballs available: Ball, Stretched Ball, Ellipsoid, and Stretched Ellipsoid. Each of these options demonstrates unique behavior when evaluated (figure 5-20). The Ellipsoid Options provide two methods of generating elliptical Metaballs. The Weight option determines the strength the sphere of influence will have on other Metaballs, a positive or negative value indicating an additive or subtractive.

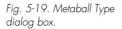

Fig. 5-19. Metaball Type dialog box.

Fig. 5-20. Metaball types created with the Metaball tool.

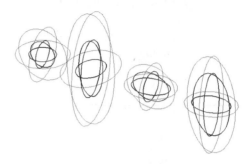

☛ **NOTE:** *Refer to the Metaformz section of this chapter for additional information on using Metaballs.*

Object Type Tool Palette

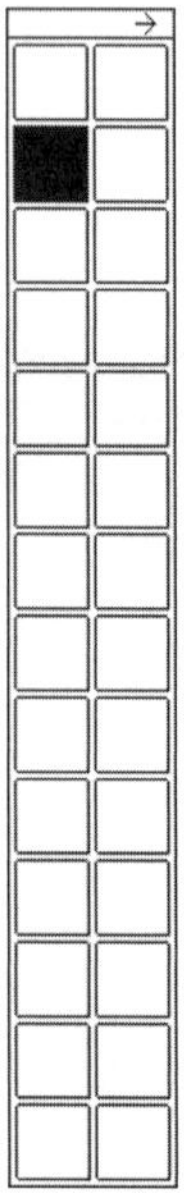

Fig. 5-21. Screen location of Object Type tool palette.

The Object Type tool palette (screen location, figure 5-21; palette, figure 5-22) contains modifier tools that allow you to directly generate 2D and 3D objects in form•Z.

Fig. 5-22. Object Type tool palette.

Tools 1: 2D Surfaces and Enclosures

LEARNING OBJECTIVE 5-5: *Creating 2D objects with lines, curves, and other geometry.*

TOOLS COVERED:

- 2D Surface
- 2D Enclosure

Fig. 5-23. Tools covered.

2D Surface

Fig. 5-24. Object made with 2D Surface tool.

The 2D Surface and 2D Enclosure tools are modifiers that create 2D objects from lines, splines, polygons, circles, and ellipses. The 2D surface tools generate 2D areas or lines based on a shape you draw. Whenever a closed line is drawn, the 2D Surface tool creates a surface that can be rendered (figure 5-24). An open shape will result in a line that will not be shown when the scene is rendered.

2D Enclosure

The 2D Enclosure tool generates 2D surfaces consisting of "walls." The walls are two parallel lines offset from lines you draw. This tool is ideal for creating walls in plan or section to later be extruded with the Derivative 3D Extrusion tool. The 2D Enclosure tool always generates a surface that can be rendered, even if the original line you draw is open (figure 5-25).

Fig. 5-25. Object made with 2D Enclosure tool.

The 2D Enclosure Options dialog box allows you to define a wall's Width and Justification. The wall Width is the separation between parallel wall lines generated from the line you draw (figure 5-26).

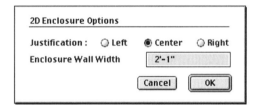

Fig. 5-26. 2D Enclosure Options dialog box.

The Justification option under the 2D Enclosure tool determines which direction the parallel wall lines are offset from the original line you draw. Figure 5-27 shows 2D enclosures generated with Left, Center, and Right justification for the same original line, which was drawn in a counterclockwise direction.

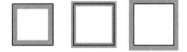

Fig. 5-27. 2D enclosures made with same source line but with different Justification settings: Left, Center, and Right.

TIPS AND STUDY AND APPLICATION NOTES

1. 2D surfaces are not solid objects and can be used in Boolean functions only with other surface objects.

2. Open shapes that are drawn with the 2D Surface tool will be visible in wireframe views but will not show up in a rendering.

3. 2D Surface and Enclosure objects can be derived from any object after the object has been created using the Derivative 2D Surface and Derivative 2D Enclosure tools in the Derivatives tool palette.

Tools 2: 3D Solids and Enclosures

LEARNING OBJECTIVE 5-6: Producing 3D extruded forms with modified drawing tool functions.

TOOLS COVERED:

- 3D Extrusion
- 3D Converged
- 3D Enclosure

Fig. 5-28. Tools covered.

These tools modify the function of the drawing tools (Polygons, Circles, Lines, Arcs, and Splines) to produce 3D extruded forms. When used in conjunction with the drawing tools, the 3D Extrusion tools generate 3D solids and surfaces by extruding all of the points of an entity upward to a prescribed height.

The three objects shown in figure 5-29 have been produced using the 3D Extrusion tool. The first object is an extruded circle, the second is a rectangle, and the third is a closed Vector Line. The heights of these solids were determined by the settings in the Heights pull-down menu.

Fig. 5-29. 3D extruded solid objects.

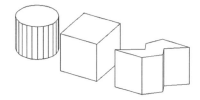

3D Extrusion

The 3D Extrusion and 3D Converged tools have the same options. The extruded form can be extruded perpendicular to the reference plane or perpendicular to the surface you are modeling. The Perpendicular to Surface setting is useful for modeling objects that are not parallel to any reference plane, such as roofs and ramps. The tools also have an option to automatically close any open shapes to produce a 3D solid (figure 5-30).

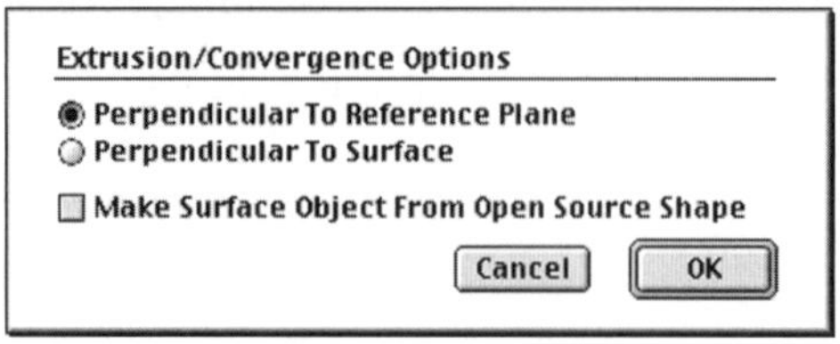

Fig. 5-30. 3D Extruded/ Converged Solid options.

3D Converged

The 3D Converged tool produces 3D objects from elements created with the drawing tools. However, unlike the 3D Extrusion tool (which extrudes all of the points of an object directly upward to a prescribed height), the Converged Solid tool extrudes the points of an object upward and inward, converging them at a point at the prescribed height. The resulting objects are cones and pyramids. Figure 5-31 shows the same source shapes as in figure 5-29, but this set is of solids generated using the Converged Solid tool.

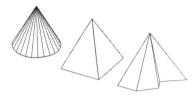

Fig. 5-31. 3D converged solids.

3D Enclosure

The 3D Enclosure tool, similar to the 2D Enclosure tool, creates walls by offsetting the line drawn to create a thickness. The 3D Enclosure tool goes one step beyond the 2D Enclosure tool to extrude the wall shapes, creating enclosed spaces. Figure 5-32 shows the same source shapes again, but this time modeled using the 3D Enclosure tool.

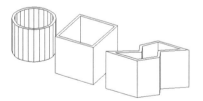

Fig. 5-32. 3D enclosure solids.

There are several options for the 3D Enclosure tool. These options set the wall thickness of the enclosure (Enclosure Wall Width) and determine whether the enclosure will include a top (Top) and bottom (Base) cap (figure 5-33).

Fig. 5-33. Options for 3D Enclosure.

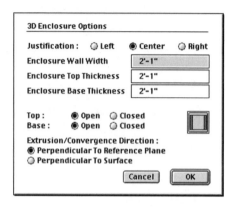

TIPS AND STUDY AND APPLICATION NOTES

1. All of the 3D solid tools are also available as derivative object operations (in the Derivatives tool palette) that can be performed on 2D lines, circles, polygons, and arcs.

2. The heights of 3D objects are determined by setting a height or by selecting Graphic/Keyed from the Heights pull-down menu. The Graphic/Keyed option allows you to graphically set the height of each object as you draw it, or to type a numerical height in the Prompts window.

3. To create a 3D solid using Lines or Arcs it is necessary to triple click the mouse to close the line; otherwise, the tool will produce a surface object.

Insertions Tool Palette

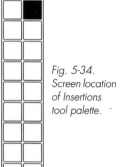

Fig. 5-34. Screen location of Insertions tool palette.

The Insertions tool palette (screen location, figure 5-34; palette, figure 5-35) contains modifier tools that allow you to add detail and definition to existing solids and surfaces by inserting lines, faces, openings, and holes.

Fig. 5-35. Insertions tool palette.

Tools 1: Insertions

LEARNING OBJECTIVE 5-7: Adding or subtracting simple volumes using modifier insertion tools.

TOOLS COVERED:

- Insert Outline
- Insert Face/Outline
- Insert Hole
- Insert Opening

Fig. 5-36. Tools covered.

These modifier tools generate new outlines, faces/volumes, holes, or openings in objects and are useful for adding to, or subtracting from, an original object. All Insertions tools are modifiers and have generative capabilities only when used in conjunction with the appropriate drawing tool, such as 2D Rectangle or Vector Line.

Insert Outline requires that you preselect a surface face on which the outline will be drawn (figure 5-37). Insert Outline produces a new outline on an object. An outline is a series of connected segments that are not 3D but instead are flush with the surface. This differs from Insert Face/Volume in that the latter generates an extruded face or volume within an object according to a given height, whereas the Insert Outline only inscribes a line on its surface.

Fig. 5-37. The insertion of outlines of segments or closed shapes using Insert Outline.

The height of a Face/Volume insertion can also be negative, which will "carve" away from an object (figure 5-38). Both operations will keep the latest generated face selected for further modification. The "height" of the Face/Volume insertion (as with the other insertion tools) is relative to the face of the object, not to the current reference plane.

Fig. 5-38. A Face/Volume insertion with positive height and negative height.

With the Insert Hole tool, a shape can be used to create either a cavity or recess in an object (figure 5-39), whereas Insert Opening (figure 5-40) always generates a hole that completely penetrates the object. When using the Insert Hole tool pay careful attention to the Heights setting; if the Height is set to a size that is bigger than the depth of the object, the Insert Hole tool will not work.

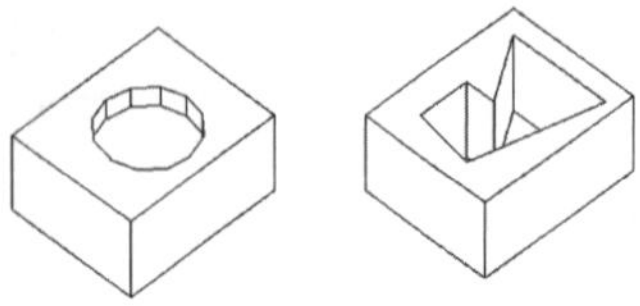

Fig. 5-39. A hole partially inserted and fully inserted into an object.

Fig. 5-40. Inserting openings through a 3D enclosure.

TIPS AND STUDY AND APPLICATION NOTES

1. All insertion operations require the preselection of a face to be modified. Faces can be preselected by selecting the Topological Level of Face and using the Pick tool to select two segments.

2. Using the Insertion tools on a controlled object (such as the parametric solids) will clear its control parameters. This means that a "smart" parametric cube will become a regular cube.

3. The difference between a hole and an opening in form•Z is that a hole does not necessarily go all the way through an object, although it can, whereas an opening by definition passes entirely through an object.

4. The Insert Hole tool, unlike the Insert Opening tool, will permit you to insert a hole that overlaps the edge of a face, generating a groove or notch in a corner.

Using the Tools

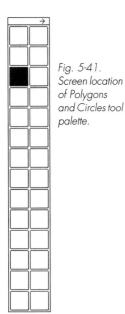

Fig. 5-41. Screen location of Polygons and Circles tool palette.

Polygons and Circles Tool Palette

The Polygons and Circles tool palette (screen location, figure 5-41; palette, figure 5-42) contains drawing tools used to draw closed shapes. When used with the tools from the Object Type tool palette, the template can be used to produce 2D and 3D objects.

Fig. 5-42. Polygons and Circles tool palette.

Tools 1: Rectangles

LEARNING OBJECTIVE 5-8: *Producing 2D and 3D rectilinear surfaces, objects, holes, and openings.*

TOOLS COVERED:

- Rectangle
- Rectangle, 3 Point

Fig. 5-43. Tools covered.

The Rectangle and Rectangle, 3 Point tools create 2D or 3D rectilinear forms. These rectangle tools must be used in conjunction with a modifier tool from the Object Type or Insertions tool palette. When used with the appropriate modifier, the rectangle tools produce surfaces, objects, holes, and openings.

Rectangle

The Rectangle tool uses two points to define a four-sided figure. The two points define opposite corners of a rectangle with its sides parallel to the reference plane axes. Objects generated with this tool do not necessarily have to lie within the reference plane (by using the Point snap and snapping to other 3D objects, you can create a tilted plane), but will always be generated with their sides parallel to the reference plane axes (figure 5-44).

Fig. 5-44. A 2-point rectangle.

Rectangle, 3 Point

The Rectangle, 3 Point tool uses three mouse clicks to define a rectangle on the reference plane. The three points define three corners of the rectangle. Unlike the Rectangle tool, the sides of the 3-point rectangle are

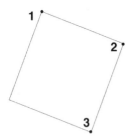

Fig. 5-45. A 3-point rectangle.

Fig. 5-46. A variety of objects modeled with the rectangle tools.

not necessarily parallel to the reference plane axes. Instead, their angle is determined by the first two mouse clicks (figure 5-45).

As previously mentioned, the Rectangle and Rectangle, 3 Point tools are used in conjunction with the Object Type tools to produce 2D and 3D solid forms, as well as voids and openings. Figure 5-46 shows a range of objects created with the rectangle tools.

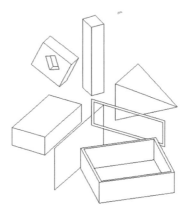

TIPS AND STUDY AND APPLICATION NOTES

1. The Rectangle and Rectangle, 3 Point tools draw objects on the active reference plane. To draw a rectangular object with a different orientation, you change the reference plane using Window Tools.

Tools 2: Polygon

LEARNING OBJECTIVE 5-9: *Generating a polygon of a specified number of sides.*

TOOL COVERED:

- Polygon

Fig. 5-47. Polygon tool

This tool generates a polygon with a predefined number of sides from two clicks of the mouse and is used in conjunction with one of the Object Type tools such as 2D Surface or 3D Extrusion. Polygons are defined with two mouse clicks (one defining the center of the polygon and a second that defines the radius). The shape of the polygon is set with the # of Segments option.

Figure 5-48 shows a series of solid objects created by inputting various values in the # Of Segments in Polygon option.

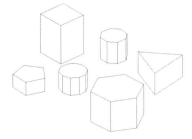

Fig. 5-48. Examples of polygons.

Figure 5-49 shows the Polygon Options dialog box. You can select one of the predefined polygons or custom create one by entering a value in the # of Segments in Polygon field. You can also add polygon patterns to each segment of the polygon by selecting one of the predefined sets of patterns.

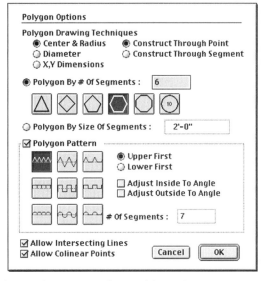

Fig.5-49. Polygon Options dialog box.

Figure 5-50 shows the range of possible polygon patterns that can be generated. The process of creating a patterned polygon is the same as a nonpatterned one. You first select the Polygon Pattern option to have form•Z automatically generate the pattern on each segment, then define two points to determine the center and the edge of the polygon. You can define the pattern's depth with a third mouse click. Polygon Patterns also allows you to generate cog wheels (figure 5-51).

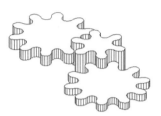

Fig. 5-51. Cog wheel examples.

Fig. 5-50. Polygon pattern types.

TIPS AND STUDY AND APPLICATION NOTES

1. You define the number of sides for a polygon in the Polygon Options dialog box.

2. The two clicks you use to draw the polygons can be defined via the Center and Radius, Diameter, or X,Y Dimensions options. The click points can be established at polygon vertices or at the center of a segment.

3. Polygon Patterns allows you to create patterns around the periphery of a polygon by selecting one of the predefined patterns.

Tools 3: Circles and Ellipses

LEARNING OBJECTIVE 5-10: *Drawing 2D and 3D circular and elliptical objects.*

TOOLS COVERED:

Fig. 5-52. Tools covered.

- Circle, Center and Radius
- Circle, Diameter
- Circle, 3 Point
- Ellipse, Major and Minor Radius
- Ellipse, Diameter and Radius

The Circle and Ellipses tools are used in conjunction with the modifier tools in the Object Type tool palette for drawing 2D and 3D circular and elliptical objects. Three methods are available for drawing circles: Circle, Center and Radius; Circle, Diameter; and Circle, 3 Point.

Using the Tools

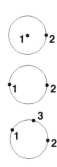

Fig. 5-53. Circle, Center and Radius; Circle, Diameter; and Circle, 3 Point.

Fig. 5-54. Ellipse, Major and Minor Radius and Ellipse, Diameter and Radius.

Fig. 5-55. Arc/Circle/ Ellipse Resolution dialog box.

Fig. 5-56. Circular objects created with various numbers of segments.

Circle, Center and Radius creates a circle from a center point and radius that you define using the mouse or by typing values into the Prompts window. Circle, Diameter describes a circle by its diameter, giving you the ability to create a circle that is touching two lines at once. The 3-Point method uses three points on the circumference of a circle to describe it. This tool is useful when it is necessary to create a circle that is touching multiple edges (figure 5-53).

Two tools are available to create ellipses. Ellipse, Major and Minor Radius uses three points to describe an ellipse. The first point locates the edge, and the second locates the center. These first two points define one dimension of the ellipse. The third point defines the dimension of the other axis of the ellipse. Ellipse, Diameter and Radius uses three points on the circumference of the ellipse to define its form (figure 5-54).

The options for arcs, circles, and ellipses determine the number of sub-divisions the curved object will have. Subdivisions can be defined by their number (# of Segments) or dimension (Max Size of Segments). The number of subdivisions determines the "smoothness" of a form (figures 5-55 and 5-56).

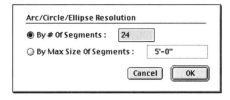

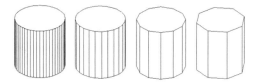

TIPS AND STUDY AND APPLICATION NOTES

1. The default # Of Segments setting of 24 is adequate for most purposes. However, for special cases (a close-up detail or a

curved object that is important to your model) you may want to increase the # Of Segments to 48 or 64, or even 96.

2. As with most tools that produce curvilinear forms, it is necessary to be conscious of how many polygons are produced when modeling. Increasing the number of segments in a circle or an ellipse will increase the total number of faces and polygons, consequently increasing rendering time.

Lines, Splines, and Arcs Tool Palette

The Lines, Splines, and Arcs tools (screen location, figure 5-57; palette, figure 5-58), like the Polygons and Circles tools, are used in conjunction with the Object Type tool palette to model objects. However, unlike the Polygon tools, these tools can generate open shapes as well as closed shapes. It is possible to use multiple tools in the same operation by switching between them.

Fig. 5-58. Lines, Splines, and Arcs tool palette.

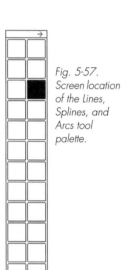

Fig. 5-57. Screen location of the Lines, Splines, and Arcs tool palette.

Tools 1: Points and Lines

LEARNING OBJECTIVE 5-11: *Using point and line tools to create objects and surfaces.*

TOOLS COVERED:

- Point
- Segment
- Vector Line
- Stream Line

Fig. 5-59. Tools covered.

When used in conjunction with the Object Type modifiers, the Point, Segment, Vector Line, and Stream Line tools create objects and surfaces of almost any shape. The four tools generate different types of objects. The Point tool draws single points defined by a mouse click or typed as coordinates in the Prompts palette. The Segment tool draws a single line segment defined by two mouse clicks or typed coordinates. The Vector

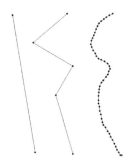

Fig. 5-60. Lines created with Segment, Vector Line, and Stream Line tools.

Line tool draws polygonal objects by connecting a series of points defined by the mouse or typed coordinates. The Stream Line tool allows you to sketch freely by holding down the mouse and drawing on the reference plane. This tool leaves behind a string of connected segments as the cursor is moved across the screen (figure 5-60).

When the Segment, Vector Line, and Stream Line tools are used with the 3D modifiers (Extrusion, Enclosure, and Converged Solid), open lines create surface objects and closed lines result in solids. The Point tool, when used with the Extrusion and Converged Solid modifier, produces a single extruded point, and when used with the Enclosure modifier, produces an extruded square (figure 5-61).

Fig. 5-61. A variety of objects made with the line and point tools.

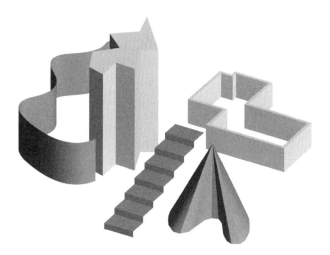

The Vector Line and Stream Line tool options have a toggle for Allow Intersecting Lines and Allow Collinear Points. Turning these off prevents accidentally producing poorly formed objects (figure 5-62). The option Stream Distance for the Stream Line tool determine the resolution of a curve and consequently the number of polygons in an object.

Fig. 5-62. Vector Line Options dialog box.

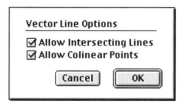

Chapter 5: Modeling Tools

TIPS AND STUDY AND APPLICATION NOTES

1. The option Allow Intersecting Lines permits you to model an object that intersects itself.

☞ **WARNING:** *The Allow Intersecting Lines option can produce a poorly formed object that will not work with the Boolean tools, as well as create unexpected results in rendering.*

2. To automatically close a Vector Line or Stream Line, triple click the mouse to define the last point.

3. When drawing with Stream Line, pay careful attention to the Stream Distance. These affect the resolution of the line and the number of polygons in the objects created.

4. When using the Line tools, it is possible to undo a mouse click while drawing. Simply select Undo from the Edit pull-down menu or use the keyboard shortcut: Mac (Cmd-C), Windows (Ctrl-C).

Tools 2: Splines

LEARNING OBJECTIVE 5-12: *Using spline tools to create objects and surfaces.*

TOOLS COVERED:

- Spline, Quadratic Bezier
- Spline, Cubic Bezier
- B-Spline Cubic
- Spline Sketch

Fig. 5-63. Tools covered.

The four Spline tools located in the Lines, Splines, and Arcs tool palette are used to generate curves in form•Z. Each spline tool allows for a different generation method and curve behavior. These tools will generate both parametric and controlled curves, which can be edited with the C-Curve and Edit Controls tools. Figure 5-64 shows the four spline curves, each generated with the same sequence of clicks. Note in the illustration that the same point position and sequence was used to generate the four spline curves.

Fig. 5-64. The four spline curves from left to right: Spline, Quadratic Bezier; Spline, Cubic Bezier; B-Spline Cubic; and Spline Sketch.

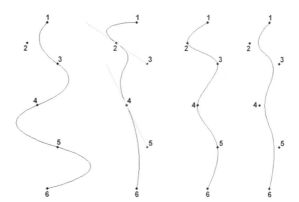

The Spline, Quadratic Bezier tool allows you to pick points that lie on the curve after the second point. form•Z then interpolates the segments between the points to create a Bezier curve.

The Spline, Cubic Bezier tool provides more control over your spline curve by allowing you to define the tangency for each curve segment. After the first point is defined, pairs of points are entered. As in figure 5-64, point 2 defines a point on the curve and point 3 establishes the direction in which the curve passes through that point. Subsequent pairs of points are entered until the curve is terminated with a double click.

The B-Spline Cubic tool creates a curve that passes through each point that is entered. form•Z then interpolates the segments between the points to create a third-degree B-Spline curve.

The Spline Sketch tool actually provides two methods of generating a spline curve. Individual points can be entered similar to the other spline curves and form•Z will interpolate the segments between the points, as in figure 5-64. Note, however, that this type of spline curve does not actually pass through the entered points, but is rather an "average" of these points. By holding down the mouse button and dragging the cursor, a stream of points are defined similar to the Stream Line tool except that form•Z interpolates the segments between these points to smooth out the line (figure 5-65).

Fig. 5-65. Spline Sketch tool used to stream points by holding down the mouse button while dragging the cursor across the screen.

All spline tool options have a toggle for Allow Intersecting Lines and Allow Collinear Points, as well as a Smooth Interval that determines the resolution of the curve and consequently the number of polygons in an object (figure 5-66).

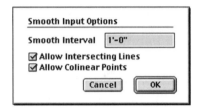

Fig. 5-66. Smooth Input Options dialog box.

TIPS AND STUDY AND APPLICATION NOTES

1. Pay careful attention to the Smooth Interval. This effects the resolution of the spline curve and the number of polygons in the objects created.

Tools 3: Arcs

LEARNING OBJECTIVE 5-13: *Creating arcs with various location methods.*

TOOLS COVERED:

- Arc, Clockwise, Endpoint-Last
- Arc, Counterclockwise, Endpoint-Last
- Arc, Clockwise, Center-Last
- Arc, Counterclockwise, Center-Last
- Arc, 3-Point, Endpoint-Last
- Arc, 3-Point, Midpoint-Last

Fig. 5-67. Tools covered.

These tools create arcs in various ways, either by locating center, beginning, and end points or by defining three points on the arc. The Arc, Clockwise, Endpoint-Last and Arc, Counterclockwise, Endpoint-Last tools allow you to draw arcs by first clicking the initial point, then the center, and then the end point of the arc (figure 5-68).

Fig. 5-68. Arcs created with the Arc, Clockwise, Endpoint-Last and Arc, Counterclockwise, Endpoint-Last tools.

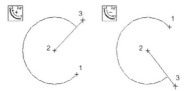

The Arc, Clockwise, Center-Last and Arc, Counterclockwise, Center-Last tools allow you to draw arcs by first defining the initial point, then the end point, and then the center of the arc (figure 5-69).

Fig. 5-69. Arcs created with the Arc, Clockwise, Center-Last and Arc, Counterclockwise, Center-Last tools.

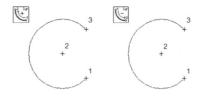

The 3-Point Arc, Endpoint-Last and 3-Point Arc, Midpoint-Last tools allow arcs to be drawn by clicking three points that defines the perimeter of the arc. The two tools differ only in regard to the order in which the perimeter points are entered (figure 5-70).

Fig. 5-70. Arcs created with the 3-Point Arc, Endpoint-Last and 3-Point Arc, Midpoint-Last tools.

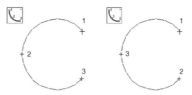

Arcs can be limited By # Of Segments or By Maximum Size Of Segments. Figure 5-71 shows the Arc/Circle/Ellipse Options dialog box.

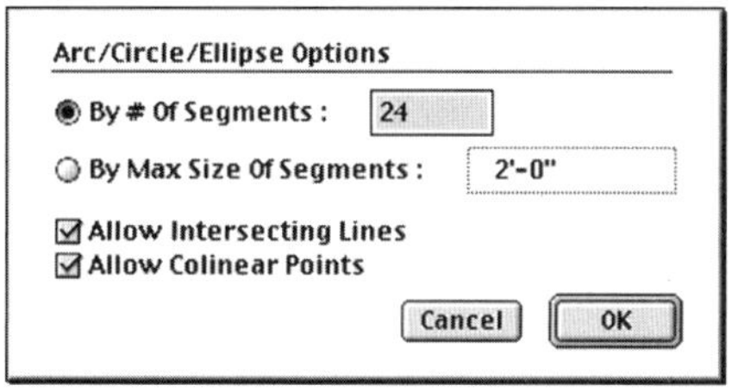

Fig. 5-71. Arc/Circle/Ellipse Options dialog box.

Chapter 5: Modeling Tools

MacOS/PC NOTE: *You can define an arc by the center point first, initial point second, and end point last by using the Ctrl key (Mac) or Ctrl + Alt keys (Windows) when clicking the first point.*

TIPS AND STUDY AND APPLICATION NOTES

1. Double clicking at the end point of an arc will end the arc; triple clicking will close it. You can also use the E key to end and C key to close the arc.

2. The Allow Intersecting Lines option permits you to create arcs that cross over each other. However, this option usually produces unexpected results and poorly formed objects.

Topological Levels Tool Palette

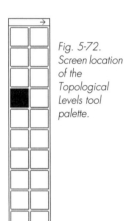

Fig. 5-72. Screen location of the Topological Levels tool palette.

The tools within the Topological Levels tool palette (screen location, figure 5-72; palette, figure 5-73) modify the function of many of the other tools, such as the Pick tools and Transformation tools. These tools determine which part of an object (points, line segments, and so on) will be affected by a tool.

Fig. 5-73. Topological Levels tool palette.

Tools 1: Modeling Components

LEARNING OBJECTIVE 5-14: *Using Topological Levels to modify the behavior of modeling tools.*

TOOLS COVERED:

- Point
- Segment
- Outline
- Face
- Object
- Group
- Hole/Volume
- Automatic

Fig. 5-74. Tools covered.

Topological levels are a classification system for elements in space. The form•Z topological classification describes levels from simplest to most complex: Point, Segment, Face, Object, and Group. The palette also

Using the Tools

includes two special types of levels: Outline and Hole/Volume. The final tool in the palette, Automatic, is not a Topological Level but is a tool that combines the functions of the other Topological Levels tools. Levels modify the behavior of many of the tools. For example, the Move tool can be used to transform the shape of an object by moving its points (Point Topological Level), or to change the object's position by moving the entire object (Object Topological Level).

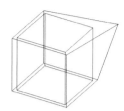

Fig. 5-75. A single point has been chosen and moved to distort the original object.

Point

The Point is the simplest element, needing only three coordinates (an *x*, *y*, and *z* position) to define its location in space. The point does not define a surface or volume and will not show up in a rendering. Selecting a Typological Level of Point allows you to manipulate a simple point, or group of points (figure 5-75).

Fig. 5-76. A segment has been chosen and moved to distort the original

Segment

In form•Z, a segment is a line between two points, and is defined by six coordinates: an *x*, *y*, and *z* value for each endpoint of the line. A segment is essentially an extruded point (figure 5-76).

Face

In form•Z, a face is a 2D element defined by at least three points, each determined by a set of *x*, *y*, and *z* coordinates. A face is a surface and is capable of reflecting light and producing shade and shadow (figure 5-77).

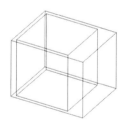

Fig. 5-77. The face selected was moved to produce one thicker wall.

Outline

An outline is similar to a face, except that it includes only the segments around the defined plane, not the actual 2D surface (figure 5-78).

Hole/Volume

Hole/Volume is unique to form•Z and allows you to select of number of outlines that constitute a hole or a volume on a face. A hole or volume on a face (figure 5-79) is made using the Insert Hole tool in the Insertions tool palette. For example, selecting the Typological Level of Hole/Volume will allow you to move, scale, or duplicate a window opening in a wall.

Fig. 5-78. The "inside top" outline was selected and scaled to produce slanted interior walls.

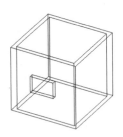

Fig. 5-79. Example of Hole/Volume use.

Fig. 5-80. An object selected and moved.

Fig. 5-81. A collection of 3D Enclosure objects constitutes this group.

Object

The Typological Level of Object is the default setting, and allows you to perform operations on entire objects when using the Geometric Transformation tools such as Move and Scale (figure 5-80).

Group

A group consists of objects, defined as a collection or set, using the Group tool. For more information about groups, see the discussion of the Group tool later in this chapter (figure 5-81).

Automatic

The Automatic tool can be used instead of the Point, Segment, Face, and Object Topological Levels tools. The Automatic tool is an "intelligent" tool and is designed to select which topological entity is desired based on the location of the cursor. For example, if the cursor is over a segment, the tool assumes you want to select a segment; if you are over a point, it assumes you want to select a point.

TIPS AND STUDY AND APPLICATION NOTES

1. form•Z defines objects by their Topological Level. Topology is the study of the properties of geometric objects that are preserved under deformations. No matter what the shape, form•Z interprets everything you draw as elements in these categories: Points, Segments, Outlines, Faces, Objects, Groups, Holes, and Volumes. You work with these categories by selecting the appropriate Topological Level.

2. The Option key (Mac) or Ctrl + Shift keys (PC) will allow you to select an object instead of a segment, face, or point when using the Automatic tool.

Using the Tools

Pick Tool Palette

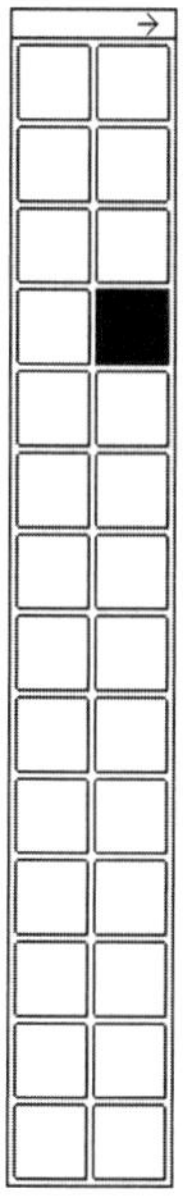

Fig. 5-82. Screen location of the Pick tool palette.

The Pick tools (screen location, figure 5-82; palette, figure 5-83) are used to select entities to be edited, as well as provide some basic editing functions. When an entity is "picked," it is highlighted in red (the default color).

Fig. 5-83. Pick tool palette.

Tools 1: Pick

LEARNING OBJECTIVE 5-15: *Selecting entities for performing modeling operations.*

TOOL COVERED:

- Pick

Fig. 5-84. Pick tool.

The Pick tool is used to activate entities to be manipulated using the other tools in the Model Tools palette. Entities can be picked individually by clicking on them with the mouse, or by drawing a window around them. Face and outlines must be picked by clicking on two edges of the respective face or outline. Once two edges are selected, the remaining edges belonging to the face or outline will automatically be highlighted. (See figure 5-85.)

When picking outlines, faces, holes, and volumes, selecting the Inside Boundaries option allows you to select entities by clicking anywhere inside their boundaries. This is an alternative to the default Clicking On Edges option.

The Click And Drag option adds editing functionality to the Pick tool by allowing you to move, rotate, or scale entities by clicking on the entity and then dragging or otherwise manipulating it. If you want, you can use modifier keys to rotate and scale the object when using the Click and Drag option.

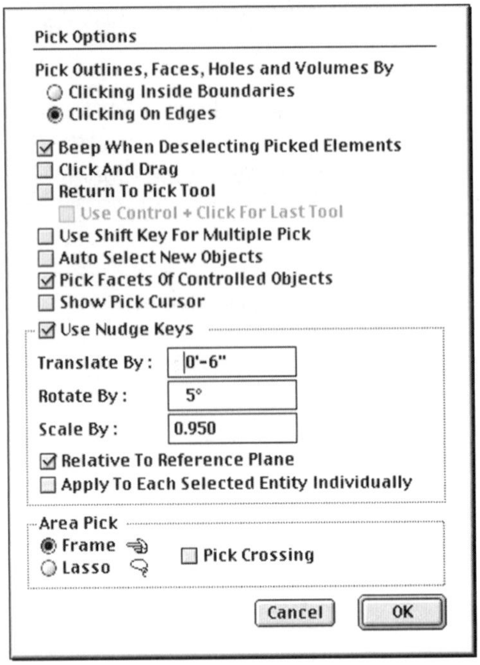

Fig. 5-85. Pick Options dialog box

Checking the Nudge Keys option allows you to incrementally move, rotate, or scale entities by using the arrow keys in conjunction with modifier keys. Default modifier keys are listed in the Key Shortcuts at the bottom of the Edit pull-down menu. The Area Pick options, Frame and Lasso, allow you to pick entities by drawing a freeform outline around them (Lasso) or a box (Frame). The Frame and Lasso options shown in figures 5-86 and 5-87, respectively. The Pick Crossing option allows you to pick objects by crossing the Lasso or Frame across them instead of having to surround them.

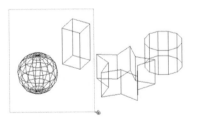

Fig. 5-86. Picking multiple objects using the Frame option.

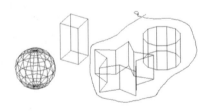

Fig. 5-87. Picking multiple objects using the Lasso option.

Using the Tools

TIPS AND STUDY AND APPLICATION NOTES

1. You can unpick (deselect) entities by clicking on them a second time. To unpick everything, click on an empty space.
2. Pick Parade can be used in instances where it is difficult to pick an entity because multiple objects are overlapping. Holding down the Shift key with picking will "parade" you through each selection.

Tools 2: Edit Controls

LEARNING OBJECTIVE 5-16: Editing the controls of parametric objects.

TOOL COVERED:

- Edit Controls

Fig. 5-88. Edit Controls tool..

The Edit Controls tool allows you to manipulate the attributes of form•Z's primitive, Nurbz, and Patch objects. This includes objects generated with tools from the Primitives tool palette and NURBS and Patches tool palette.

Each of the objects made with tools from the Primitives tool palette have control attributes such as Height, Width, and Depth for the Cube. If you select the Edit Controls tool and click on a Primitives object, its control handles will be displayed. By clicking on one or more of these control handles, you can edit the object's attributes (figure 5-89).

Fig. 5-89. A parametric cube with its edit controls.

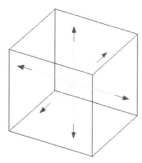

Objects generated with Nurbz and Patch tools have control points and these points can be edited using the Edit Controls tool. When working with a Nurbz or Patch surface, Edit Controls activates the control points on the object and allows them to be edited (figure 5-90).

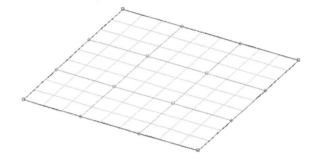

Fig. 5-90. A Patch object with its control points.

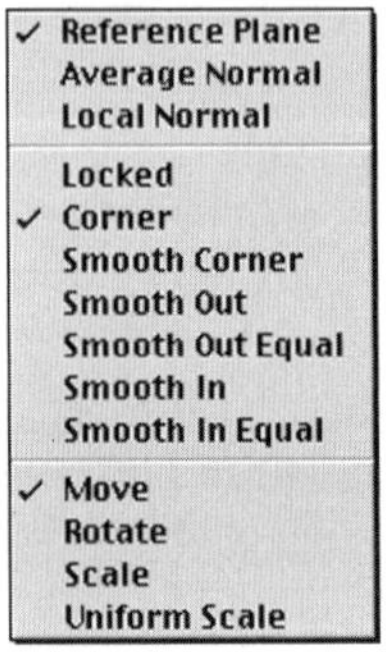

When editing Nurbz or Patch objects with the Edit Controls tool, there is a contextual options menu that controls the type of transformation, the reference plane, and the smooth and lock control points. This contextual menu (figure 5-91) pops up at the cursor whenever the Edit Controls tool is active. The mouse button is held down and the Option key (Mac) or Shift + Alt keys (PC) are depressed.

Fig. 5-91. Edit Controls contextual menu.

Parametric objects in form•Z can be automatically converted to Nurbz or Patch objects and deformed or shaped using the Edit Controls tool. Selecting the Convert To option in the Options dialog box (or Tool Options palette) will add control points to an object, allowing its surface to be shaped and deformed with the Edit Controls tool. (See figure 5-92.)

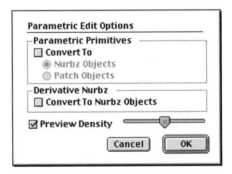

Fig. 5-92. Parametric Edit Options dialog box.

TIPS AND STUDY AND APPLICATION NOTES

1. The Edit Controls tool does not work with plain polygon objects (those generated with the drawing tools and tools from the Object Type tool palette) because they have no control parameters.

Tools 3: Edit Surface

LEARNING OBJECTIVE 5-17: *Editing the surface controls of Parametric, Patch, or Nurbz.*

TOOL COVERED:

- Edit Surface

Fig. 5-93. Edit Surface tool..

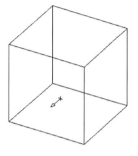

Fig. 5-94. Editing the surface of a Primitive object.

Using the Edit Surface tool, you can manipulate the surfaces of a Parametric, Nurbz, or Patch object. When working with a Primitive object, the Edit Surface tool will allow you to move faces of the object, changing its length or width dimensions. When using the Edit Surface tool to move the faces of a Primitive object, the cursor changes from a cross to an arrow. The arrow indicates in which direction the surface is being pulled or pushed when editing (figure 5-94).

If you are working on a Nurbz object, you can use the Edit Surface tool to pull and push at points on the surface to sculpt the form of the object (figure 5-95). When editing a surface with the Edit Surface tool, the point at which you click on the object is the point that will be moved. The Edit Surface tool can be used with the Topological Levels of Object, Face, and Segment to change the shape of an entire object, or portions of it.

Fig. 5-95. Editing the surface of a Patch object.

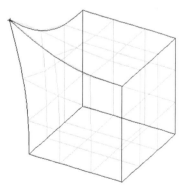

When sculpting a surface, the Edit Surface tools move points on the surface in a direction perpendicular to the Reference Plane or at a Normal to the surface. The Normal to a surface is the direction exactly perpendicular to a surface at any point. To change this option, use the contextual Edit Surface menu (figure 5-96). This menu pops up at the cursor

Fig. 5-96. Edit Surface contextual menu.

when the Edit Surface tool is active, that is, the object is active, the mouse button is held down, and the Option key (Mac) or the Shift + Alt keys (PC) are depressed.

TIPS AND STUDY AND APPLICATION NOTES

1. The options dialog box for the Edit Surface tool is identical to the Edit Controls tool. The options allow you to automatically convert an object to a Nurbz object when you select it with the Edit Surface tool.

2. The Edit Surface tool does not work on plain polygon objects, that is, objects generated with tools from the Object Type tool palette.

Derivatives Tool Palette

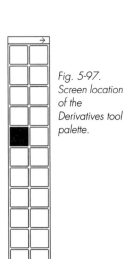

Fig. 5-97. Screen location of the Derivatives tool palette.

The tools in the Derivatives tool palette (screen location, figure 5-97; palette, figure 5-98) are used to create 3D objects; that is from 2D objects through processes such as extrusion and skinning. Some of the tools perform a converse function of creating 2D objects from 3D objects through actions such as sectioning.

Fig. 5-98. Derivatives tool palette.

Tools 1: 2D Surface Object and 2D Enclosure

LEARNING OBJECTIVE 5-18: *Generating 2D surface objects from source objects or parts of source objects.*

TOOLS COVERED:

- 2D Surface
- 2D Enclosure

Fig. 5-99. 2D Surface and 2D Enclosure tools.

2D Surface and 2D Enclosure are derivative tools that can generate 2D surface objects from objects or parts of objects. These tools produce objects similar to those produced with the 2D tools in the Object Type tool palette; however, they can be used on objects that have already been created.

The 2D Surface tool generates surfaces from lines, faces, or objects. From a line, the 2D Enclosure tool generates two parallel lines offset from the original. It also automatically creates beginning and end segments, which always results in a surface that is closed (figures 5-100 and 5-101). 2D surface objects can be generated from any source object, face, or line. For example, if you apply the 2D Surface tool to a 3D solid, it will be "exploded," and all of faces will be converted into individual surfaces.

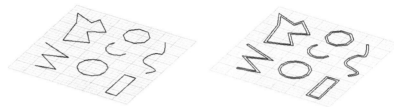

Fig. 5-100. Objects generated with the 2D Surface Object tool.

Fig. 5-101. Objects generated with the 2D Enclosure tool.

The 2D Enclosure tool produces objects similar to those generated by the 2D Enclosure tool in the Object Type tool palette. However, unlike its twin in the Object Type tool palette, this 2D Enclosure tool can be used on objects other than lines. For example, it will derive a 2D frame from a solid object. If this tool is applied to a 3D solid, all edges will be interpreted as source objects and the entire object will be turned into a 2D frame, as demonstrated in the Geodesic Dome exercise in Chapter 7.

TIPS AND STUDY AND APPLICATION NOTES

1. The 2D Enclosure tool contains options identical to the tool of the same name in the Object Type tool palette. The options allow you to set the Enclosure Wall Thickness and the Justification (figure 5-102).

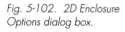

Fig. 5-102. 2D Enclosure Options dialog box.

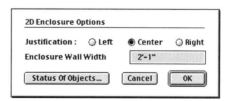

2. 3D objects can be derived from a 2D surface object or 2D enclosure by using tools in the Derivatives tool palette.

Tools 2: 3D Extrusion, 3D Converged, and 3D Enclosure

LEARNING OBJECTIVE 5-19: *Creating 3D objects from lines, faces, and objects.*

TOOLS COVERED:

- 3D Extrusion
- 3D Converged
- 3D Enclosure

Fig. 5-103. Tools covered.

These three derivative tools produce 3D objects from lines, faces, objects, and parts of objects. These tools perform functions similar to their twins, the modifiers in the Object Type tool palette. However, these tools can generate 3D objects from a wide variety of source objects and entities.

The 3D Extrusion tool extrudes lines and surfaces to produce solid and surface objects by extruding all of the points of the source object directly upward to a prescribed height. In figure 5-104, the 3D Extrusion tool has been applied to a 2D surface object. The form was extruded with the Perpendicular to the Reference Plane option. Objects can also be extruded with the Perpendicular to the Surface option.

Fig. 5-104. A 2D surface object can be transformed into a 3D solid using the 3D Extrusion tool.

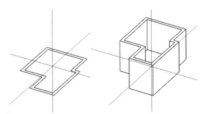

NOTE: *For more information on the 3D Extrusion tool, see its twin (also named 3D Extrusion) in the Object Type tool palette.*

The 3D Converged tool also extrudes the points of the source object upward, but also moves them inward to converge at a single point at the prescribed height. The resulting objects are cones and pyramids. This tool, like the other 3D derivative tools, can be applied to lines, surfaces, and faces of existing objects. In figure 5-105, the 3D Converged tool has been applied to a single face of a rectangular solid to create a pyramidal top.

Fig. 5-105. The top face of an object is used as a source for the 3D Converged tool to create a pyramidal top.

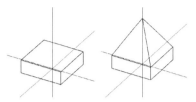

The 3D Enclosure tool produces "walled" objects from lines, surfaces, and objects. This tool is similar to the 2D Enclosure tool, producing walls from edges and lines, but it also extrudes walls into 3D solids. Unlike the 3D Enclosure tool in the Object Type tool palette, this tool can be used on solid objects as well as lines.

Figure 5-106 shows the 3D Enclosure tool applied to a face of a 3D solid. Edges are interpreted as lines and thickened to create walls. The height of the face was set in the Heights menu to be the same dimension as the wall thickness, creating a uniform frame.

Fig. 5-106. A 3D frame generated from the face of a 3D object using the 3D Enclosure tool.

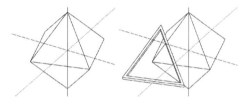

➥ **NOTE:** *For more information on the 3D Enclosure tool, see the Geodesic Dome exercise in Chapter 7.*

TIPS AND STUDY AND APPLICATION NOTES

1. The height of objects generated with the 3D Extrusion, 3D Converged, and 3D Enclosure tools is set by selecting a height, or the Graphic/Keyed option, in the Heights pull-down menu.

Tools 3: Parallel

LEARNING OBJECTIVE 5-20: *Generating parallel objects from pre-selected objects.*

TOOL COVERED:

- Parallel

Figure 5-107. Parallel tol.

The Parallel tool generates parallel objects from preselected objects. Parallel objects are objects of similar shape but different size or location. The original object can consist of lines, surfaces, or solids. You can create single or double parallel objects. Figure 5-108 shows the Parallel Options dialog box. Here you have the option of creating a Single Parallel or Double Parallel object.

Fig. 5-108. Parallel Options dialog box.

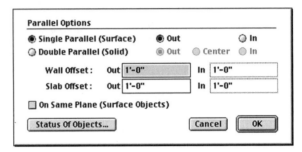

Single Parallel (figure 5-109) will offset the surface of an object by a distance predefined in the dialog box. When applied to a solid, the surface of the solid is moved either In or Out by the predefined distance, but the object remains a solid (figure 5-110).

 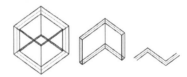

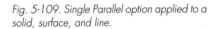

Fig. 5-109. Single Parallel option applied to a solid, surface, and line.

Fig. 5-110. Double Parallel with Center option applied to a solid, surface, and line.

Figure 5-111 shows the Double Parallel option. Here, Center was selected. The objects are offset by the predefined distance using the original object as the new object's center. When In or Out is selected, the original object is retained and a copy of the surface or vector line is offset in the previously designated direction (either in or out). This tool is useful for architectural applications such as adding wall or slab thickness to complex objects that would otherwise be difficult to create. In figure 5-111, three primitive forms were combined using Boolean Union. The Double Parallel object option in the Out direction was then applied.

Fig. 5-111. Parallel tool applied to more complex objects.

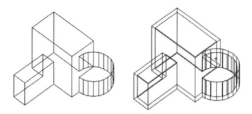

TIPS AND STUDY AND APPLICATION NOTES

1. In the options, Slab is considered any horizontal surface with a tilt of 5 degrees or less. All other entities are considered Walls.

2. In the case of a horizontal surface, Out is up and In is down when using the Single Parallel or Double Parallel options.

Tools 4: Projection and Unfold

LEARNING OBJECTIVE 5-21: *Producing 2D objects from 3D shapes using geometric projection or unfolding.*

TOOLS COVERED:

- Projection
- Unfold

Figure 5-112. Tools covered.

The Projection tool produces 2D objects from 3D shapes either by geometric projections or by unfolding the object. The Projection tool contains two main options: Projection Of View and Orthographic Projection (figure 5-113).

Fig. 5-113. Projection Options dialog box.

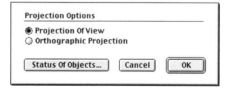

The Projection Of View option produces a 2D object on the reference plane that is a projection of the 3D object in the current view. When this tool is used, the derivative object will not look correct. This is because the object has been created in a reference plane. In figure 5-114, the object was projected onto the *xy* reference plane; therefore, it does not appear correct until viewed in the top view.

Fig. 5-114. Projection of view.

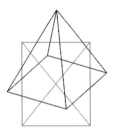

The Orthographic Projection tool creates 2D projections using the current reference plane to determine the projection. This is ideal for creating orthographic drawings from a 3D object (figure 5-115).

Fig. 5-115. Orthographic projection.

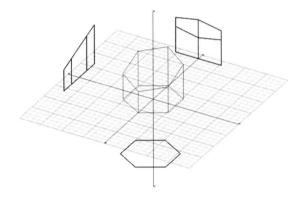

Unfold is used to derive a 2D construct of a 3D solid. The solid object in figure 5-116 has been unfolded several ways to produce the 2D objects shown in figure 5-117. The unfolded object is derived differently based on where you click on the 3D object. In the dialog box (figure 5-118), you can select from the various modes of unfolding.

TIPS AND STUDY AND APPLICATION NOTES

1. To ensure you get only one 2D object per object when you project, instead of one object for every face, toggle Store All Volumes as One Object under the Status of Objects in the Unfold Options.

2. The Unfold tool calculates every face in an object; therefore, the computation time is proportional to the number of polygons in an object. A very complicated object may take a considerable amount of time

Using the Tools

Fig. 5-116. The original object.

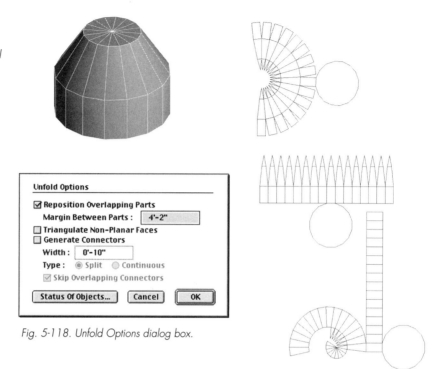

Fig. 5-118. Unfold Options dialog box.

Fig. 5-117. Unfolded projection of the original object.

Tools 5: Terrain

LEARNING OBJECTIVE 5-22: *Creating a 3D solid from 2D contour lines.*

TOOL COVERED:

- Terrain

Fig. 5-119. Terrain tool.

The Terrain tool creates a 3D solid from a series of 2D contour lines. The tool is specifically designed for creating models of landscape and sites. The Terrain tool contains three primary options that define the type of terrain model that will be generated: Mesh, Stepped, or Triangulated.

The Mesh option generates a polygonal meshed solid that smoothly interpolates transitions between contours. The Stepped terrain model extrudes each contour until it reaches the starting height of the next contour, producing a model that looks like stacked layers of cardboard. The Triangulated option generates a solid by generating faces between contour lines and triangulating them to make them planar. A triangulated terrain object usually contains fewer polygons than a mesh terrain, and is normally smoother than a Stepped model.

There are two options for placing contour heights: Use Contour Heights and Adjust Heights At Intervals Of. The former option requires you to move all contours to their proper height in the *z* dimension. The latter option will place the contours at even intervals based on a given increment as specified in the Terrain Model Options dialog box (figure 5-120).

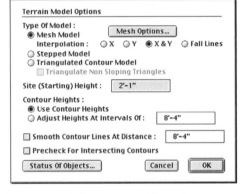

Fig. 5-120. Terrain Model Options dialog box.

Fig. 5-121. 2D lines for the Terrain tool.

The Terrain tool creates a 3D solid landform from a series of contour lines. 2D contours should be drawn with the 2D Surface object tool. Along with the contours, you will need to supply a boundary object (a 2D Surface object that will define the extents of the solid landform). The 2D contours must be closed lines, or if they are open, both ends must lie outside the boundary of the form (figure 5-121).

Mesh Terrain creates a polygonal meshed object from contours (figure 5-122). Stepped Terrain produces a model that looks as if it were cut out of layers of cardboard (figure 5-123). Triangulated creates a mesh by creating faces between contours and then triangulating them (figure 5-124).

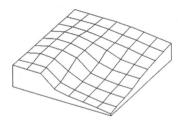

Fig. 5-122. A Meshed Terrain model.

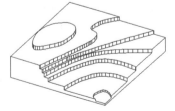

Fig. 5-123. A Stepped Terrain model.

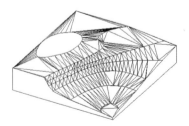

Fig. 5-124. A Triangulated Terrain model.

Using the Tools

TIPS AND STUDY AND APPLICATION NOTES

1. If you are using the Adjust Heights At Intervals Of option, it is necessary to select contour lines in order from bottom to top (lowest to highest elevation).

2. If you are using the Adjust Heights At Intervals Of option, and you have two separate lines at the same height, join them with the Join tool first, which will enable you to select them as one object.

3. Before using your terrain model in Boolean operations, you may need to triangulate the model using the Triangulate tool.

Parametric Derivatives Tool Palette

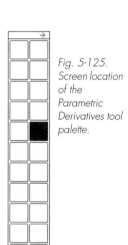

Fig. 5-125. Screen location of the Parametric Derivatives tool palette.

The tools in the Parametric Derivatives tool palette (screen location, figure 5-125; palette, figure 5-126) are used to generate objects of complex shape. Tools such as Revolved Object (Revolve), Sweep and Skin generate 3D objects from simple 2D lines while the Bolt/Screws and stair tools allow you to generate complete 3D objects of great complexity in only a few mouse clicks.

Fig. 5-126. Parametric Derivatives tool palette.

Tools 1: Revolved Object

LEARNING OBJECTIVE 5-23: *Producing "lathed" forms.*

TOOL COVERED:

- Revolved Object

Fig. 5-127. Revolved Object tool.

The Revolved Object (Revolve) tool produces lathed objects; that is, forms that appear as if they have been produced by spinning on a woodworkers lathe or potter's wheel. These forms are "turned" or revolved from a single 2D profile.

The default settings of the Revolve tool will produce a 3D solid or surface object from a 2D profile. To revolve a profile, select the Revolved Object tool, the profile or source shape, and then an axis of revolution (x, y, or z). The resulting object will be a solid if both ends of the profile are touching, but do not cross, the axis. Figure 5-128 shows a profile that has been revolved to produce a solid.

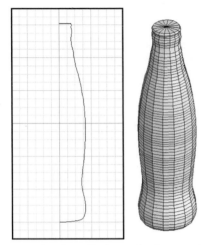

Fig. 5-128. The profile that touches the axis at both ends generates a solid object when revolved.

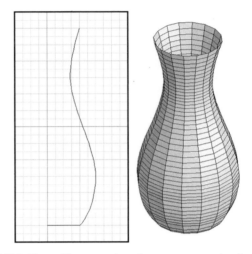

Fig. 5-129. The profile, not touching the axis on one end, will produce a surface object, open at the top like a vase.

If both ends of the source profile do not touch the axis of revolution, a surface object will be generated. Figure 5-129 shows a surface object that has been generated using the Revolved Object tool. Surface objects cannot be used in Boolean operations, unlike a solid object produced with the Revolved Object tool.

The Revolved Object dialog box (figure 5-130) contains options that allow you to set the Revolution Angle, Direction of Revolution, # of Steps, as well as the type of surface or object created: Faceted or Nurbz (figure 5-131). The revolution angle is the number of degrees the object is revolved around the axis. The default revolution angle is 360, which is a full revolution producing a round object. Any number less than 360 for Revolution Angle will produce an object that has a sectional slice missing, like the apple shown in figure 5-132. When working with open source shapes (as in figure 5-129) on which you intend to perform Stitch and Trim operations, toggle on the Surface Object From Open Source option.

Fig. 5-130. Revolved Object dialog box.

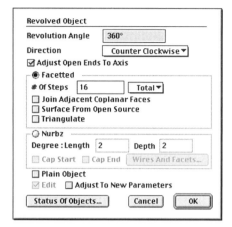

Fig. 5-131. The exact same profile was used to create both a Faceted object (left) and a Nurbz object (right). Notice that although the two objects have approximately the same complexity wireframe, the Nurbz object appears smoother when rendered.

Fig. 5-132. By changing Revolution Angle to 270 in the Revolved Object options dialog box, an apple is generated with a slice missing.

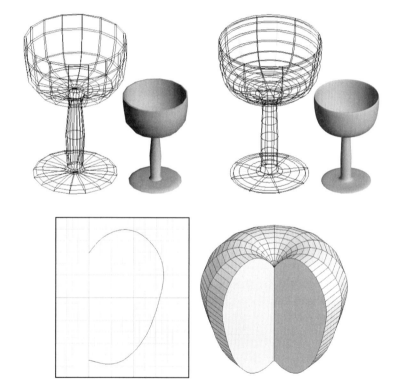

Chapter 5: Modeling Tools

TIPS AND STUDY AND APPLICATION NOTES

1. When forming solid objects to be used later with Boolean operations, it is important that your source shapes terminate on the axis of revolution.

2. See the Goblet exercise in Chapter 7 for additional information about the Revolved Object tool.

Tools 2: Helix and Screws

LEARNING OBJECTIVE 5-24: Creating complex spiral forms.

TOOLS COVERED:

- Helix
- Bolt/Screw

Fig. 5-133. Tools covered.

form•Z has provided built-in methods of generating many complex objects. Using the Helix tool, you can create a variety of helical and spiral shaped objects that would be very difficult to model otherwise. The Helix dialog box (figure 5-134) provides you with a number of generation methods and options.

Fig. 5-134. Helix dialog box.

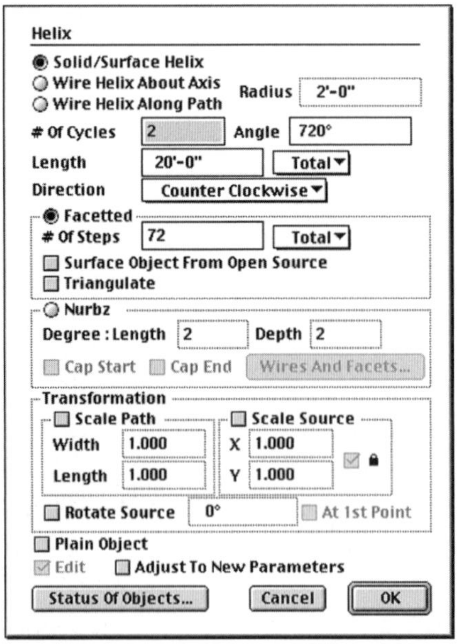

Using the Tools

By adjusting the various options, helix objects can be generated from both open and closed shapes. The Solid/Surface Helix option allows you to generate both Faceted and Nurbz objects whereas the Wire Helix About Axis and Wire Helix Along Path options allow you to generate 2D spirals that can be used as paths for Sweep operations (figure 5-135).

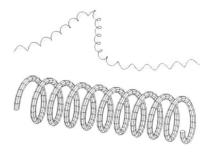

Fig. 5-135. Spiral objects generated with the Helix tool.

The Bolt/Screw tool is another built-in method of easily generating complex shaped objects. The Bolts/Screws dialog box (figure 5-136) provides a very intuitive graphic interface for adjusting the settings to create a limitless number of different screws and bolts. In addition to adjusting numeric dimensions, you can also choose between plain, slotted, and Phillips-head type screws and bolts (figure 5-137). Unchecking the Tip, Neck, or Head box allows you to eliminate these parts from your objects. You can adjust the number of segments in your objects in the Points Per Cycle field.

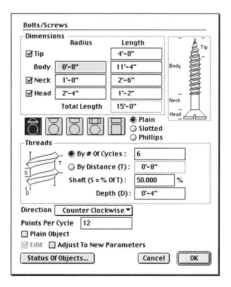

Fig. 5-136. Bolts/Screws dialog box.

218   *Chapter 5: Modeling Tools*

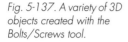
Fig. 5-137. A variety of 3D objects created with the Bolts/Screws tool.

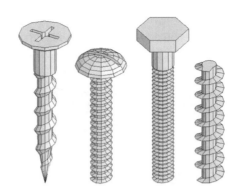

TIPS AND STUDY AND APPLICATION NOTES

1. These tools can easily generate thousands of polygons in only a few clicks. It is important to keep the number of segments and faces in your objects as low as possible to prevent unnecessarily long rendering times.

2. See the five-minute Bolt exercise in Chapter 7 and the Telephone exercise in Chapter 9 for addition information.

Tools 3: Stairs

LEARNING OBJECTIVE 5-25: *Generating 3D staircases.*

TOOLS COVERED:

- Spiral Stair
- Stair From Path

Fig. 5-138. Tools covered.

form•Z has two tools dedicated specifically to the modeling of 3D staircases. The Spiral Stair tool is used for creating a staircase that spirals about an axis of revolution, whereas the Stair From Path tool is used to create runs of stairs from a path line (figure 5-139).

Using the Tools

Fig. 5-139. Stairs generated with the Spiral Stair and Stair From Path tools.

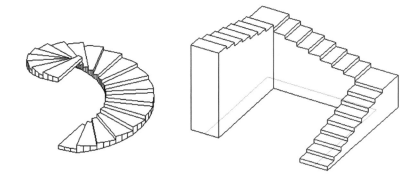

The Spiral Stair dialog box (figure 5-140) contains all options needed to adjust every aspect of your spiral stair. The Warn When Exceeds Limits option, when active, will notify you when your stairs exceed certain criteria that can be defined by the user. This allows you to verify that your stairs meet certain building code requirements and other such restrictions.

Fig. 5-140. Spiral Stair dialog box.

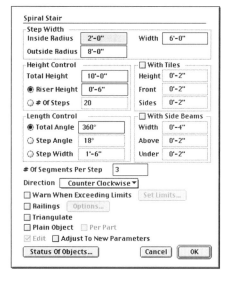

In addition to adjusting the dimensions and type of stairs created, the Railings options offer you a number of methods of easily generating handrails and railings for your steps. Figure 5-141 shows some of the possible results.

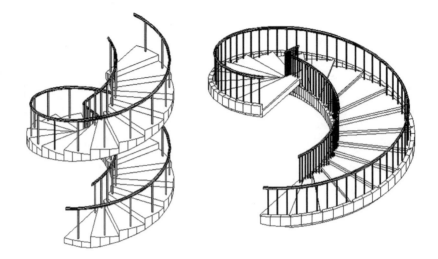

Fig. 5-141. By adjusting the Spiral Stair options, you can create many types of spiral stair and railing combinations.

The Stair From Path tool generates stairs from any 2D path, including straight lines, splines, and arcs. The path line will yield more predictable results when drawn on the *xy* plane or parallel to it, but it is possible for the path line to lie anywhere in 3D space. The path line can also be open or closed, but a closed path is required to generate repeating, multiple-flight stairs. Figure 5-142 shows the Stair From Path dialog box.

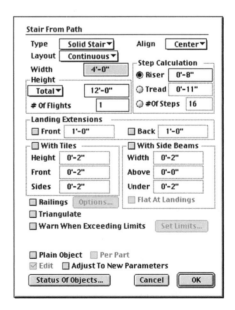

Fig. 5-142. Stair From Path dialog box.

Using the Tools

Many of the settings, including those related to Railings and Warn When Exceeding Limits, function in the same manner as the Spiral Stair tool. The Type pull-down menu allows you to choose between six different stair types, as shown in figure 5-143. The Align pull-down menu determines how the stairs are generated in relation to the 2D path line, whereas the Layout pull-down menu allows you to determine how the steps of your stairs are configured and if landings are included.

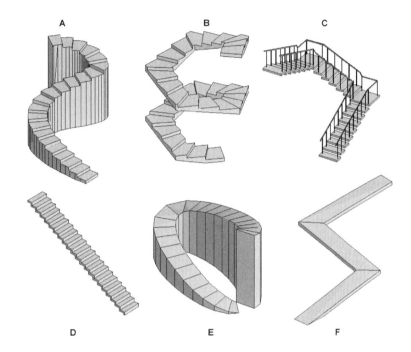

Fig. 5-143. The six different stair types available with the Stair From Path tool generated with different stair options from various 2D path lines: (a) Solid Stair, (b) Beam Stair, (c) Steps Only, (d) Steps and Risers, (e) Solid Ramp, and (f) Beam Ramp.

TIPS AND STUDY AND APPLICATION NOTES

1. See the five-minute Stair exercise in Chapter 7 for additional information.

Tools 4: Sweep

LEARNING OBJECTIVE 5-26: *Creating organic and curvilinear forms with a sweep.*

TOOL COVERED:

- Sweep

Fig. 5-144. Sweep tool.

The Sweep tool is one of the many tools in form•Z for creating organic and curvilinear forms. Like the C-Mesh and Skin tools, Sweep uses a combination of 2D objects to produce a 3D form. The Sweep tool extrudes, or "sweeps," one 2D profile (referred to as a source) along another 2D profile (referred to as a path).

Axial Sweep

Axial Sweep combines a single open or closed 2D source and single 2D path into a swept object with a uniform cross section (figures 5-145 and 5-146). The source and the path can be located anywhere and do not have to touch. Avoid path bends that are too tight or the swept object will fold over itself, producing a poorly formed object.

Fig. 5-145. A sweep path.

Fig. 5-146. A 3D object created with a single source sweep.

2 Source Sweep

The 2 Source Sweep command sweeps a source along a path. During the sweep, the source shape gradually transforms from one source shape into another. If the two source objects do not have the same number of points, the function will add points to make a smooth transition between the two (figures 5-147 and 5-148).

Fig. 5-147. The same sweep path.

Fig. 5-148. Merging from a circular source shape to a square source shape along a single path.

2 Path Sweep

The 2 Path Sweep command sweeps a single source between two paths, compressing or stretching the source to keep it in contact with both paths. Preserve Source Height maintains the proportions of the original source as it gets stretched (figures 5-149 and 5-150).

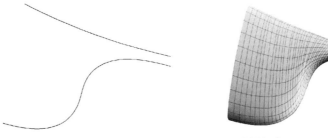

Fig. 5-149. Two sweep paths.

Fig. 5-150. One source object swept between two paths.

Boundary Sweep

A boundary sweep requires one open source and one path. This operation sweeps the source profile around the path. The Boundary Sweep command closes the form by placing a flat face on its top and bottom (figures 5-151 and 5-152).

Fig. 5-151. A 2D sweep path.

Fig. 5-152. A single source swept around a path to produce a boundary.

TIPS AND STUDY AND APPLICATION NOTES

1. The preview window in the Sweep Edit dialog box (figure 5-153) allows you to view the final form before it is produced. This window also controls parameters of the operation and attributes of the elements. These options let you change the direction of both the source and the path, as well as rotate, translate, and scale the source to alter the outcome. The Sweep Edit dialog box opens automatically when the Sweep tool is used.

Chapter 5: Modeling Tools

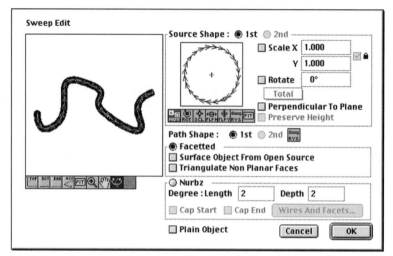

Fig. 5-153. Sweep Edit dialog box.

2. For more information about the Sweep tool, see the Chair exercise in Chapter 8 and the Telephone exercise in Chapter 9.

Tools 5: Skin

LEARNING OBJECTIVE 5-27: *Generating 3D surfaces by sweeping source shapes along 2D paths.*

TOOL COVERED:

- Skin

Fig. 5-154. Skin tool.

The Skin tool is used to generate a 3D surface by means of source shapes swept along 2D paths. It is similar to the Sweep tool but is not limited to a set number of source shapes and paths.

The Skin tool uses paths to guide interpolations between the sequentially positioned source shapes. It is important to align the source shape and the paths properly for the tool to function properly. It is a good idea to use Point Snap when placing the source shapes on the paths so that the corresponding points are on top of each other. Figure 5-155 shows a telephone handset with two sources shapes placed at the ends of four paths.

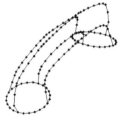

Fig. 5-155. Source shapes and paths for the skinning of a telephone handset.

There are many options available under the Skin dialog box (figure 5-156), the most important being the # of Paths and # of Sources options,

Using the Tools

which determine how many paths and source objects you can use in a skin. It is useful to spend some time with this command, to understand just how many possibilities exist.

Fig. 5-156. Skin dialog box.

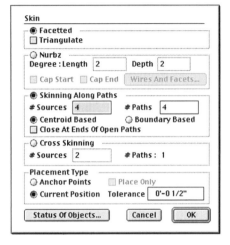

Fig. 5-157. The completed skinned model.

In the example shown in figure 5-157, # Sources is 2 and # Paths is 4. In the Placement Type, By Current Position is used because the source shapes and the paths are in positions they will occupy in the final model.

The Close At Ends command of the Open Paths option in the Skin Options dialog box is crucial in determining whether the resulting object is to be open or closed at its ends. When this option is not selected, the skin operation will create a surface with open ends, whereas when the option is selected, the object is generated as a solid. In figure 5-158, the figure on the left is a surface, whereas the figure on the right is a solid as a result of the Close At Ends command.

Fig. 5-158. Close At Ends command of the Open Paths option.

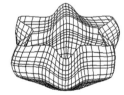 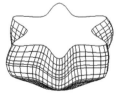

Chapter 5: Modeling Tools

TIPS AND STUDY AND APPLICATION NOTES

1. When using the Skin tool it is visually helpful to have the Show Points option selected in the Wire Frame Options dialog box in the Display Options menu.

2. In the Placement Type options, By Anchor Points allows you to have the source shapes and paths located anywhere. The Skin tool will assemble them based on predefined anchor points. However, By Current Position should be used whenever possible because it produces more reliable results; you position the entities rather than letting the computer calculate their positions.

3. The C-curve tool can be used to smooth out paths and source shapes. One of the Tangent Quick Curves should be used because they are the only C-curve option that will preserve the locations of the current points along the line.

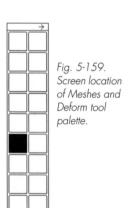

Fig. 5-159. Screen location of Meshes and Deform tool palette.

Meshes and Deform Tool Palette

The tools in the Meshes and Deform tool palette (screen location, figure 5-159; palette, figure 5-160) are a powerful set of procedures for advanced shape editing, including meshing, smoothing, triangulating, and deforming.

Fig. 5-160. Meshes and Deform tool palette.

Tools 1: Mesh

LEARNING OBJECTIVE 5-28: *Generating meshes from surfaces, solid objects, or selected faces.*

TOOL COVERED:

- Mesh

Fig. 5-161. Mesh tool.

Meshes, or subdivided surfaces, can be generated from surfaces, solid objects, or selected faces of objects. Generating meshes lends malleability to a surface and allows for operations to be performed on the meshed portions that would otherwise lack the points and segments required. The Mesh Options dialog box is shown in figure 5-162. The resolution of the mesh is controlled by the values entered in the *x*, *y*, and *z* fields. A mesh can be generated for All Directions or a specific direction.

Fig. 5-162. Mesh Options dialog box.

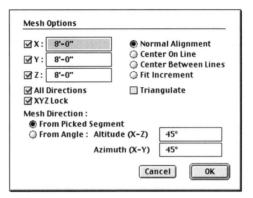

Meshing adds points and segments to an object or surface and allows for further operations (such as deformations). These points and segments increase the "flexibility" of an object. The finer the mesh, the more flexible and smooth an object will be when deformed (figure 5-163).

Fig. 5-163. An unmeshed object versus a meshed object.

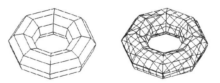

Meshes can be performed recursively, meaning that a meshed object can be meshed again to increase its smoothness. Figure 5-164 shows a mesh generated on a selected portion of a model. First, the top face of the object was meshed. Then selected faces of the mesh were further meshed with a finer resolution so that the deformation could be performed on a specific part of the model.

Fig. 5-164. Meshing faces of objects and faces within meshed surfaces.

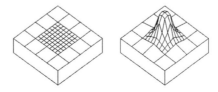

The mesh will vary depending on which segment you click on when generating it. As a default, the mesh grid is created parallel to the first segment picked. This can be changed by deselecting the From Picked Segment option in the Mesh options dialog box (figure 5-165).

Fig. 5-165. Mesh direction determined by the first mouse click.

TIPS AND STUDY AND APPLICATION NOTES

1. The Mesh tool can also generate meshes relative to the reference plane rather than relative to a picked segment. This is done using the From Angle option under Mesh Direction in the Meshes option box. Altitude is the angle measured from the $+x$ axis on the zx plane. Azimuth is the angle measured from the $+x$ axis on the xy plane.

2. The Mesh tool creates a mesh on an object with forced rectangular dimensions, which often result in nonplanar faces (which means that the four points that define a face are not in the same plane). This may become a problem later on when performing other operations on the object.

✓ **TIP:** *To avoid this problem, it is a good idea to have the Triangulate option checked when generating the mesh.*

Tools 2: Smooth Mesh

LEARNING OBJECTIVE 5-29: *Smoothing mesh edges to change the resolution of an object.*

TOOL COVERED:

- Smooth Mesh

Fig. 5-166. Smooth Mesh tool.

This tool changes the resolution of an object by subdividing its faces while at the same time smoothing its edges by adjusting the positions of the points of the newly generated faces. The Smooth Mesh Options dialog box is shown in figure 5-167.

Using the Tools

Fig. 5-167. Smooth Mesh Options dialog box.

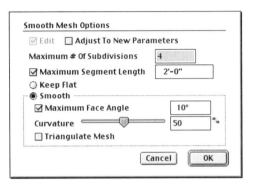

Toggling the Edit option will invoke the Smooth Mesh Edit dialog box (figure 5-168) when the Smooth Mesh tool is used on an object. This preview box contains the parameters for the smoothing of the object. If the Edit option is not selected, the smoothness of an object being meshed is determined by the # of Subdivisions, the Maximum Segment Length, the Maximum Face Angle, and the Curvature.

Fig. 5-168. Smooth Mesh Edit dialog box.

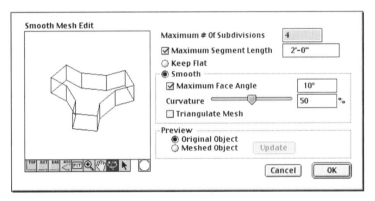

Figure 5-169 shows results of the Keep Flat option, which preserves the shape of the object as it meshes the object. The Smooth Mesh algorithm produces subdivisions based on an object's edge conditions. This option may be desirable over the regular Mesh tool when further operations will be performed on the object.

Fig. 5-169. Smooth Mesh with Keep Flat option result.

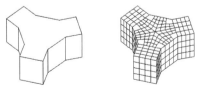

To generate the object in figure 5-170, the Smooth option was selected in the Smooth Mesh Options dialog box. Maximum Face Angle parameters are used to generate denser meshes around portions of a model that have greater curvature. The lower the value in the box, the higher the resolution of the mesh that will result.

Fig. 5-170. Smooth Mesh with Smooth option.

Values for Curvature options determine the amount of curvature in the mesh. Smaller values add more curvature to the edges. Greater values add more curvature to faces, resulting in a more rounded model.

TIPS AND STUDY AND APPLICATION NOTES

1. If you find that Smooth Mesh is creating too high a resolution, you can edit the object, supplying new Smooth Mesh parameters, because the tool keeps in memory the original object used to generate the smoothed object. To view the original object, select Original Object in the Preview section of the Smooth Mesh Edit dialog box.

Tools 3: Reduce Mesh

LEARNING OBJECTIVE 5-30: *Simplifying the resolution of a generated mesh surface or object.*

TOOL COVERED:

- Reduce Mesh

Fig. 5-171. Reduce Mesh tool.

This tool changes the resolution of an object by joining adjacent faces that fall within a selected degree tolerance set by the user. The default value set by the program is a 5-degree difference between adjacent faces. The reduction can also be selected to only affect similar surfaces, such as those with identical color or texture, by selecting the appropriate options in the Reduce Mesh Options dialog box (figure 5-172).

Using the Tools

Fig. 5-172. Reduce Mesh Options dialog box.

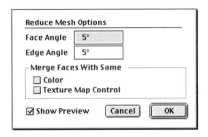

Figure 5-173 and figure 5-174 show the Reduce tool applied to a simple curving mesh.

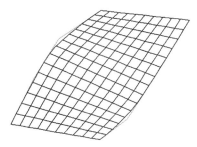

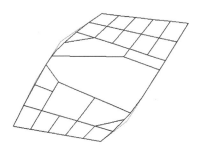

Fig. 5-173. The mesh before reduction.

Fig. 5-174. The same mesh after the Reduce tool.

TIPS AND STUDY AND APPLICATION NOTES

1. The Reduce tool, like the Smooth tool, has a Keep Flat (or Smooth) option, which will determine whether your object will actually change shape or simply change resolution. Attention should be paid to this option, as it can alter the object significantly.

2. The Reduce tool can be useful for simplifying, and speeding up, rendering times. Its main advantage is that it reduces the number of polygons in a model, and that in turn makes rendering quicker.

Tools 4: Define Profile, Move Mesh, Disturb, and Deform

LEARNING OBJECTIVE 5-31: *Deforming meshed objects randomly, interactively, or with defined profiles.*

Chapter 5: Modeling Tools

TOOLS COVERED:

- Define Profile
- Move Mesh
- Disturb
- Deform

Fig. 5-175. Tools covered.

This group of tools is used to deform meshed objects randomly, interactively, or through the use of defined profiles. The Define Profile tool is used to create parametric profiles to use for deformations. There are three tools for deforming objects: Move Mesh, Disturb, and Deform.

The Move Mesh method of deforming objects uses a predefined profile to describe the shape an object will take when deformed. These profiles can be chosen from those available in the Profiles palette (figure 5-176), or can be defined from a 2D line using the Define Profile tool.

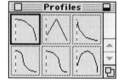

Fig. 5-176. Profiles palette.

➻ **NOTE:** *The Profiles palette is discussed in Chapter 4.*

The Move Mesh method uses predefined profiles to deform a mesh, as illustrated by the deformed surface shown in figure 5-177. The profile can be applied to an area or along a line. The method in which a profile is applied can be set under Movement Shape in the Move Mesh options dialog box (figure 5-178).

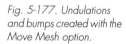

Fig. 5-177. Undulations and bumps created with the Move Mesh option.

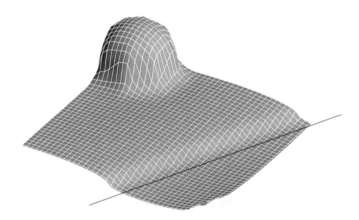

Using the Tools

Fig. 5-178. Move Mesh Options dialog box.

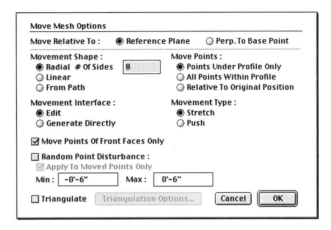

The Disturb tool randomly moves the points of a mesh to create a disturbed surface. This is useful for instantly creating rough terrain (figure 5-179).

Fig. 5-179. Using the Disturb tool on a surface.

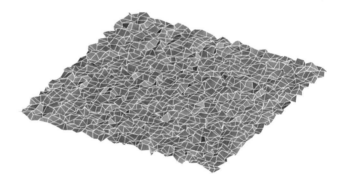

The Deform tool deforms meshed objects through operation commands such as Twisting, Tapering, Bending, Bulging, and Shearing (figure 5-180). To use this tool, select it from the tool bar and click on a meshed object. A bounding box will appear. You can then click on either the control points or the vertices to establish the deformation of the bounding box in relation to the chosen base reference plane. Depending on the type of deformation selected, the object can be bent and twisted at will (figure 5-181). Other features of the Deform Options allow the object to be symmetrically deformed by selecting the Through Center option, instead of the default, which typically deforms along one edge.

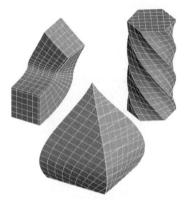

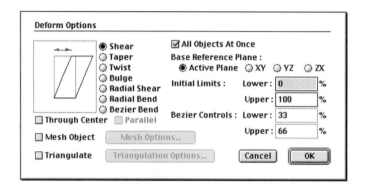

Fig. 5-180. Objects deformed through bending, twisting, and tapering.

Fig. 5-181. Deform Options dialog box.

TIPS AND STUDY AND APPLICATION NOTES

1. The Deform tool contains a convenient option (Initial Limits) for localizing the effect of your chosen type of deformation by setting percent limits for the beginning and end of the object, thus offering better control and the option of more localized deformations.

Tools 5: Displacement

LEARNING OBJECTIVE 5-32: *Using displacement maps.*

TOOL COVERED:

- Displacement

Fig. 5-182. Displacement tool.

The Displacement tool displaces the geometry of an object based on the lightness and darkness of a referenced grayscale image. This referenced image is called a displacement map. Figure 5-183 shows a displacement map that will be used in a Displacement operation.

Using the Tools

Fig. 5-183. Displacement map.

A simple cubic object can be created and its top face used for the Displacement operation. To do this, first select Face as the Topological Level. Then click on the Displacement tool and select the top face of the object by clicking on the two edges that define the face. This invokes the Displacement Map Edit dialog box (figure 5-184). A previously prepared displacement map can be loaded by double clicking on the image in the lower left-hand corner of the dialog box.

Fig. 5-184. Displacement Map Edit dialog box.

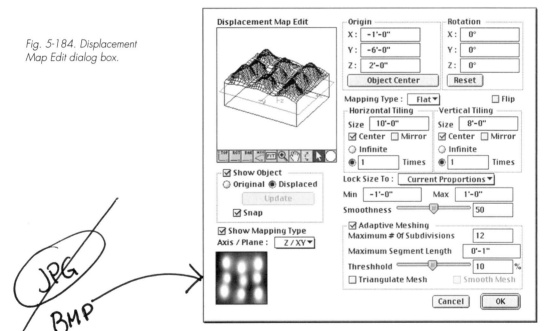

The top face needs to be subdivided or meshed into smaller faces for the Displacement tool to be able to displace the surfaces in reference to the selected displacement map. In order to do this, the Adaptive Mesh option needs to be selected to automatically generate a mesh for the surface. This automatically meshes the model. The three options under

Adaptive Meshing affect the resolution of the mesh. Select an appropriate resolution to fit the scale of your model.

Depending on the geometry of the model, you would choose from among Flat, Cylindrical, or Spherical options under Mapping Types. Because the previous example is a flat surface, you would select the Flat option. Map types are further discussed with the Texture Map tool later in this chapter.

A preview of the displaced object can be viewed by selecting Displaced under Show Options located under the preview box. The circular icon located at the lower right corner of the preview box can also be selected to view a rendered preview (figures 5-185 and 5-186).

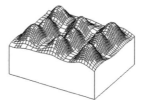

Fig. 5-185. A Displacement Mapped object.

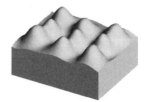

Fig. 5-186. A Displacement Mapped object rendered.

TIPS AND STUDY AND APPLICATION NOTES

1. The Displacement tool retains a memory of the original object so that further editing of the displacement mapped object is possible.

MacOS/PC NOTE: *Image formats that can be loaded as displacement maps are PICT on the MacOS, BMP and Metafile on Windows, and TIFF, Targa, PNG, and JPEG for both platforms.*

Tools 6: Triangulate

LEARNING OBJECTIVE 5-33: *Converting nonplanar faces to planar faces.*

TOOL COVERED:

- Triangulate

Fig. 5-187. Triangulate tool

Three points by definition describe a plane. If a surface has a fourth point that is not in plane with the first three, the surface is nonplanar (not flat). Nonplanar surfaces usually create difficulties when additional operations are performed on them, such as when an object containing the surface is used as a Boolean operand. The Triangulate tool is used to force nonplanar faces into planar states by subdividing them into triangles.

The Triangulate tool creates triangles by inserting segments in a model to convert nonplanar faces to planar faces. In figure 5-188, nonplanar faces were created when a cube was twisted. The Triangulate tool was used to convert the nonplanar faces to triangles.

Fig. 5-188. A twisted cube before and after triangulation.

As a default, C-Mesh and other derivative surface tools do not have triangulated faces. To use objects created with the C-Mesh, Revolve, Skin, and Sweep tools in Boolean operations it is usually necessary to triangulate them. In figure 5-189, a C-Meshed object has been triangulated.

Fig. 5-189. A C-Meshed object before and after triangulation.

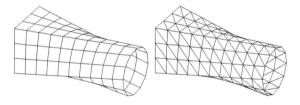

The Triangulate tool options allow you to specify the Triangulation Method, whether to Triangulate All Faces, and whether to perform a Strict Planarity Test (figure 5-190).

Fig. 5-190. Triangulate tool options.

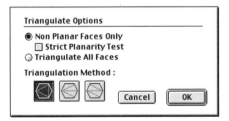

Triangulating all faces is unnecessary for Boolean operations. In figure 5-191, a C-Mesh is first triangulated using Triangulate Non Planar Faces Only, then triangulated using the Triangulate All Faces option.

Fig. 5-191. A surface before and after triangulation with the Triangulate Non Planar Faces Only option and the Triangulate All Faces option.

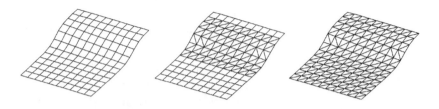

TIPS AND STUDY AND APPLICATION NOTES

1. Error messages that begin "Non planar faces encountered..." can be remedied by triangulating the object.

2. Because the Triangulate tool converts faces, it will not work if you have the Topological Level below the level of Face. Single faces can be converted to planar triangles if the Topological Level is set to Face.

3. Triangulating an object will increase the number of polygons and consequently the rendering time.

✓ **TIP:** *Do not triangulate an object if it is not necessary because it can drastically increase the number of polygons in your model.*

Rounding and Draft Angles Tool Palette

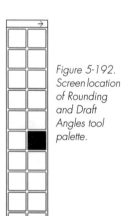

Figure 5-192. Screen location of Rounding and Draft Angles tool palette.

The Rounding and Draft Angles tool palette (screen location, figure 5-192; palette, figure 5-193) contains tools that allow you to smooth and bevel the edges of objects, as well as apply tapers to entire objects.

Fig. 5-193. Rounding and Draft Angles Tool Palette.

Tools 1: Round

LEARNING OBJECTIVE 5-34: *Rounding corners and edges.*

TOOL COVERED:

- Round

Fig. 5-194. Round tool..

Using the Tools

Fig. 5-195. Rounded object.

The perfectly straight edge is theoretical and does not exist in reality. Edges and corners are rounded by ergonomic design, by manufacturing processes, or by just plain wear. In order to replicate this state, form•Z includes the Round tool, which allows you to round corners and edges directly or through a preview dialog box. Rounding adds more polygons to your model, but is usually worth it for the added detail and realism (figure 5-195).

The Plain Rounding option in the Round Options dialog box (figure 5-196) applies the rounding directly to the model, either after you pre- or post-select points, segments, or an entire object to be rounded.

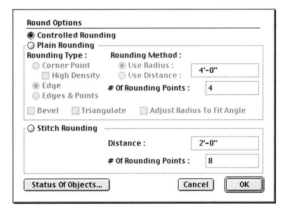

Fig. 5-196. Round Options dialog box.

The Controlled Rounding option opens the Rounding Edit dialog box, where you can test and preview a variety of corner and edge rounding options before applying any given selection to a model. The cursor control in the Rounding Edit dialog box will allow selection of points and vertices individually, and the cursor will change depending on which topological level it will affect (figures 5-197 and 5-198).

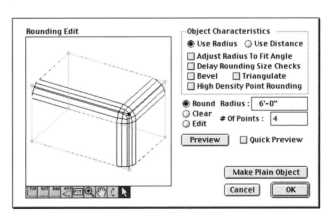

Fig. 5-197. Rounding Edit dialog box.

Fig. 5-198. Object rounded using various Rounding Type options: Corner Point, Edge, and Edge and Point.

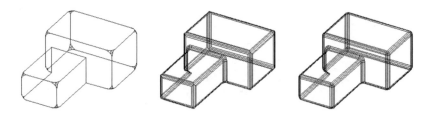

Rounding includes two major options: Use Radius and Use Distance. The Use Radius option rounds angles using an arc that has the same radius as specified in the options box. The Use Distance option uses the fixed distance specified in the option box to create a rounding that is not dependent on the local angle (figure 5-199).

Fig. 5-199. Rounding options Use Radius (top) and Use Distance (bottom) applied.

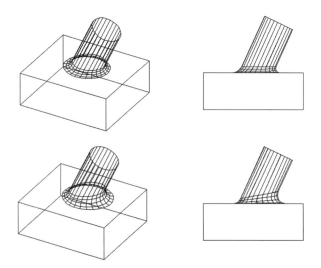

Stitch Rounding is applied to a closed series of segments that have resulted either from the Stitch or Boolean operation.

TIPS AND STUDY AND APPLICATION NOTES

1. As a general rule, when rounding several edges, it is easier to use Plain Rounding.

Tools 2: Draft Angle

LEARNING OBJECTIVE 5-35: *Creating draft angles of a designated inclination.*

Tool Covered:

- Draft Angle

Fig. 5-200. Draft Angle tool

Draft angle is a subtle taper on the face of objects and is normally used in the design of objects that need to be easily removed from molds. This tool creates draft angles of a designated inclination, either to a selected portion of an object or the entire object.

Figure 5-201 shows the Draft Angle Options dialog box. Base Reference Plane allows for the selection of the plane from which the draft angle calculation will be made. In addition to the regular Cartesian planes, Arbitrary Reference Planes can be selected as the Base Reference Plane by using the Active Plane option.

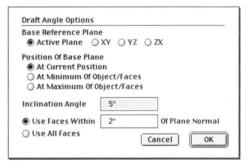

Fig. 5-201. Draft Angle Options dialog box.

The Position Of Base Plane option can be set to At Current Position, which projects the draft angle from the current reference plane. The At Minimum Of Object/Faces option selects the base plane in reference to the lowest point on the object. At Maximum Of Object does the opposite and selects the highest point.

Inclination Angle is the angle of the taper on faces. Use Faces Within performs the Draft Angle operations only on those faces with less than the predefined degree of inclination. When this option is selected, the Draft Angle operation is performed on one face at a time (figure 5-202). When the Use All Faces option is selected, draft angles are generated for all faces (figure 5-203).

Chapter 5: Modeling Tools

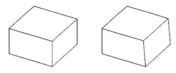

Fig. 5-202. Draft Angle tool applied to a single face.

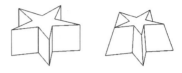

Fig. 5-203. Draft Angle tool applied to an entire object.

TIPS AND STUDY AND APPLICATION NOTES

1. The Draft Angle tool is useful when trying to create faces with a specific inclination angle.
2. Draft angles can also be created for individual faces by using the Topological Level of Face option and selecting portions of the model that need to have a draft angle.

NURBS and Patches Tool Palette

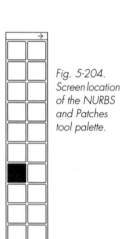

Fig. 5-204. Screen location of the NURBS and Patches tool palette.

Tools from the NURBS and Patches tool palette (screen location, figure 5-204; palette, figure 5-205) are used to create and edit parametric surface objects. These objects are used for sculpting complex 3D forms such as industrial design objects and organic forms.

Fig. 5-205. NURBS and Patches tool palette.

Tools 1: C-Curve

LEARNING OBJECTIVE 5-36: *Deriving controlled curves with the Vector Line tool.*

TOOL COVERED:

- C-Curve

Fig. 5-206. C-Curve tool.

C-curves (controlled curves) are curves derived from simple lines by using the C-Curve tool. The Vector Line tool is used to define a framework of control points that will be used to manipulate the C-curve.

Using the Tools

These control points are used to push and pull a curve to achieve subtle curvatures.

The Controlled Curve Options dialog box is shown in figure 5-207. The options in this dialog box determine the level of control and complexity of the generated C-curve. Arranged in order of complexity, NURBS are the most complex and Tangents the least.

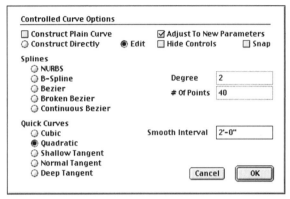

Fig. 5-207. Controlled Curve Options dialog box.

A C-curve is generated from a rough shape that can be drawn using 2D Vector Line, as shown in the first image of figure 5-208. Depending on the Controlled Curve options, a curve is generated from the line by interpolating a curve between the points of the line. The points of the line remain and can be used as control points. The smoothness of the curve is determined by the # Of Points option or the Smooth Interval option.

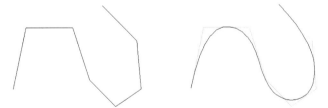

Fig. 5-208. Before and after C-curve generation.

Controls for Spline curves and Quick curves are shown in figure 5-209. The NURBS (Nonuniform Rational Bezier Spine) option gives you the most control over the shape of a curve. The NURBS option in the C-curve settings is not to be confused with Nurbz (explained later in this section).

Fig. 5-209. Controls in NURBS and a Quick Quadratic curve.

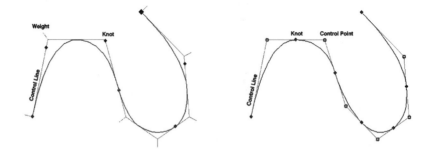

In the Splines section of the Controlled Curve Options dialog box (figure 5-207), Degree determines the degree of smoothing or interpolation used to generate the curve. The higher the degree, the more subtle are the curves created. # Of Points determines how many points there will be in the final curve. The more points, the smoother the curve and the more calculations needed to generate the curve. In Quick Curves, the Smooth Interval option determines the size of the segments in the curve. Figure 5-210 shows various ways Splines and Quick Curves behave in relation to the original generating control lines.

Fig. 5-210. Various types of C-curve behavior in relation to original control lines.

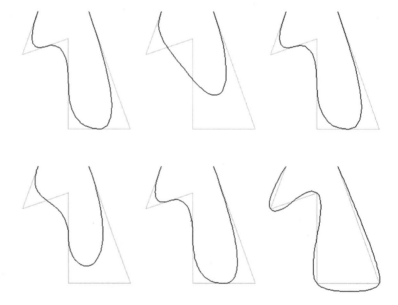

Because the C-curve function retains memory of original control lines, the control lines can be edited at any time by selecting the C-Curve tool or the Edit Control tool and clicking on the respective C-curve. The controls will reappear and the line can be edited. To adjust the parameters on an existing C-curve, change the settings in the Controlled Curve options or Tool Palette options, select the Adjust to New Parameters options, and click on the C-curve.

TIPS AND STUDY AND APPLICATION NOTES

1. As a rule, the higher the level of a curve, the more computation time is required to generate and manipulate it.

2. Despite the fact that form•Z software calls these elements curves, they are in fact a series of very small segments. Keep this in mind when generating 3D objects from these curves because each segment will become a face and add to rendering time.

3. Most of the C-Curve options generate curves that do not pass through control points. The Tangent Curves option, however, creates a curve that intersects the points of the original control line. This is especially useful when performing a Sweep, C-Mesh, or Skin operation for which you want certain points on the object to be in specific and predictable locations.

4. The Hide Controls option in the Controlled Curve options can be selected so that the curves can be seen without the sometimes distracting controls.

5. Behavior similar to the C-Curve tool can also be achieved using the Spline Sketch tool in the Lines, Splines, and Arcs tool palette in conjunction with the Edit Controls tool.

MacOS/PC NOTE: In Quick Curves, the Point Type option can be changed interactively by pressing Cmd (Mac) or Ctrl (Windows) and clicking the knot point that needs to be edited. This gives you the Point Type dialog box.

Tools 2: C-Mesh

LEARNING OBJECTIVE 5-37: *Generating meshed surfaces containing control points from 2D lines.*

TOOL COVERED:

- C-Mesh

Fig. 5-211. C-Mesh tool.

C-Mesh generates meshed surface objects containing control points from 2D lines. C-Meshed objects are objects with a set of internal controls for manipulation and deformation. These controls can be edited using the C-Mesh tool again or using the Edit Controls tool. The object generated is dependent on the shape of the generating curves, their direction, the order in which the curves are chosen, and C-Mesh option settings.

To create a C-Mesh from more than two curves, the curves must be pre-picked before selecting the C-Mesh tool. Figure 5-212 shows a C-Meshed object generated from two lines using the Create C-Mesh tool. The C-Mesh tool has a behavior similar, yet inferior to, the Nurbz tool covered later in this section. In most situations it is preferable to use the Nurbz tool instead of the C-Mesh tool because it offers additional control and editing of the resultant surface.

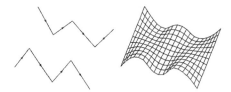

Fig. 5-212. A C-Meshed surface generated from two lines.

Figure 5-213 shows the Controlled Mesh Options dialog box. The first four options apply to the entire mesh, whereas the other options affect either the length or depth of the mesh. The Construct Plain Mesh option creates a plain object, which contains no control lines. The Edit option (versus Construct Directly) causes the tool to prompt you to edit a mesh directly after it is created. The Adjust to New Parameters option for Edit C-Mesh reconfigures a C-Mesh based on new settings for Length and Depth of net.

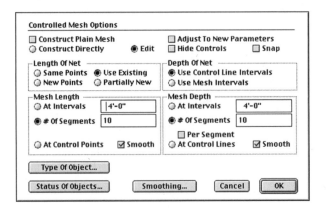

Fig. 5-213. Controlled Mesh Options dialog box.

In figure 5-214, a C-Mesh has been created from two lines, each with a different number of points. The Same Points option will not allow an object to be created from these sources because the sources do not contain the same number of points. The Use Existing, New Points, and Partially New Points options generate different sets of control lines from the same sources.

Fig. 5-214. C-Meshes generated using the Use Existing, New Points, and Partially New Points options.

Figure 5-215 shows the effects of the Mesh Length and Mesh Depth settings on the generation of a meshed surface. In figure 5-215, three meshes have been made from identical generation lines, but # of Segments for Mesh Length has been set at a different value for each: 5, 10, and 20.

Fig. 5-215. Three C-Meshed objects created with the same generation lines but different settings in the # of Segments option along their mesh length.

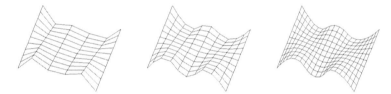

Meshed objects and surfaces generated with the C-Mesh tool can also be edited by pulling and stretching the control lines and points. You activate and manipulate a C-Meshed object's control lines using the C-Mesh tool again on the same object, or using the Edit Controls tool. When the process is complete, double click to finish editing (figure 5-216).

Fig. 5-216. Using the C-Mesh tool to manipulate the control lines of a mesh.

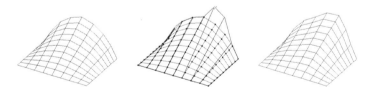

TIPS AND STUDY AND APPLICATION NOTES

1. It is possible to perform Boolean and Trim/Stitch operations on C-Meshes, but the result is a plain mesh object. Plain meshes

have none of the internal control parameters of C-Mesh surfaces.

2. The type of curves used for control lines in a C-Mesh are set in the Controlled Mesh Smoothing options (figure 5-217), accessed by clicking on Smoothing in the Controlled Mesh options dialog box. The various curve types are discussed in the previous section on C-curves.

Fig. 5-217. C-Mesh Smoothing options.

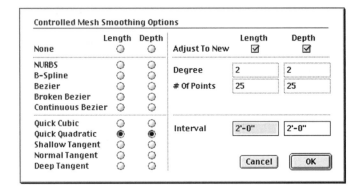

Tools 3: Nurbz

LEARNING OBJECTIVE 5-38: *Generating parametric surfaces and objects from 2D lines.*

TOOL COVERED:

- Nurbz

Fig. 5-218. Nurbz tool.

The Nurbz tool generates a Nurbz surface from a number of control lines. A Nurbz surface is a parametric surface that can be shaped and edited after its generation by means of control points. Nurbz (usually spelled NURBS in 3D modeling software) stands from Non-Uniform Rational Bezier Spline. This name refers to the type of mathematical equation used to calculate the surface geometry. The Nurbz tool can also be used to edit the parameters of an existing Nurbz surface. The Nurbz tool seems similar at first to the C-Mesh tool in the same palette. However, it offers much more control over the surface through the Edit Controls and Edit Surface tool.

Nurbz surfaces are generated from two or more lines or curves. Figures 5-219 and 5-220 show a set of generation curves and the resulting surface. The number of lines required to generate a surface is the number of the Depth Degree of the surface (discussed in material that follows) plus one. If there are more than the minimum number of lines used to generate the surface, they must be pre-picked and the surface will be generated based on the order they are selected. The number of points in the generation lines does not need to be equal, but the number of points in each line must be one more than the number of the Length Degree.

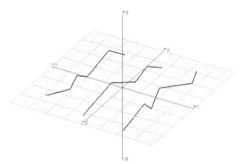

Fig. 5-219. Generation curves for producing a Nurbz surface.

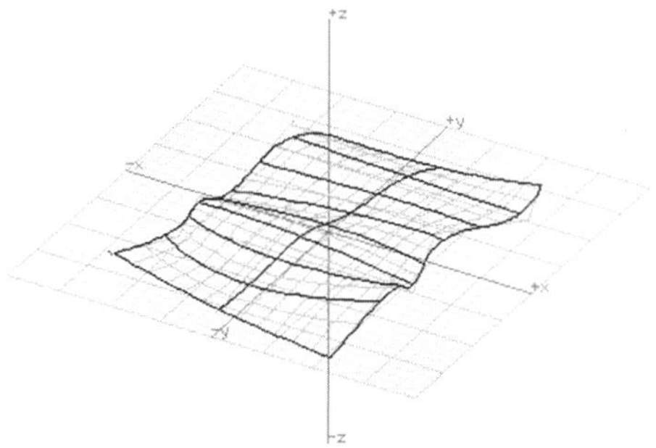

Fig. 5-220. The resulting Nurbz surface.

When Nurbz surfaces are generated, a set of control points is generated. These control points can be edited using the Edit Controls tool to shape the Nurbz surface. See Edit Controls under the Pick Tool palette for more discussion on editing controls. Figure 5-221 shows a Nurbz surface being edited, with its control points.

Fig. 5-221. Nurbz control points.

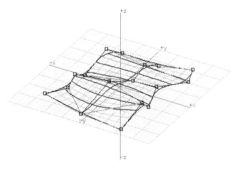

The Length Degree and Depth Degree of a Nurbz surface, set in the Nurbz Options dialog box (figure 5-222), determine how many control points, and consequently the degree of control over a surface's shape.

The "flexibility" of a surface and its ability to curve and bend is directly related to the degree of the surface. The higher the degree, the more control points and consequently a higher resolution surface with more control over shape. Keep the degree of the surface as low as possible (but usually above 1 because a 1-degree surface is polygonal). With a higher degree of surface comes more computational overhead, making rendering and navigating your model slower. In addition, with a high degree surface, editing may become tedious, as there are more points to manipulate.

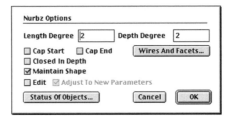

Fig. 5-222. Nurbz tool options.

The Cap Start and Cap End options in the Nurbz options will close the ends of a surface generated from closed generation lines. If Cap Start is selected, a surface will be created as a face to "cap" the open start of a Nurbz surfac. Cap End will create a face to cap the end of a Nurbz surface.

To edit a Nurbz surface parameter of an existing Nurbz object, use the Adjust to New Parameters option. This option can be used to change the degree of a surface. When this is used to change the surface degree, modeling information is lost or created and the shape of the surface will be affected. With the Maintain Shape option selected, the software attempts to preserve the shape of the object being adjusted by inserting control vertices.

TIPS AND STUDY AND APPLICATION NOTES

1. The resolution of the Wires and Facets for a Nurbz surface can be set by clicking on the Wires and Facets button in the Nurbz options. For information about Wires and Facets, see the introduction to this chapter.

2. The number of points in the generation lines do not need to be equal, but the number of points in each line must be one more than the number of the Length Degree setting in the Nurbz options.

3. For practice using Nurbz surfaces, see the Umbrella example in Chapter 9.

Tools 4: Patches

LEARNING OBJECTIVE 5-39: Creating and manipulating patch surfaces.

TOOLS COVERED:

- Patch Derive
- Patch Grow
- Patch Divide
- Patch Attach

Fig. 5-223. Tools covered.

The Patch tools are used to create and modify patch surface models. Patch surfaces are, as the name implies, small flexible surfaces combined and shaped to create complex 3D surface models. Patches are shaped and sculpted using the Edit Surface and Edit Controls tools. Figure 5-224 shows a patch surface being edited with the Edit Surface and Edit Controls tools. Modeling with patches is usually less exact than working with Nurbz surfaces but much easier to work with in a sculptural manner.

Fig. 5-224. Editing a Patch with Edit Surface (left) and Edit Controls (right).

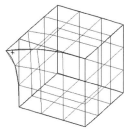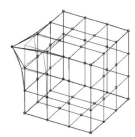

The method of modeling a Patch surface is an additive process, usually beginning by deriving a single patch, which is then grown, shaped, and combined with other patches to create an entire system. Patches can be derived from existing 3D objects, such as extruded objects and parametric solids, as well as 2D surfaces using the Patch Derive tool. To derive a patch object from a 2D or 3D object, select the Patch Derive tool and click on the object. (See figure 5-225.)

Fig. 5-225. Patch Derive options.

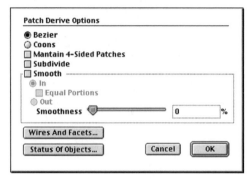

When deriving a patch, the object can be smoothed in the process by selecting Smooth in the Patch Derive options. Within the Smooth section of the Patch Derive options, you can select the option of Smooth In or Smooth Out. The direction of in or out determines whether the resulting patch object is reduced (In) during the smoothing process or enlarged (Out). Figure 5-226 shows a cube that has been converted to a patch object using Patch Derive. The left side of the figure shows a patch derived without the Smooth option, the middle shows a patch derived with the Smooth option set to In, and the right side of the figure shows a patch derived with the Smooth option set to Out.

Fig. 5-226. A patch derived without the Smooth option, with Smooth set to In, and with Smooth set to Out.

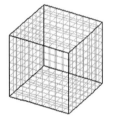
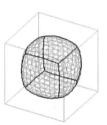
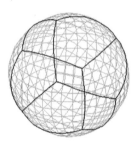

Once a single patch object exists, you can use the Patch Grow tool to generate additional patches adjacent to the first patch. Patch Grow will "grow," or create, a new patch attached to the edge of an existing patch surface. An important note about Patch Grow is that it only works on 2D patches, not on patch objects derived from 3D solids. New patches can also be generated between two existing patches to stitch them together. To generate a patch between two patches, select the Between Edges option in the Patch Grow options.

Figure 5-227 shows new patches grown from a single existing patch, and a new patch generated between two existing patches. The Patch Grow options also include several options for how new patches are

Using the Tools

generated in relation to the original patches. The options of Align Patch Normals, Perpendicular to Patch, Parallel to Reference Plane, and Perpendicular to Reference Plane determine the direction in which new patches are grown.

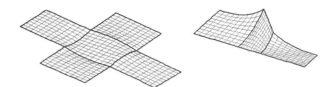

Fig. 5-227. Patches grown from a single patch (left) or between two patches (right).

Patches generated with the Derive Patch tool or Patch Grow tool can always be further subdivided with the Patch Divide tool. The Patch Divide tool will increase the resolution of a patch surface, making it more flexible and allowing it to be sculpted to a finer level of detail. If a patch has already been subdivided using the Patch Divide tool, the subdivision process can be reversed by selecting the Undivide option in the Patch Divide options.

Sets of patch surfaces can be joined together to create a single continuous patch surface using the Patch Attach tool. The Patch Attach tool will join the edges of two patches. If the two edges are not touching, the first patch selected will be modified to join the second patch, depending on the options selected in the Patch Attach options (figure 5-228). The Move Edge option will adjust the edge of the first patch to align with the edge of the second patch, stretching the first patch. The Move Patch option will move the entire first patch to align with the edge of the second patch, resizing the first patch. Figure 5-229 illustrates the attach options. The left of the figure shows the original patches, the middle shows the patches attached using the Move Edge option, and the right side shows the patches joined using the Move Patch option.

Fig. 5-228. Patch Attach options.

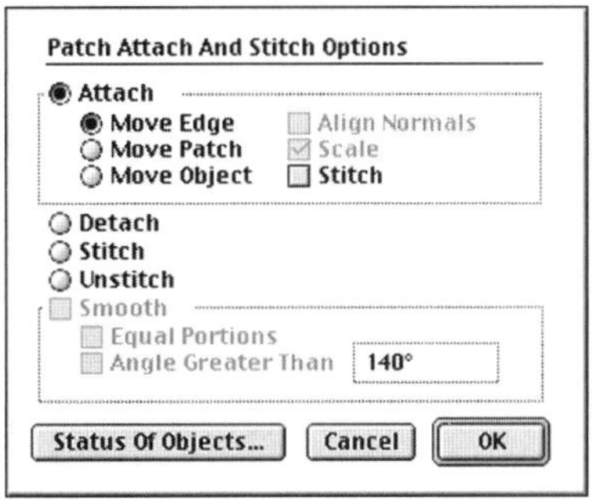

Fig. 5-229. Attaching patches using the Move Edge options (middle) and the Move Patch options (right).

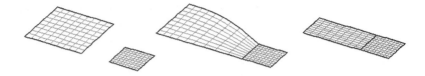

TIPS AND STUDY AND APPLICATION NOTES

1. Patches can be sculpted and edited using the Edit Surface and Edit Controls tools. For more information on these tools, see the Pick tool palette section earlier in this chapter.

2. For a working example using Patches, see the Sandal example in Chapter 9.

Metaformz Tool Palette

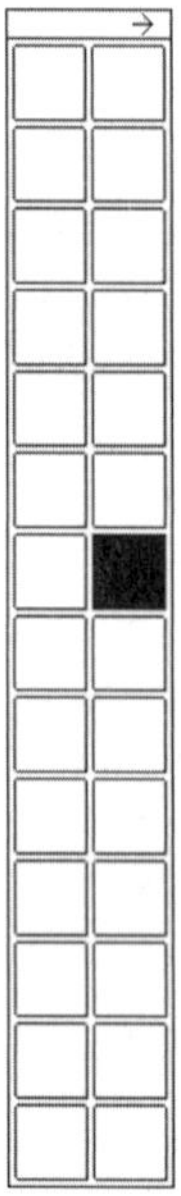

Fig. 5-230. Screen location of Metaformz tool palette.

This palette (screen location, figure 5-230; palette, figure 5-231) contains tools for working with Metaballs that are created with the Metaballs tool located in the Balls tool palette. Metaformz are controlled (or parametric) objects with "mass" and a "region of influence" designed for creating organic forms. When these objects are combined, a smooth blended surface results. The result does not have an exact size, but is a negotiation between the metaform's mass and region of influence and its proximity to other metaforms.

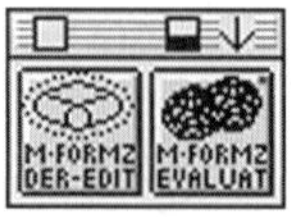

Fig. 5-231. Metaformz tool palette.

Tools 1: Metaformz

LEARNING OBJECTIVE 5-40: *Creating organic-shaped objects using Metaformz.*

TOOLS COVERED:

- Metaformz Derive/Edit
- Metaformz Evaluate

↝ **NOTE:** *Refer to the Metaballs tool in the Balls tool palette for additional information on creating metaballs.*

Using the Tools

In addition to being generated with the Metaballs tool, metaballs can also be created by deriving them directly from any 2D or 3D objects by using the Metaformz Derive/Edit tool. This method of deriving metaforms provides additional types of metaballs outside the Ball and Ellipsoid. These are Tube, Conic Tube, Torus, Sheet, and Polyhedron (figure 5-232).

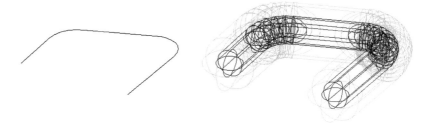

Fig. 5-232. Metaforms created with the Metaformz Derive/Edit tool.

The combining of metaforms is called evaluation. This process converts a collection of metaforms into a solid that can be rendered. The evaluation of a single metaform is performed using the Metaformz Evaluate tool; multiple metaforms are evaluated by grouping them using the Group tool. Once evaluated, a metaform can be reevaluated at any time using the Metaformz Evaluate tool. In figure 5-233, three metaforms are grouped to create a surface. In the last image, the metaforms group has been reevaluated using the Metaformz Evaluate tool, and the threshold has been set at a greater value.

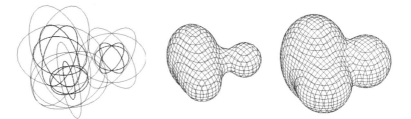

Fig. 5-233. Metaballs evaluated using the Group tool to make metaforms, then reevaluated using Metaformz Evaluate.

TIPS AND STUDY AND APPLICATION NOTES

1. Metaforms have no surface and, subsequently, will not render until they have been evaluated. Metaforms are evaluated with the Metaformz Evaluate tool or by grouping a collection of metaforms using the Group tool.

2. The plus or minus value for the Weight parameter of a metaform gives the metaform an additive or subtractive quality that determines its effect on the adjacent metaforms in the group evaluated.

3. Refer to the Pear exercise in Chapter 7 for additional information on working with metaforms.

Booleans and Intersections Tool Palette

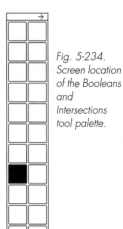

Fig. 5-234. Screen location of the Booleans and Intersections tool palette.

Tools from the Booleans and Intersections tool palette (screen location, figure 5-234; palette, figure 5-235) are used to combine and separate objects. The palette includes tools for creating composite shapes, such as the Union and Difference tools for solid objects and Trim/Split and Stitch tools for surface objects. The palette also contains analytical tools for working with your models, such as the Section and Contour tools.

Fig. 5-235. Booleans and Intersections tool palette.

Tools 1: Boolean Operators

LEARNING OBJECTIVE 5-41: *Adding and subtracting solid objects.*

TOOLS COVERED:

- Union
- Intersection
- Difference
- B-Split

Fig. 5-236. Tools covered.

The Boolean operations of Union, Intersection, Difference, and B-Split are tools for combining objects or subtracting one object from another. Boolean operations can only be performed on solid 3D objects.

The Status Of Objects option box is a part of the first three Boolean tools and is opened by double clicking on the Boolean operation icon. Status of Operand Objects determines if form•Z will Keep, Ghost, or Delete

Using the Tools

the original objects after performing the operation (figure 5-237 and figure 5-238).

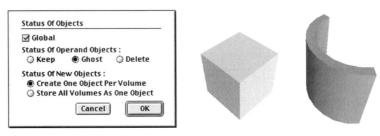

Fig. 5-237. Status Of Objects option box.

Fig. 5-238. Solid A (left) and solid B (right).

The Union operation takes solids A and B and combines their volumes into one new solid (figure 5-239).

Fig. 5-239. Boolean Union of A and B.

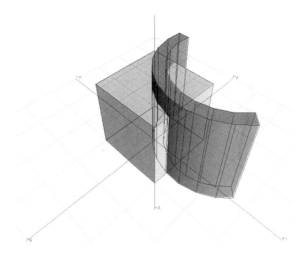

The Intersection operation creates a new solid representing the volume common to solids A and B (figure 5-240).

Fig. 5-240. Boolean intersection of A and B.

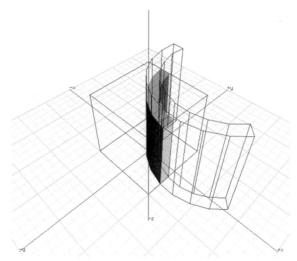

The Difference operation subtracts the volume of the second object from the first. You can subtract solid A from B or B from A (figure 5-241).

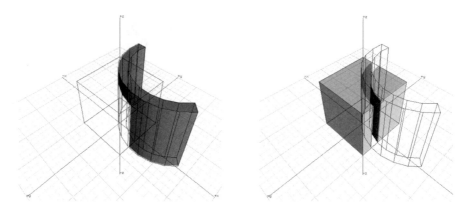

Fig. 5-241. Boolean difference of A from B (left) and B from A (right).

The B-Split (Boolean Split) tool splits the intersecting objects into distinct volumes. With One Way Split, the first object is split by the second. With Two Way Split, the objects split each other. (See figures 5-242 through 5-244.)

Fig. 5-242. Boolean Split Options dialog box.

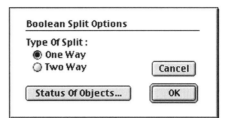

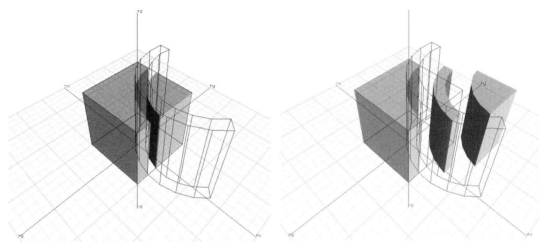

Fig. 5-243. One-way Boolean split.

Fig. 5-244. Two-way Boolean split.

TIPS AND STUDY AND APPLICATION NOTES

1. George Boole (1815-1864), an English mathematician, wrote several classical works on mathematical logic. He developed means of expressing logical processes using algebraic symbols and later created a branch of mathematics known as symbolic logic. In The *Mathematical Analysis of Logic and Investigation of the Laws of Thought*, he established formal logic and a logic based on "and, or, and not" called Boolean algebra. In *Treatise on Differential Equations*, he illuminated parallels between differential operators and rules of algebra.

Tools 2: Trim, Stitch, and Intersection

LEARNING OBJECTIVE 5-42: *Combining and subtracting 3D surface objects.*

TOOLS COVERED:

- Trim/Split
- Line of Intersection
- Stitch

Fig. 5-245. Tools covered.

Whereas Boolean tools are used for adding and subtracting solid objects, Trim, Stitch, and Intersection tools are used for combining and subtracting 3D Meshed Surface objects and Solids. It is important to note that these operations will not work on Surface Solid objects.

The Trim/Split tool contains two main options: Trim and Split. These options are similar to the Difference and B-Split operations, whereas the Stitch option, when toggled on, combines surfaces much like the Union tool. The Stitch option also has a rounding function that will round a shared edge as it stitches two surfaces together (figure 5-246).

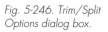

Fig. 5-246. Trim/Split Options dialog box.

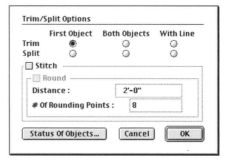

The Trim/Split operations are performed most of the time between surfaces, but lines can also be used as operands. When using surfaces, you can apply the operation to one surface or both. Figure 5-247 shows two surfaces being combined using the Trim option. The first image shows the original objects placed in position and intersecting each other. The second image shows one object being trimmed using the First Object option. The third image shows both objects being trimmed using the Both Objects option.

Fig. 5-247. Two surfaces combined using the Trim option.

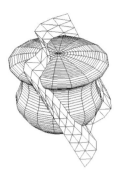  

Figure 5-248 shows the Split option being used on two surface objects. The first image shows the original objects. The second image shows the result of a Split operation performed with the First Object option. The third image shows the resulting object using the Both Objects option.

Fig. 5-248. Two surfaces and the Split option.

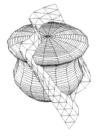 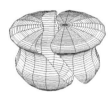 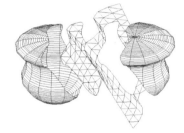

In figure 5-249, the Trim option with the Stitch option toggled on has been applied to two surfaces. Portions of both surfaces have been trimmed away and the remains joined to produce a solid. The First Object and Both Objects options determine whether one or two objects are produced as a result.

Fig. 5-249. Two surfaces combined using the Trim option with Stitch selected.

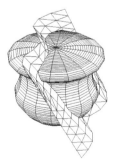  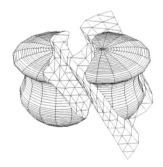

A line can be used instead of a surface object to Trim, Split, and Stitch another surface. The line in the first image of figure 5-250 has been used to first Trim, then Split, and finally Trim and Stitch, the revolved surface object.

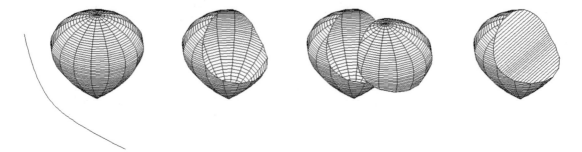

Fig. 5-250. Trim and Split tool using a line and surface.

The Line of Intersection tool is used to generate a separate object for the intersection between two objects. This line can then be edited and manipulated independently from the original objects (figure 5-251).

Fig. 5-251. Line of Intersection tool used to generate a line where the two objects intersect.

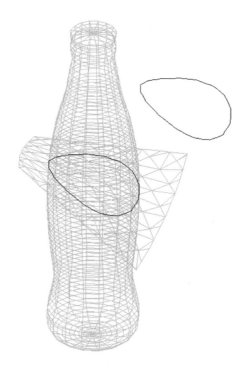

Using the Tools

The Stitch tool performs independently in much the same manner as the Stitch option within the Trim/Split tool. It is used to stitch ends of surface objects together whose ends coincide. Through the Stitch Options dialog box, the intersection between stitched surfaces can also be rounded (figure 5-252).

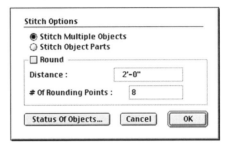

Fig. 5-252. Stitch Options dialog box.

TIPS AND STUDY AND APPLICATION NOTES

1. The portions of surfaces trimmed away or kept are determined by where on the object you click when selecting it to be trimmed. Be careful of the order in which you select objects because, depending on where on an object you click, very different results are possible.

2. Although a Trim performed with the Both Objects option may appear to have produced a solid object (such as in figure 5-247), it does not actually do so; it has produced two surfaces with shared edges. To produce a solid object from two surfaces, use the Stitch option.

3. Refer to the Make One/Two Sided tool in the Topologies tool palette for information on converting Surface Solid objects to One-Sided Surface objects so that Trim and Stitch operations can be performed.

Tools 3: Section/Contour of Solid

LEARNING OBJECTIVE 5-43: *Cutting for analysis.*

TOOLS COVERED:

- Section
- Contour

Fig. 5-253. Tools covered.

The Section tool allows you to analyze a 3D model by cutting it into two distinct parts or by producing a 2D profile with the cutting plane.

2D Section

The 2D option for the Section tool produces a 2D profile from a 3D model and a 2D cutting plane or line. The plane or line can be positioned anywhere in 3D space. The one stipulation is that if it were extended infinitely, it must intersect the object to be sectioned. To perform a 2D Section, select the 2D option under the Section tool and pick the 3D object to be cut, then the cutting plane or line (figures 5-254 through 5-256).

Fig. 5-254. The original object.

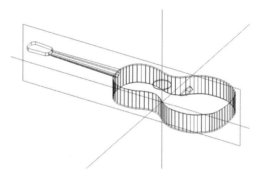

Fig. 5-255. Position a plane to use as a cutting blade in 2D Section.

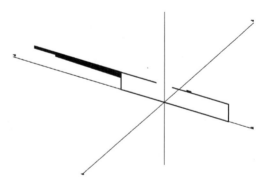

Fig. 5-256. A 2D section of the object.

3D Section

The 3D Section option uses a 2D surface to slice an object in half, producing two solid objects. Like the 2D Section option, the 3D option requires a face or line to be selected as a cutting plane. To make a 3D Section, select the 3D Section option, click on the 3D object, and then on the cutting plane or line to split your object (figure 5-257).

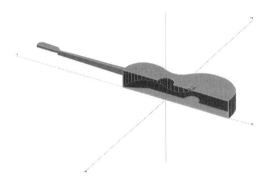

Fig. 5-257. A 3D section of the object.

The Contour tool produces 2D profiles from a 3D object, resembling a topographical contour map. The tool makes 2D contours, either at regular intervals or of a predefined number of even intervals. The contours can be later used to fabricate a physical version of your form•Z model. The Contour tool uses planes parallel to the reference plane to cut the object, not a defined cutting plane. If you want to cut sections at other angles, change the reference plane via the Reference Plane tool in the Window Tools palette (figure 5-258).

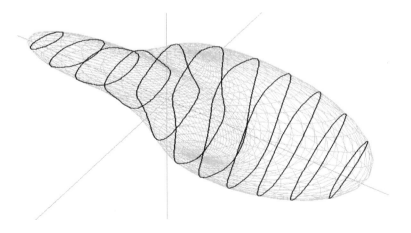

Fig. 5-258. Contours of an object derived with the Contour tool.

Chapter 5: Modeling Tools

TIPS AND STUDY AND APPLICATION NOTES

1. The Section tool contains various options that allow it to work either as a 3D modeling tool or as a 2D analytical tool. The 3D Section option for the Section tool allows you to cut a model into 3D pieces using a plane as a virtual knife. This function is practical both as a modeling tool, allowing you to chip away at a simple solid to produce a complex form, or as a presentation tool, allowing you to split in half and fold open hollow forms to reveal their insides.

Tools 4: Cage

LEARNING OBJECTIVE 5-44: *Reducing object complexity and generating low-resolution "stand in" objects.*

TOOL COVERED:

- Cage

Fig. 5-259. Cage tool.

As the size and complexity of your form•Z scenes increase, they can become increasingly difficult and time consuming to manipulate and adjust. This is particularly true when setting up animations. The Cage tool allows you to build temporary, very low-resolution (thus quicker to manipulate) duplicates of the objects in your scene. These objects can then be used as "stand-in" objects while lighting your scenes and setting up animations and renderings. The cage is generated by projecting the profile of the object in two planes, extruding them until they cover the other projection, and then generating a Boolean Intersection object from the results. The Cage Options dialog box provides a number of methods of generating these low-resolution objects (figure 5-260).

Fig. 5-260. Cage Options dialog box.

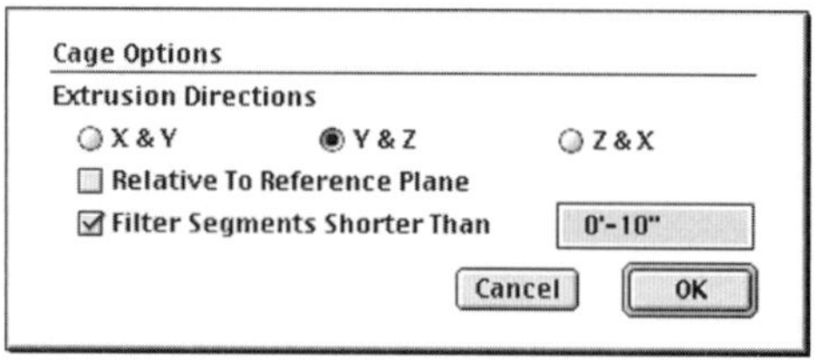

The three radio buttons located in the Cage Options dialog box determine the two planes from which the profiles are projected (figures 5-

261 and 5-262). The Relative To Reference Plane option allows the profiles to be projected from an Arbitrary Reference Plane, and the Filter Segments Shorter Than option sets the resolution of the Cage.

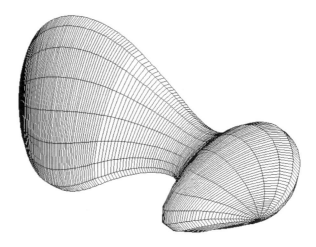

Fig. 5-261. The original object.

Fig. 5-262. Cages extruded in different planes produce different results.

TIPS AND STUDY AND APPLICATION NOTES

1. Cages will not include holes or voids from original objects.
2. If you are using Cages as "stand-in" objects, be sure to keep your originals so that you can replace them before final rendering.
3. You may find the Reduce Mesh tool in the Meshes and Deform tool palette more useful for permanently reducing the resolution of objects or generating more accurate "stand-in" objects.

268 Chapter 5: Modeling Tools

Fig. 5-263. Screen location of the Join and Group tool palette..

Join and Group Tool Palette

The tools in the Join and Group tool palette (screen location, figure 5-263; palette, figure 5-264) provide various methods of gathering objects into simple entities, sets, and hierarchies in order to facilitate subsequent operations.

Fig. 5-264. Join and Group tool palette.

Tools 1: Join and Separate

LEARNING OBJECTIVE 5-45: *Combining and splitting objects.*

TOOLS COVERED:

- Join
- Separate

Fig. 5-265. Tools covered.

The Join tool binds objects as a single object, whereas the Separate tool splits a joined object into its constituent parts. The Join tool combines selected objects and defines them as a single object. However, unlike the Boolean Union tool, Join does not recalculate the shapes into a combined object. In addition, unlike Boolean Union, objects that are not touching in any way can be joined into a single object. The Join tool also differs from the Group tool, which simply organizes multiple objects into a named group and does not redefine them all as one object.

The Separate tool can be used to separate objects that have been joined, as well as to extract elements from objects. Figure 5-266 shows the Separate Options dialog box. The Volumes option is used to create independent objects from objects containing more than one volume, such as those combined with the Join tool. Along Stitch separates Objects that have previously been attached using the Stitch operation. Along Segment cuts an object down preselected segments, and At Boundary Of Selected Faces cuts out part of an object that is defined by the boundary of preselected faces (figure 5-267).

Using the Tools

Fig. 5-266. Separate Options dialog box.

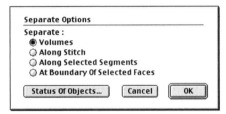

Fig. 5-267. Separate options: Along Segments and At Boundary Of Selected Faces.

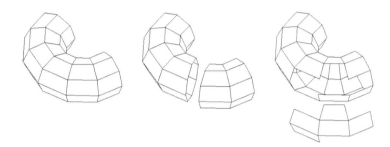

TIPS AND STUDY AND APPLICATION NOTES

1. The Join command is especially useful when you perform Boolean operations on more that one object simultaneously, because joined objects will behave like a single object during Boolean operations. However, overlapping joined objects should be avoided because this may in some cases cause the Boolean operation to fail.

Tools 2: Group and Ungroup

LEARNING OBJECTIVE 5-46: *Creating a hierarchy of objects.*

TOOLS COVERED:

- Group
- Ungroup

Fig. 5-268. Tools covered.

The Group tool organizes multiple objects into groups, creating a hierarchy among the objects in a model. The Ungroup tool is used to separate objects from the groups in which they have been placed.

The Group command creates hierarchical organizations of objects in your model. Objects can be selected and added to groups, and groups can be selected and nested into higher-order groups. This process of nesting objects within groups simplifies the task of selecting and performing operations on sets of related objects. The object names may also be displayed in the color of their current surface style for easier reference.

Grouped objects can be collectively manipulated by using the Topological Level of Group with other tools. The Objects palette also displays and permits the direct manipulation of the hierarchy and organization of grouped objects (figure 5-269).

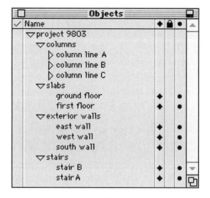

Fig. 5-269. Objects palette showing a nested hierarchy of groups.

The Ungroup command removes objects from a group. The operation can be performed on several levels. The Dismantle One Level option ungroups only one step within a hierarchy of groups. Dismantle All Levels will recursively ungroup the selected group, as well as all groups within its hierarchy. The Remove Object From Group option can be used to remove a single object or group from a higher-level group (figure 5-270).

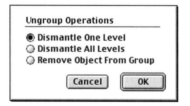

Fig. 5-270. Ungroup Operations dialog box.

Using the Tools

TIPS AND STUDY AND APPLICATION NOTES

1. As shown in figure 5-269, groups can be contained within other object groups, and organizational hierarchies can be created in the Objects palette. For more information on object groups, see Chapter 4.

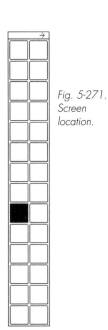

Fig. 5-271. Screen location.

Text Tool Palette

The Text tool palette (screen location, figure 5-272; palette, figure 5-272) displays the tools that allow you to create and edit text within your 3D project.

Fig. 5-272. Text tool palette.

Tools 1: Text Manipulations

LEARNING OBJECTIVE 5-47: Producing and manipulating text.

TOOLS COVERED:

- Text Place
- Text Edit
- Text Line Edit

Fig. 5-273. Tools covered.

form•Z provides three tools for producing and manipulating text: Text Place, Text Edit, and Text Line Edit. These tools allow you to produce text at a designated point or place it in relation to a given set of lines. After the text has been created, it can be manipulated like a normal object with the Scale, Move, Rotate, and Mirror commands. It can also be re-edited later using the Text Edit and Text Line Edit tools.

The Text Place tool contains options for text type and text placement (figure 5-274). The Text Placement dialog box, which is invoked by double clicking on the Text Place tool, determines whether Plain Text or Text As Object is produced. Plain text is flat text that cannot be manipulated using any tools in form•Z except the Text Edit tool. Plain text will also not show up in a rendering. The Text As Object option produces text that can be manipulated with all tools in form•Z. It can also be used in Boolean functions.

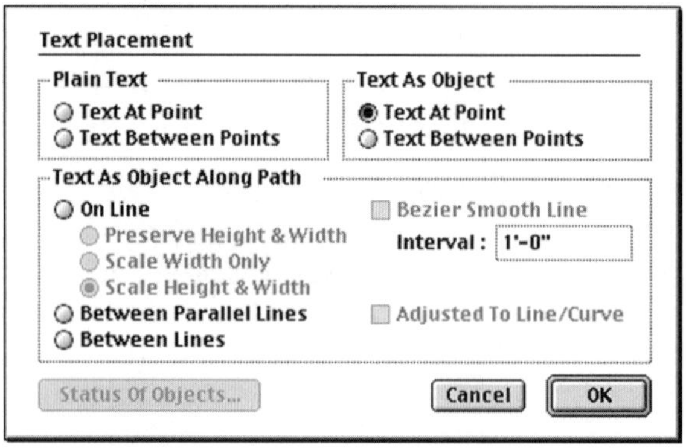

Fig. 5-274. Text Placement dialog box.

Figure 5-275 shows Text As Object being placed using the At Point and Text Between Points commands. The standing text has been placed with the Standing (Perp. to Plane) option available in the 3D Text Editor.

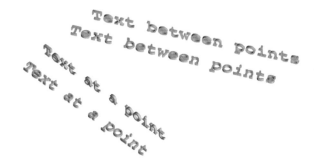

Fig. 5-275. Placing text at a point and between points.

Text in figure 5-276 has been placed using the Text On A Line, Text Between Parallel Lines, and Text Between Lines options.

Fig. 5-276. Placing text on or between lines.

Using the Tools

When placing text you will be presented with the 3D Text Editor. This option box allows you to set the size of your type and its resolution, and to enter the text to be placed. The words of the text, as well as the text paths, can be edited using the Text Edit tool (figure 5-277).

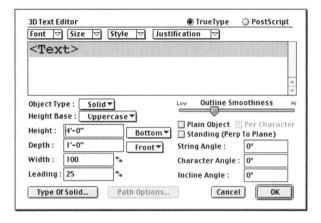

Fig. 5-277. 3D Text Editor.

TIPS AND STUDY AND APPLICATION NOTES

1. Text and text lines can be edited, even after transforming the text using transformation tools such as Move, Scale, Rotate, and Mirror. However, many functions in form•Z will change the status of text from a text object to a regular object, making it impossible to edit using Text Edit tools. Functions such as the Boolean tools and Derivatives tools will change text objects to plain objects.

2. For more information on the use of text in form•Z, see the Logo exercise in Chapter 7.

Symbols Tool Palette

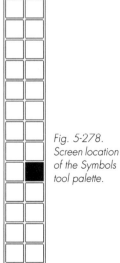

Fig. 5-278. Screen location of the Symbols tool palette.

The Symbols tool palette (screen location, figure 5-278; palette, figure 5-279) contains all the tools for building and editing object libraries for repeated use throughout your projects.

Fig. 5-279. Symbols tool palette.

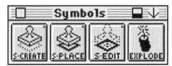

Tool 1: Symbols

LEARNING OBJECTIVE 5-48: *Changing all instances of an object using Symbol Library.*

TOOLS COVERED:

- Symbol Create
- Symbol Place
- Symbol Edit
- Symbol Explode

Fig. 5-280. Tools covered.

The tools in the Symbols tool palette allow you to store models in libraries and then repeat them symbolically, as references to the originals. These repeated symbol instances are representations of originals that are stored in the library and that allow for easy repetition and placement of common models, not only in a single project but across projects.

Global changes to these models are facilitated through these symbols, allowing you to change all instances of an object by changing the original in the Symbol Library. A global change is effected by altering the original. Symbols also economize file size and memory use and are ideal for such repetitious entities as furniture, architectural elements, and trees.

Symbols first need to be modeled before they can be defined as symbols. In the following illustrations, a desktop computer, monitor, and keyboard were modeled at varied levels of detail and stored in the Symbol Library.

You define a model as a symbol with the Symbol Create tool. Preselect the model, click on the Symbol Create tool, and click anywhere inside the modeling window. form•Z asks you to Name New Definition. When the name is entered, the new symbol is displayed in the Symbols palette.

Each definition can have up to three levels of detail, as indicated by the numbers 1, 2, and 3 in the lower portion of the Symbols palette (figure 5-281). Symbols created with the Symbols Create command are placed in the active Detail Level. You can place more symbols into other levels of the same symbol by selecting another Detail Level and then defining the symbols using Symbol Create but naming it with the same name. The grayed-out boxes in the palette simply mean that a symbol has not been defined yet for that Detail Level.

Using the Tools

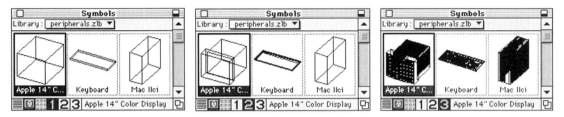

Fig. 5-281. Symbols palettes showing the same symbols with different Detail Levels.

Each symbol has an Origin and Handle that define how it is placed and manipulated. When you double click on a symbol in the Symbols palette, the Symbol Definition dialog box is opened (figure 5-282). In the Preview box, the red coordinate axes are the Origin, and the small, gray crosshairs represent the Handle. Origin is used to place a symbol, and Handle is used to scale, rotate, and transform a symbol when placing it.

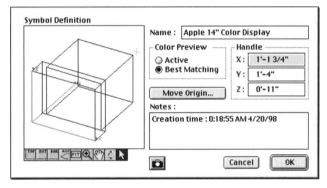

Fig. 5-282. Symbol Definition dialog box.

Handle and Origin can be manipulated in the Preview box. The locations of the Origin and Handle can also be assigned when creating the symbol definition by selecting the Pick Origin And Handle option in the Symbol Creation dialog box. This dialog box is opened by double clicking on the Symbol Create tool (figure 5-283).

Fig. 5-283. Symbol Creation dialog box.

The Symbol Place tool is used to place symbol instances into a project. Double clicking the tool opens the Symbol Instance Placement dialog box (figure 5-284). In the dialog box, you can choose to directly or manually adjust the scale and rotation of the symbol instance while it is placed. Pick Origin And Uniform Scale and Pick Origin and X, Y, And Z Scale options allow for manual scaling where the scale is adjusted interactively while it is placed. The first click places the symbol instance in relation to its Origin. The symbol instance can then be scaled and rotated using its Handle. The second click ends the operation.

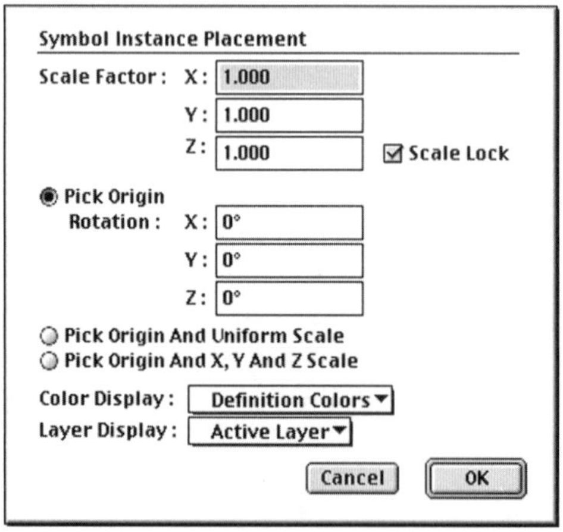

Fig. 5-284. Symbol Instance Placement dialog box.

Symbol Edit allows you to edit all properties of symbol instances individually or simultaneously. These properties include Scale, Rotation, and Detail Level. The Edit Individually or Simultaneously option is located in the Symbol Instance Edit dialog box, which is opened by double clicking on the Symbol Edit tool. When a symbol instance is selected with the Symbol Edit tool, the Symbol Instance Parameters dialog box (figure 5-285) is opened. Here, you can perform new adjustments to symbol instances.

Fig. 5-285. Symbol Instance Parameters dialog box.

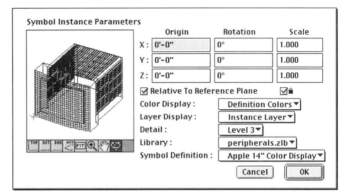

The Symbol Explode tool replaces symbol instances with the original models the symbol references. Figure 5-286 shows symbol instances rendered.

Fig. 5-286. Rendering of symbol instances.

TIPS AND STUDY AND APPLICATION NOTES

1. One symbol instance can be replaced by another, resulting in a global change of all instances that reference the original. This is done by creating a symbol and assigning it the same name as the symbol it is replacing.

2. Symbols can be nested hierarchically within each other and one another. In other words, symbols can contain within themselves other symbols or groups of symbols. An example of nesting is a keyboard button symbol called multiple times in a keyboard model, which itself is a symbol.

3. Symbols can liberate a complex project from time-consuming model regenerations by keeping all symbol instances at a relatively low level of detail until a higher level of detail is needed.

4. Symbol Libraries and objects can be edited easily by opening the library from the File pull-down menu and selecting form•Z Library as the file type.

5. See Chapter 4 for more information on the Symbols tool palette.

Line Editing Tool Palettes

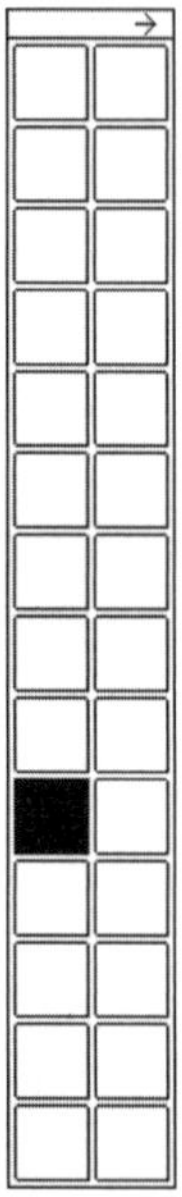

Fig. 5-287. Screen location of the Line Editing tool palette.

The Line Editing tool palette (screen location, figure 5-287; palette, figure 5-288) is for editing 2D lines and splines created with the 2D Surface tool or 2D Enclosure tool in conjunction with tools from the Lines, Splines, and Arcs tool palette and the Polygons and Circles tool palette. With the exception of Insert Point and Insert Segment, these tools can be applied to 2D surface objects only.

Fig. 5-288. Line Editing tool palette.

Tools 1: Break Line

LEARNING OBJECTIVE 5-49: *Creating a break in a line segment.*

TOOL COVERED:

- Break Line

Fig. 5-289. Break Line tool.

The Break Line tool is used for creating a break in 2D line segments, thus creating two segments. This tool can be used on any type of 2D element. To create a break in a segment with the Break Line tool, select the tool and click on a 2D object. A break will be created at the point at which you click on the object. The Break Line tool can also be used to create a break at an intersection of two lines by selecting the Break With Line option in the Break Line dialog box. To create a gap in the segment at the point it is broken, toggle the Break Distance option. Figure 5-290

Using the Tools

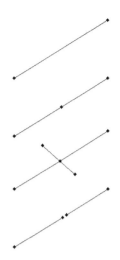

Fig. 5-290. The results of the Break Line tool: the original segment (top), the segment broken (second from top), a segment broken with the Break With Line option (second from bottom), and the Break Distance option (bottom).

shows a single segment broken using the default settings, the Break With Line option and the Break Distance option.

TIPS AND STUDY AND APPLICATION NOTES

1. When using the Break Line tool it is easier to see the effects of the tool if you turn on the Show Points option in the Wire Frame options under the Display pull-down menu.

Tools 2: Close Line

LEARNING OBJECTIVE 5-50: *Closing an open sequence of segments.*

TOOL COVERED:

- Close Line

Fig. 5-29. Close Line tool.

The Close Line tool will close an open sequence of segments either as a Trim, Join, or Connect operation. The method of closing lines is set using the Close Line dialog box. If the ends of a line sequence intersect, the Trim option will trim the two ends and cleanly join them. If the two ends do not intersect, they will be extended until they meet and are joined. The Join option joins two segments into one entity by taking the endpoint of the line sequence and moving it to be coincident with the first point in the sequence. The Connect option will draw an additional line segment between the beginning and endpoint in the sequence to create a closed series. Figure 5-292 shows a line sequence closed with each of the Close Line options.

Fig. 5-292. The Close Line tool used to close a line. From left to right: original line, Trim option, Connect option, and Join pption.

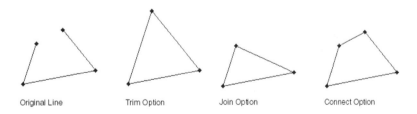

Tools 3: Trim Lines

LEARNING OBJECTIVE 5-51: *Trimming pairs of segments or multiple segments.*

Chapter 5: Modeling Tools

TOOL COVERED:

- Trim Lines

Fig. 5-293. Trim Lines tool.

The Trim Lines tool will trim intersecting lines at their point of intersection, or if the lines do not intersect, the Trim Lines tool will extend them to their point of intersection. The Trim Lines tool has several different behaviors when joining segments. These are set in the Trim Segment Options.

The default option for the Trim Lines tool is Trim Pairs of Segments. The Trim Pairs of Segments option is used for trimming two lines at a time. The two lines used in a trim operation with the Trim Pairs of Segments option will both be affected, each one trimmed to the point of intersection with the other line. The Trim Pairs of Segments option for the Trim tool has several options itself. The most important of these options are Fit Fillet and Bevel.

With Fit Fillet toggled in the Trim Pairs of Segments section of the Trim Lines dialog box, the pair of lines will be trimmed and a radius will be generated at the point of intersection to create a fillet. The Bevel option will create a Bevel at the point of intersection of the lines trimmed. The results of the various Trim Pairs of Segments option are illustrated in figure 5-295.

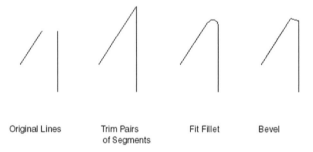

Fig. 5-295. The results of the various Trim Pairs of Segments options in the Trim Lines dialog box. From left: original line, Trim Pairs of Segments, Fit Fillet, and Bevel.

The Trim Segments With Line option in the Trim Lines dialog box will allow you to trim multiple lines to a single intersecting line. The tool can be used for aligning multiple lines, as shown in figure 5-296.

Fig. 5-296. Using the Trim Segments With Line option to trim multiple lines to a single line.

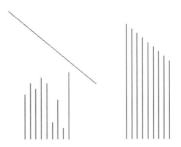

Tools 4: Connect Lines and Join Lines

LEARNING OBJECTIVE 5-52: *Connecting a sequence of line segments.*

TOOLS COVERED:

- Connect Lines
- Join Lines

Fig. 5-297. Tools covered.

The Connect Lines and Join Lines tools will connect two or more line segments to create a continuous sequence of segments. The Connect Lines tool will join lines that have coincident endpoints or endpoints that are not touching. If the two lines chosen for the connect operation have coincident endpoints, they will be joined into a single continuous line sequence. If the two lines chosen for the connect operation are not touching, a segment will be generated between the endpoints to create a continuous line sequence.

There are two options for the Connect Lines tool: Connect Segments and Connect Lines. The Connect Segments tool will join a pair of line segments to create a single continuous sequence, an example of which is shown in figure 5-298. The second option, Connect Lines, is used to connect multiple lines into a single sequence. To connect multiple lines, pre-pick the lines using the Pick tool, select the Connect Lines tool and the Connect Lines option, and click any open area of the modeling window. If you select the Close Lines Sequence option under the Connect Lines option, the sequence of lines will be closed in the process of the Connect Lines operation.

Fig. 5-298. Using the Connect Lines tool with the Connect Segments option to connect two lines.

The Join Lines tool will join a series of lines that have coincident endpoints into a single continuous line sequence. This tool is similar to the Connect Lines tool, but will only work on lines that have coincident endpoints. If the endpoints of lines are not touching, they will not be joined. The definition for how close two endpoints must be to be considered "touching" is set in the Tolerance option in the Join Lines dialog box. If the Join All option is selected in the Join Lines dialog box, several lines (all having coincident endpoints) can be joined with a single click.

By clicking on one line in a series of segments with the Join Lines tool, with the Join All option selected, all of the lines in the series will be joined. Similar to the Connect Lines tool, the Close Line Sequence option will close an open sequence of lines during the process of Joining.

TIPS AND STUDY AND APPLICATION NOTES

1. The Connect Lines tool will connect the endpoint of the first line selected with the starting point of the second line selected. Because of this behavior, it is important to pay close attention to the direction of lines in the model. To see the direction of lines in your model, turn on the Show Direction option in the Wire Frame options under the Display pull-down menu.

Tools 5: Fillet/Bevel Lines

LEARNING OBJECTIVE: *5-53 Creating fillets or bevels on line sequences.*

TOOL COVERED:

- Fillet/Bevel Lines

Fig. 5-299. Fillet/Bevel tool.

The Fillet/Bevel Lines tool generates filleted radii or bevels on the corners of line sequences. If you have a 2D line sequence generated with one of the polygon tools or Vector Line tool, you can smooth the corners using the Fillet/Bevel Lines tool. The Fit Fillet option in the Fit Fillet/Bevel Lines dialog box will generate rounded corners based on the

Using the Tools

number of edges and radius or distance specified. The Bevel option will cut each corner of a line sequence with a bevel base on the Offset or Length set in the Fit Fillet/Bevel Lines dialog box. Figure 5-300 shows an original line sequence that has been filleted using Fit Fillet and beveled using the Bevel option.

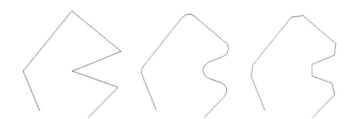

Fig. 5-300. A line sequence show in its original state (left), filleted using the Fit Fillet option (middle), and beveled using the Bevel option.

Tools 6: Insert Point/Insert Segment

LEARNING OBJECTIVE 5-54: *Inserting points and segments in objects.*

TOOLS COVERED:

- Insert Point
- Insert Segment

Fig. 5-301. Tools covered.

The Insert Point and Insert Segment tools are used to insert points and segments into 2D and 3D objects. It is not necessary to preselect the face of an object before inserting points or segments. Unlike the other tools from the Line Editing tool palette, the Insert Point and Insert Segment tools will work on 3D objects as well as 2D objects.

The Insert Point tool will insert a point on a segment at the location of a mouse click; an example of which is shown in the middle of figure 5-302. To see the results of this operation it is necessary to turn on the Show Points option in the Wire Frame Options. The Insert Segment tool can be applied to any face by clicking on two separate segments or points. A new segment will be created between the click locations, an example of which is shown on the right in figure 5-302.

Fig. 5-302. Insert Point and Insert Segment applied.

Fig. 5-303. Screen location of the Topologies tool palette.

Topologies Tool Palette

The tools in the Topologies tool palette (screen location, figure 5-303; palette, figure 304) are used to change the topological properties of an object (such as Reverse Direction and Make First Point), as well as change a Primitive object's attributes using Drop Controls. The tools in this palette, though somewhat obscure and abstract in their function, are often necessary for making adjustments needed to perform other operations in form•Z.

Fig. 5-304. Topologies tool palette.

Tools 1: First Point/Point Marker

LEARNING OBJECTIVE 5-55: *Setting first points and point markers.*

TOOLS COVERED:

- Make First Point
- Set/Clear Point Marker

Fig. 5-305. Tools covered.

The Make First Point and Set/Clear Point Marker tools are used to assign specific properties to points on a line. The first image in figure 5-306 shows normal points on a closed curve. The second image shows a point designated as First Point, which has been set with the Make First Point tool. The third point is a Marked point, defined using the Set/Clear Point Marker tool.

Using the Tools

Fig. 5-306. Lines showing the symbol for (from left) Points, First Point, and Point Marker.

The Make First Point tool is necessary to set the First Point of a closed line. Every closed line has a First Point, the point designated as the beginning of the line sequence. Some operations, such as the C-Mesh tool, will connect points or lines to generate objects. In this process of generation, first points of each line are connected. Sometimes to create cleaner models it is necessary to move the First Point, which is done by clicking on a point in a line with the Make First Point tool.

The Set/Clear Point Marker tool is used to set or erase point markers on a line. Point Markers are sometimes used in the generation of Skins. To set a Point Marker, click on an unmarked point with the Set/Clear Point Marker tool active. To clear a marker, click on a marked point with the tool.

TIPS AND STUDY AND APPLICATION NOTES

1. Selection of the Show Point and Show Marked Points options under Wire Frame Options is essential when changing the First Point and setting Point Markers.

Tools 2: Reverse Direction

LEARNING OBJECTIVE 5-56: *Reversing direction of a line or surface.*

TOOL COVERED:

- Reverse Direction

Fig. 5-307. Reverse Direction tool.

The Reverse Direction tool is used to reverse the direction of a line or surface. This function is often necessary for Nurbz, C-Mesh creation, Boolean operations on surfaces, and other operations for which the direction of a source's curves and surfaces may determine the outcome of the generated object (figure 5-308).

Fig. 5-308. Two identical lines with opposite directions.

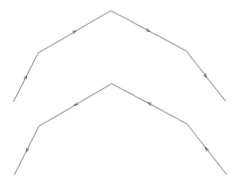

A Nurbz surface or C-Mesh might be displayed as twisted because the directions of the profiles do not have the same orientation. This is fixed by using the Reverse Direction option to modify one of the source curves (figure 5-309).

Fig. 5-309. A twisted C-Mesh remedied by changing the direction of one of the source lines.

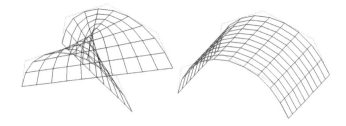

TIPS AND STUDY AND APPLICATION NOTES

1. Selection of the Show Direction options under Wire Frame is essential when changing the direction of a line or surface.

Tools 3: Make One/Two Sided

LEARNING OBJECTIVE 5-57: *Converting surface objects and surface solids.*

TOOL COVERED:

- Make One/Two Sided

Fig. 5-310. Make One/Two Sided tool

The Make One/Two Sided tool will convert a Surface Object (One Sided) to a Surface Solid (Two Sided). To convert an object to a Surface Solid or Surface Object, select the appropriate option in the Object Surface Type dialog box and, with the Make One/Two Sided tool active, click on the object. Converting an object to a Surface Solid is necessary to use some tools, such as the Boolean tools, on 2D objects. Converting an object to Surface Solid or Surface Object does not affect the shape or form of an object, only the topological definition. The conversion of an object to Surface Solid or Surface Object can always be reversed by converting it again with the Make One/Two Sided tool.

Tools 4: Drop Controls

LEARNING OBJECTIVE 5-58: *Clearing controls of parametric objects.*

TOOL COVERED:

- Drop Controls

Fig. 5-311. Drop Controls tool.

The Drop Controls tool will clear the controls of any parametric object: Primitives, Parametric Derivative, Nurbz Surfaces, Patches, and Derivative Nurbz objects. While having parametric controls adds a very useful level of control, it is often necessary to clear these controls to apply other tools. For example, objects created as Derivative Nurbz with the Revolved Object tool cannot be scaled with the Independent Scale tool. If the controls are cleared with the Drop Controls tool, the Derivative Nurbz object can be scaled.

The Drop Controls Options box has two sections: one for converting Primitives and one for converting Derivative Nurbz objects. Primitives are those objects created with tools from the Primitives tool palette. Derivative Nurbz are objects created with the Nurbz controls, such as Revolved Object, Sweeps, and Skins. The Drop Controls Options can be used to convert Derivative Nurbz objects and Primitives to plain Facetted Objects (with no controls), Nurbz Objects (with basic Nurbz controls), or Patch Objects (with Patch controls).

TIPS AND STUDY AND APPLICATION NOTES

1. The Drop Controls tool cannot be used to perform a reverse operation. Once the controls of a Primitive or Parametric Object have been dropped, they cannot be recreated.

Chapter 5: Modeling Tools

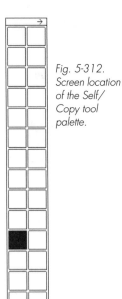

Fig. 5-312. Screen location of the Self/Copy tool palette.

Self/Copy Tool Palette

The Self/Copy tools (screen location, figure 5-312; palette, figure 5-313) modify the function of the Transformation tools. These tools control whether a transformation operation will be performed on an original object, or whether a copy of the object will be made and the transformation performed on the copy.

Fig. 5-313. Self/Copy tool palette.

Tools 1: Copying

LEARNING OBJECTIVE 5-59: *Making copies of objects.*

TOOLS COVERED:

- Self
- One Copy
- Continuous Copy
- Repeat Copy
- Multi-Copy

Fig. 5-314. Tools covered.

The Self/Copy modifiers allow you to make copies of objects. To duplicate objects, select one of the four copy modes while using one of the Transformations tools, such as Move. When the Self modifier is used (this tool is the default tool in the palette) in conjunction with a Transformation tool, the transformation will be applied to the original object instead of a copy. The One-Copy tool is for making a single copy of an object. When used with a transformation tool, the object selected is copied and the copy is transformed. The Continuous Copy, Repeat Copy, and Multi-Copy modes are useful for situations in which many duplicates are needed in some predictable geometrical arrangement.

In figure 5-315, Multi-Copy mode is active as a cone is moved 8 feet from its original position at the left. The Move command is automatically repeated seven times, the current # Of Copies setting in the Copy Options. The copies are spaced equally, according to the original Move distance. Note that the Repeat Copy mode can achieve the same result "manually" with seven mouse clicks.

Using the Tools

Fig. 5-315. Multi-Copy tool and Move.

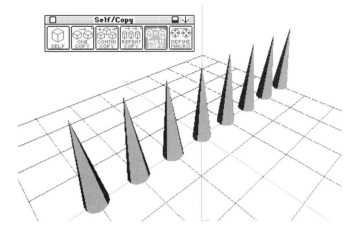

In figure 5-316, the Copy Options dialog box is opened by double clicking on the Multi-Copy icon, and its # Of Copies option is set to 7. The Even Increment setting is the default.

Fig. 5-316. Copy Options dialog box.

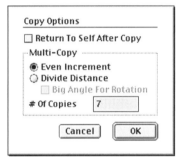

In figure 5-317, Multi-Copy is set at 11 copies, and the Rotate command is applied to a cone. A rotation value of 30 degrees is entered in the Prompts window. Because the point of rotation is outside the object (entered in the Prompt window as 0,0), the result is a circle of cones.

Fig. 5-317. Multi-Copy and Rotate.

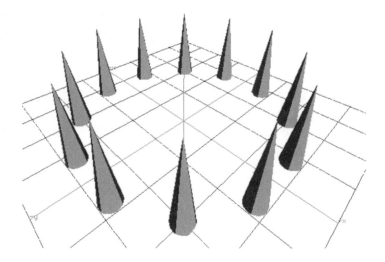

In figure 5-318, the Multi-Copy mode is set at 7 copies, and a perpendicular Move command is applied to the first floor slab. The next seven floor slabs are then created automatically, generating a building with equal floor-to-floor heights.

Fig. 5-318. Multi-Copy and Move.

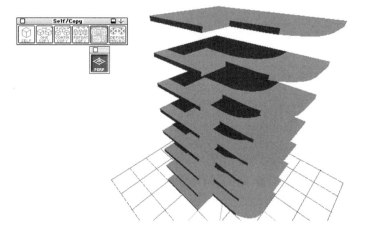

TIPS AND STUDY AND APPLICATION NOTES

1. The result of using Multi-Copy (with the settings at 7 copies) will be identical to using Repeat Copy seven times. Multi-Copy is more convenient when the precise number of desired copies is known. Repeat Copy is more useful when you need to see the result to judge how many copies are needed.

Using the Tools

Tools 2: Define Macro Transformation

TOOL COVERED:

- Define Macro Transformation

Fig. 5-319. Define Macro Transformation tool.

The Define Macro Transformation tool is discussed under "Macro Transformation Tools" in the Geometric Transformations palette.

Query Tool Palette

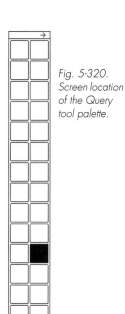

Fig. 5-320. Screen location of the Query tool palette.

The tools in the Query tool palette (screen location, figure 5-320; palette, figure 5-321) are used to display and edit the properties of an object, and can be used to measure distances between points, segments, and objects.

Fig. 5-321. Query tool palette.

Tools 1: Query Attributes

LEARNING OBJECTIVE 5-60: *Displaying and editing objects.*

TOOLS COVERED:

- Query
- Query Attributes

Fig. 5-322. Tools covered.

The Query tool displays the properties of an object, and can be used to edit and calculate certain geometric properties of objects. When the Query tool is selected and you click on an object, the Query Object dialog box (figure 5-323) is displayed. The box shows all attributes of an object and allows you to edit them using pull-down menus or by clicking on the Edit button. This dialog box includes the options for calculating the Volume, Surface Area, and # of Non Planar Faces of an object. By selecting the Query Attributes button, the Query Object Attributes dialog box is displayed (figure 5-324). This level of information can also be directly obtained by using the Query Attributes tool, which also displays the Query Object Attributes dialog box.

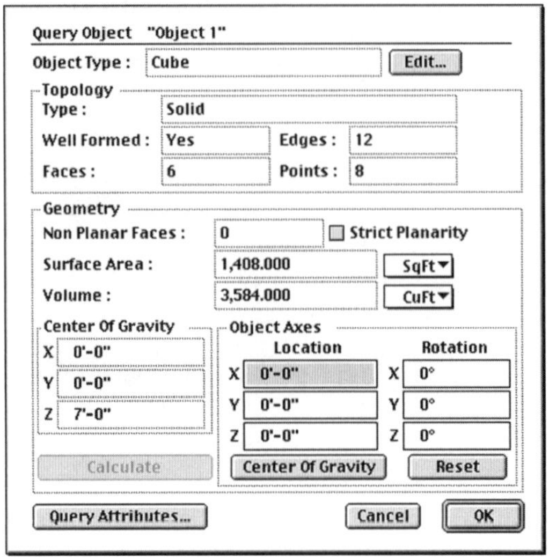

Fig. 5-323. Query Object dialog box.

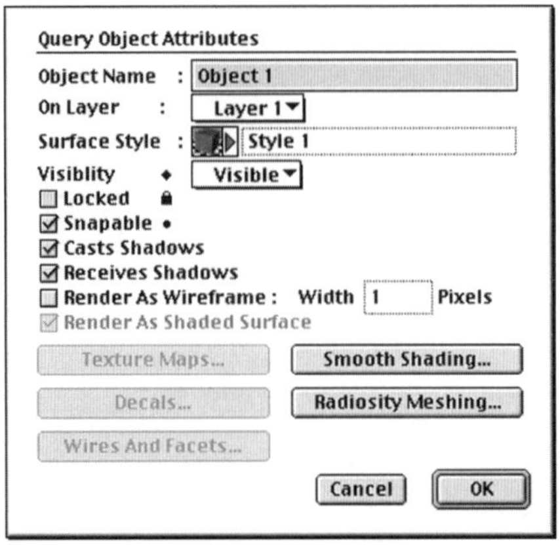

Fig. 5-324. Query Object Attributes dialog box.

If the Query option is selected and Topological Level is set to Point, Segment, Face, or Outline, a different dialog box appears for each of these geometry types that allows you to specify the attributes and position of these elements. For example, figure 5-325 shows the Query Segment dialog box.

Fig. 5-325. Query Segment dialog box.

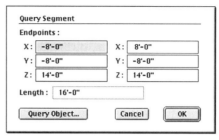

The Query tool also allows direct editing of all properties of Primitive objects. By clicking the Edit button with a primitive selected, the appropriate dialog box is displayed depending on the type of primitive selected. Figure 5-326 shows the Torus Edit dialog box, where such properties as object Origin, Rotation, and Dimensions can be directly edited.

Fig. 5-326. Torus Edit dialog box.

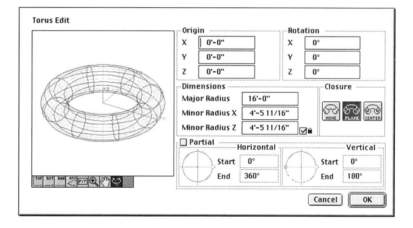

TIPS AND STUDY AND APPLICATION NOTES

1. These tools allow you to edit the attributes of objects very efficiently. In only a few clicks, you are able to edit every attribute of an object except the number of faces, points, or edges.

2. You can use the Query tool to accurately place a point, segment, or face. To do this, set Topological Level to Point, Segment, or Face and select the element. In the dialog box that appears, you can set the element's x, y, and z locations.

Tools 2: Measure

LEARNING OBJECTIVE 5-61: *Measuring distances between points, lines, segments, and surfaces.*

TOOL COVERED:

- Measure

Fig. 5-327.Measure tool.

The Measure tool is used to measure distances and lengths. These options require that you select Points, Segments, or Faces to measure between. The different options can be set in the Measurement Options dialog box (figure 5-328). When the appropriate entities have been selected and the Query is activated, the measurement appears in the Prompts window or a pop-up dialog box (figure 5-329).

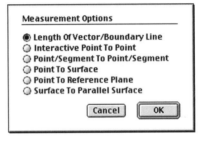

Fig. 5-328. Measurement Options dialog box.

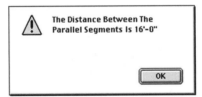

Fig. 5-329. Your requested measurements are displayed in the prompt box or a pop-up dialog box.

TIPS AND STUDY AND APPLICATION NOTES

1. The Measure tool measures distance in 3D space, which is difficult to obtain visually through comparison with the Grid Module.

2. The Measure tool's accuracy is based on the Numeric Accuracy of your Project Working Units. This can be changed in the Options menu under Working Units.

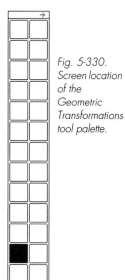

Fig. 5-330. Screen location of the Geometric Transformations tool palette.

Geometric Transformations Tool Palette

The tools in the Geometric Transformations tool palette (screen location, figure 5-330; palette, figure 5-331) are used to transform objects and topological elements in 3D space.

Fig. 5-331. Geometric Transformations tool palette.

Tools 1: Move and Rotate

LEARNING OBJECTIVE 5-62: *Changing the locations of objects and topological elements in 3D space.*

TOOLS COVERED:

- Move
- Rotate

Fig. 5-332. Tools covered.

The Move tool performs a geometric transformation on an object or topological element, changing its position in space. The Rotate tool spins an object or topological element around a point.

The Move tool translates objects or topological elements. Several objects or elements can be moved at once by pre-picking them before selecting the Move tool. Alternatively, each element can be moved, one at a time, by selecting the Move tool and then post-picking (clicking on) the element to be moved (figure 5-333).

Fig. 5-333. An object is translated using the Move tool.

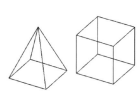
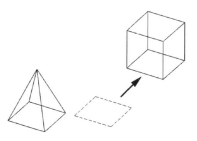

Shapes of objects can be easily transformed by moving a topological element of their geometry. To transform a topological element, select the appropriate Topological Level from the Topological Levels tool palette. In figure 5-334, the face of a cube is selected and moved.

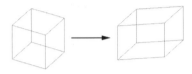

Fig. 5-334. An object transformed by moving a face.

The Rotate tool pivots an object based on a center of rotation you define. The first click determines the coordinate location of the center of rotation and the second and third clicks determine the degree of spin. The two values can also be entered manually using the Prompts window (figure 5-335).

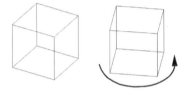

Fig. 5-335. An object is pivoted using the Rotate tool.

Shapes of objects can be transformed by selecting topological elements and rotating them. The twisted shape in figure 5-336 was produced by rotating a face.

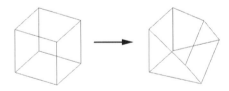

Fig. 5-336. An object transformed by rotating a face.

TIPS AND STUDY AND APPLICATION NOTES

1. Moving and rotating points, segments, and faces usually results in a nonplanar object. A nonplanar object must be triangulated before it can be used as a Boolean operand.

2. Move and Rotate operations can also be performed with the Pick tool if the Click and Drag option or the Nudge Keys option is turned on. Refer to the Pick tool for more information.

Tools 2: Independent Scale and Uniform Scale

LEARNING OBJECTIVE 5-63: *Scaling objects uniformly and individually.*

TOOLS COVERED:

- Dynamic Scale
- Uniform Scale

Fig. 5-337. Tools covered.

There are two tools for scaling objects: Dynamic Scale and Uniform Scale. Uniform Scale transforms an object uniformly in all directions, affecting its x, y, and z scales equally. The Dynamic Scale tool transforms an object in the x, y, and z directions independently.

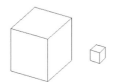

Fig. 5-338. Uniform scaling of an object.

To scale an object, select it by either pre-picking or post-picking, then define the base of scale with the first click and the percentage of scale with a second click. The base of scale is the reference point about which the object is scaled. The scale tools work by scaling all distances relative to the base of scale. Often your object will move during scaling if your base of scale is not located within the object. The Uniform Scale tool changes the scale of an object's x, y, and z dimensions uniformly (figure 5-338).

The Scale tools can be used on all Topological Levels: point, segment, face, object, and group (figure 5-339).

Fig. 5-339. Uniform scaling of a single face.

The Dynamic Scale tool is used to scale an object by different percentages in the x, y, and z dimensions (figure 5-340). Using Dynamic Scale, objects can be scaled numerically by entering separate Scale Factors for the x, y, and z dimensions in the Prompts window.

Fig. 5-340. Dynamic Scale can scale in each direction separately.

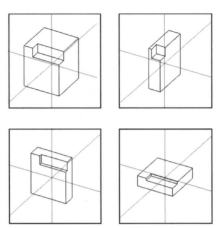

TIPS AND STUDY AND APPLICATION NOTES

1. Scaling an object affects the entire object, not just scaling the overall size of the object but proportionally affecting all of its details. Although this is the function of the tool, it is often an undesirable choice. For example, if you have a simple walled box, scaling its length will change the wall thickness along that length. The result will be that the wall thickness will no longer match.

2. To preserve the size of details while changing the size of an entire object, it is often more practical to select and move the faces of an object instead of scaling the entire object.

Tools 3: Mirror (Reflect)

LEARNING OBJECTIVE 5-64: *Reflecting entities along an axis of reflection.*

TOOL COVERED:

- Mirror

Fig. 5-341. Mirror tool.

The Mirror tool allows you to reflect entities along an axis of reflection. The axis can be designated as a point, segment, or face/surface, or designated dynamically. Figure 5-342 shows mirroring with the Dynamic Reflection option, which is the default setting for the Mirror tool. Clicking on point 1 establishes the base point of the reflection. A preview of

Using the Tools

the reflected entity and its axis of reflection is rubber-banded until point 2 is clicked, which determines the location of the mirrored object. Multiple entities can be simultaneously reflected by pre-picking the selection before the Mirror tool is selected.

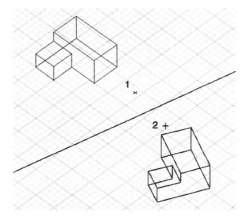

Fig. 5-342. Object during mirroring with the Dynamic Reflection option.

Figure 5-343 shows the Reflection Options dialog box, which is selected by double clicking on the Mirror tool. Entities can be reflected using the default options Dynamic Reflection, About A Point, About A Segment, or About A Surface/Plane.

Fig. 5-343. Reflection Options dialog box.

As in figures 5-344 and 5-345, when reflecting About A Point, Segment, or Surface/Plane, it is important to check whether the Relative To Reference Plane option has been selected. Figure 5-344 shows the option turned off so that the entity is reflected about the point in the x, y, and z directions.

Fig. 5-344. Mirroring about a point.

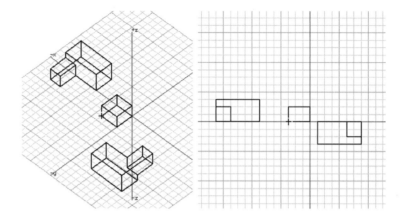

Fig. 5-345. Mirroring about a point with the Relative To Reference Plane option selected.

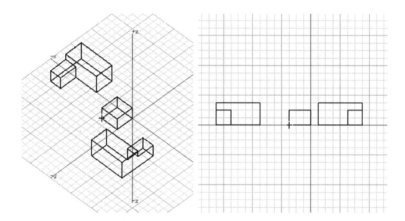

Figure 5-345 shows that when the Relative To Reference Plane option is turned on, the entity is reflected only in the two directions of the active reference plane.

TIPS AND STUDY AND APPLICATION NOTES

1. It is a good idea to use Orthogonal Snap to get perpendicular reflections. Grid Snap also helps to locate the reflected entity in a precise location.

2. When reflecting entities About A Point, Segment, or Surface/Plane, after the entity is selected, the tool automatically looks for a point, segment, or surface even when the Topological Level is not specified.

3. The Adjust Direction option allows you to adjust the direction of solid objects to normal, which is typically reversed when reflected.

Tools 4: Transformations

LEARNING OBJECTIVE 5-65: *Recording a sequence of geometric transformations and assigning macro buttons.*

TOOLS COVERED:

- Repeat Last Transformation
- Macro Transformations

Fig. 5-346. Tools covered.

The Macro Transformation tool and the Macro Define tool allow you to record a sequence of geometric transformations (i.e., Move, Rotate, Scale, and Mirror) and assign the three macro buttons to the record. This allows you to combine a frequently used sequence of transformations into a single macro button, which can then be applied to a desired object with a single action.

In this process, you first define macros by selecting a model onto which transformations will be performed and recorded. You then select Define Macro in the Self/Copy Modifier tool palette. This brings up the Name New Macro Transformation dialog box (figure 5-347), in which you enter the name of the macro transformation.

Fig. 5-347. Name New Macro Transformation dialog box.

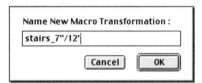

Double click on Macro Transformations. This brings up the Macro Transformations dialog box (figure 5-348). In this dialog box, you can assign macros (left side of the dialog box) to each of the three macro buttons (right side of the dialog box) by selecting the macro and then clicking on the tool you want to assign it to.

Fig. 5-348. Macro Transformations dialog box.

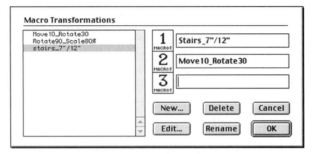

The Edit button in the Macro Transformations dialog box opens the Definition of Macro dialog box (figure 5-349). Here, you can manually input transformations by clicking on the New button, selecting the transformation Type, and assigning numeric values to the transformations. Multiple transformations can be defined in one macro simply by adding more definitions.

Fig. 5-349. Defining and editing macro transformations.

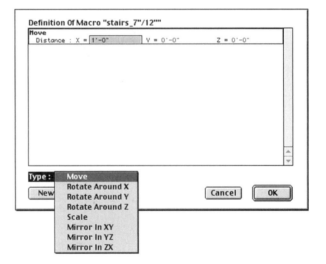

TIPS AND STUDY AND APPLICATION NOTES

1. Macro transformations can also be directly recorded while the Define Macro is selected by selecting transformation tools (Move, Rotate, Scale, and Mirror) and performing the operations. You can later edit the recorded transformations in the Macro Transformations dialog box.

2. Repeat Last Transformation (figure 5-350) is a predefined macro that, as the name suggests, repeats the last executed transformation.

Fig. 5-350. Creating a stepped pyramid with Macro Transformations.

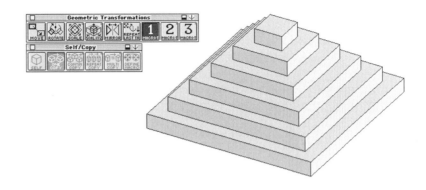

Relative Transformations Palette

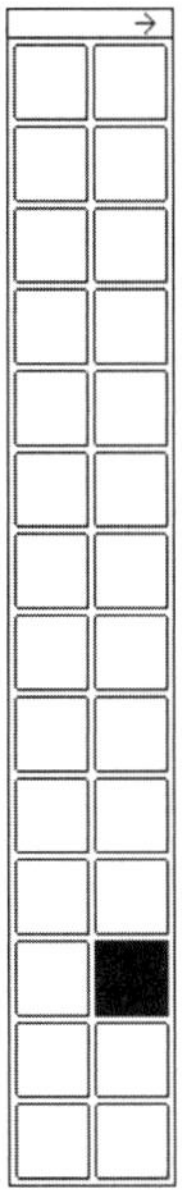

The tools of the Relative Transformations tool palette (screen location, figure 5-351; palette, figure 5-352) are used to reconfigure elements or relative distances of objects. Objects vary from points, lines, surfaces, and solids. This palette provides tools that combine multiple transformation operations and apply them to an object in a single step.

Fig. 5-351. Screen location of the Relative Transformations tool palette.

Fig. 5-352. Relative Transformations tool palette.

Tools 1: Attach

LEARNING OBJECTIVE 5-66: *Attaching one object to another.*

TOOL COVERED:

- Attach

Fig. 5-353. Attach tool.

The Attach tool is a composite transformation tool, combining a series of Move, Scale, and Rotate transformations to produce the effect of "attaching" one object to another. Attach Options determine which reference is to be used to attach one object to another, as well as whether the attachment will affect the entire object or a part of the object. The Attach Options dialog box is shown in figure 5-354.

Fig. 5-354. Attach Options dialog box.

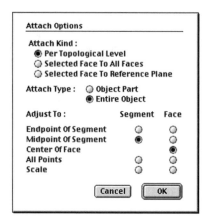

If Topological Level is set to Object, the Attach tool will perform an object-to-object attachment (figure 5-355), attaching a copy of the first object to every face of the second. The position in which the first is placed on the second is set in the Attach Options dialog box.

Fig. 5-355. Object-to-object attachment.

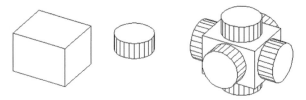

With Topological Level set to Point, Segment, or Face, it is possible to attach segments, lines, or points of one object to another. In figure 5-356, Topological Level is set to Segment. The entire object is affected because the option for Attach Type is set to Entire Object in the Attach Options dialog box.

Fig. 5-356. Attaching a segment to an object using Entire Object in the Attach Options dialog box.

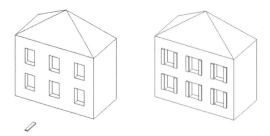

If Attach Type is set to Object Part in the Attach Options dialog box, only a portion of the object is affected instead of the entire object. In figure 5-357, Attach Type is set to Object Part and Face Adjustment is set to Scale. The Attach operation is performed on the Face level.

Using the Tools

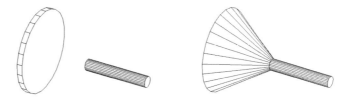

Fig. 5-357. Attaching a face with Scale set for Face Adjustment.

TIPS AND STUDY AND APPLICATION NOTES

1. Like the other Geometric Transformation tools, the way the Attach tool operates is set by the current Topological Level.

☛ **WARNING:** *Be careful when performing an object to object attachment with a curved object. The object-to-object attachment attaches one object to every face of the second object. Because curved objects usually have many faces, this operation could easily produce thousands of new objects.*

Tools 2: Align/Distribute

LEARNING OBJECTIVE 5-67: *Aligning/distributing objects according to their relative distances.*

TOOL COVERED:

- Align/Distribute

Fig. 5-358. Align/Distribute tool.

The Align/Distribute tool positions multiple objects relative to each other. The tool provides two major options, the Align and the Distribute options (figure 5-359). The Align options are used to align a set of objects randomly distributed throughout your model. The Distribute options are used for the opposite purpose, to distribute a dense set of objects throughout your model. These two option types can be used together to both align and distribute objects relative to different axes.

Fig. 5-359. Alignment/ Distribution Options dialog box.

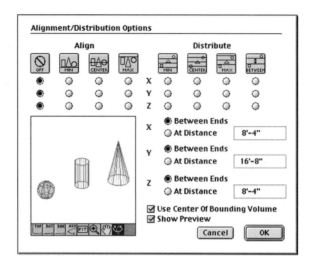

The Align/Distribute tool must be used on two or more objects. The location of the alignment and the positioning of the distribution are based on the bounding box of the group of objects. The bounding box is a rectangular solid that contains all of the objects to be aligned or distributed. The Align/Distribute tool, using the Align settings, will align all the objects to the x, y, or z edge of this bounding box. The Distribute settings will distribute the objects within this bounding box.

Figure 5-360 shows the results of different options under the Align section of the Alignment/Distribution Options. The figure shows a group of objects placed randomly in the x, y, and z directions. The Align/Distribute tool was used to align all of the objects in the x direction by selecting the MIN radio button in the X row of the Align section of the options (upper right of figure 5-360). The illustration also shows the same group of objects aligned to the y and z axes using the Align options in the Alignment/Distribution Options.

Fig. 5-360. The results of an align operation using the Align/Distribute tool. The original group of objects is located in the top left. The top right, bottom left, and bottom right show the results of aligning to the x, y, and z axes, respectively.

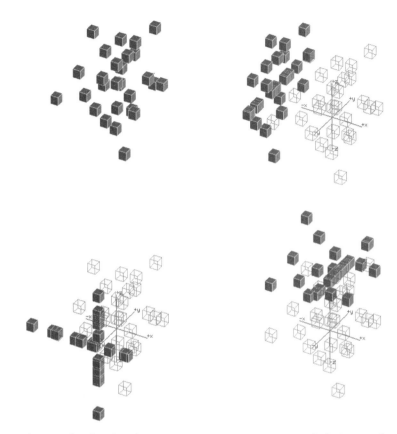

The result of a distribution operation on a group of objects is shown in figure 5-361. A cluster of objects has been distributed evenly by selecting the Between option for the x, y, and z directions under the Distribute section of the Alignment/Distribution Options. Under the x, y, and z spacing section, the option of At Distance has been selected for the x, y, and z directions. Selecting the At Distance option allows you to evenly space all of the objects as they are distributed.

Fig. 5-361. Result of the distribution operation. The Between option was selected in the Distribute section of the Alignment/Distribute options.

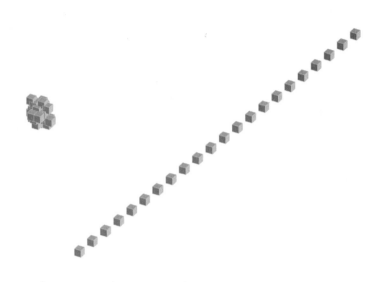

TIPS AND STUDY AND APPLICATION NOTES

1. Unlike the other Geometric Transformation tools, the Align/Distribute tool operates independently from current Topological Level.
2. At least two objects have to be selected for Align/Distribute operations.
3. Aligning a set of objects in the x, y, and z directions will usually results in all of the objects being in the same location.

Tools 3: Extend

LEARNING OBJECTIVE 5-68: *Extending a segment or group of segments to meet the face of an object.*

TOOL COVERED:

- Extend Segment

Fig. 5-362. Extend segment tool.

This tool extends a segment or group of preselected segments of an object to meet a face of another object. Figure 5-363 shows a cube and a hexagonal cylinder. The Extend Segment command will be executed on the hexagonal cylinder.

When the Extend Segment tool is selected, the Prompts window asks you to first select the segment to extend and then the face to which it will be extended. Figure 5-364 shows one segment of the hexagonal cylinder extended to the face of the cuboid.

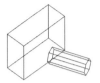 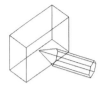 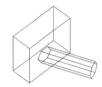

Fig. 5-363. Cube and hexagonal cylinder.

Fig. 5-364. A segment of the cylinder extended to the face of the box using the post-pick method.

Fig. 5-365. All segments in the cylinder extended by pre-picking all segments.

With the pre-pick method, all segments on the hexagonal cylinder can be preselected before the Extend Segment tool is selected to extend a group of segments simultaneously (figure 5-365).

TIPS AND STUDY AND APPLICATION NOTES

1. Segments can be extended individually or in groups to meet the face of an object.

Tools 4: Place

LEARNING OBJECTIVE 5-69: *Placing shapes perpendicular to and at selected points along a line.*

TOOL COVERED:

- Place

Fig. 5-366. Place tool.

This tool places shapes perpendicular to and at selected points along a line. When used with Self/Copy modifiers, it is especially useful for placing source objects for C-Mesh and Nurbz generation.

Figure 5-367 shows the Place Options dialog box. When using the Place tool, you must first have a source shape and a placement line (figure 5-368). The source shape can be aligned to the placement line at its Origin, which is the location of the origin (0,0,0) at the time the shape was created. The source shape can also be placed at First Point, Centroid, or Middle Of Open Ends.

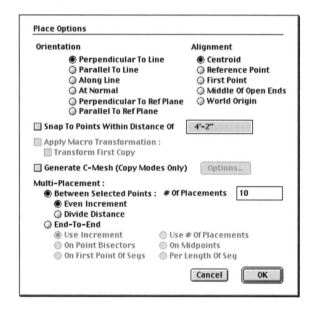

Fig. 5-367. Place Options dialog box.

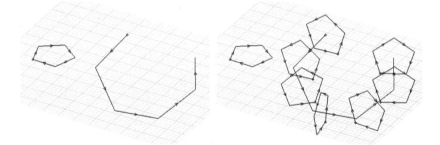

Fig. 5-368. Place on Line used with the Transformation Multi-Copy modifier.

The Place On Line tool is always used in conjunction with Self/Copy modifiers. In figure 5-368, the Multi-Copy modifier was used and the source shape was placed with the # Of Placements option set at 7, with the End-To-End option selected. This places seven copies of the source evenly along the line. When the Generate C-Mesh option is selected, a C-Mesh is created between the copied source objects (figure 5-369).

➤ **NOTE:** *Make sure one of the Multi-Copy modifiers is selected when using this option. Otherwise, multiple copies will not be generated.*

Fig. 5-369. Place on Line option used with the Generate C-Mesh option selected.

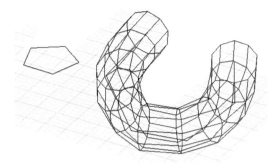

Figure 5-370 shows the use of the Generate C-Mesh option with Apply Macro Transformations. The macro for this option was predefined as having a rotation of 27, and a reductive scale of 75%. The left side of the illustration shows the placement line and the source shape. The right side of the illustration shows the rendered model.

Fig. 5-370. Use of the Generate C-Mesh option with Apply Macro Transformations.

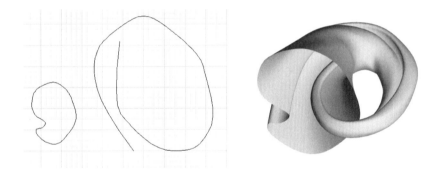

TIPS AND STUDY AND APPLICATION NOTES

1. When using the Place tool, it is helpful to have the Show First Point and Show Directions options selected in the Wire Frame Options dialog box. This provides visual cues when performing the operations.

2. The value in the Transformation Multi-Copy dialog box has no effect on the Place on Line operation. Only the value entered in the Place On Segment/Point Options dialog box is applied.

Attributes Tool Palette

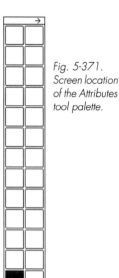

Fig. 5-371. Screen location of the Attributes tool palette.

The tools of the Attributes tool palette (screen location, figure 5-371; palette, figure 5-372) are used to define all attributes of an object other than its geometric properties. These attributes include how the object is displayed on screen and how it is rendered by each of the rendering types.

Fig. 5-372. Attributes tool palette.

Tools 1: Color

LEARNING OBJECTIVE 5-70: *Assigning the active color to an object or face of an object.*

TOOL COVERED:

- Set Color

Fig. 5-373. Set Color tool.

The Color tool assigns the active color to an object or face of an object. The active color is the color highlighted in the Surface Style palette. The Color tool contains two options in the Surface Style Options dialog box: Keep Current Color of All Faces and Clear All Face Colors.

When coloring an object with the Keep Current Color on All Faces option selected, the object color will change but the faces that have been individually colored will remain the same. The Clear All Face Colors option erases the colors of all faces of an object before assigning the new color (figures 5-374 and 5-375).

Fig. 5-374. Surface Style Options dialog box.

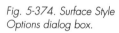

Fig. 5-375. Changing the color of an object.

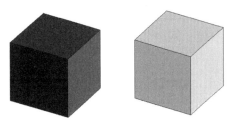

When Topological Level is set to Face, the Set Color tool can be used to change the color of individual faces of an object (figure 5-376).

Fig. 5-376. Changing the color of a face.

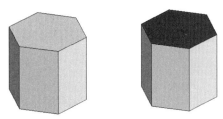

TIPS AND STUDY AND APPLICATION NOTES

1. When trying to reset the color of all faces of an object, make sure to toggle the Clear All Face Colors option in the Color Options dialog box. Keep Current Color of All Faces as the default option.
2. If you have used the Boolean functions to add or subtract objects of different colors, the new object will often have faces with colors from the operand objects. To erase the colors on all faces of an object, use the Clear All Face Colors option in the Set Color tool.

Tools 2: Shading

LEARNING OBJECTIVE 5-71: *Setting the smoothing attributes of an object for rendering.*

TOOL COVERED:

- Smooth Shade

Fig. 5-377. Smooth Shade tool.

The Smooth Shade tool sets the rendering attributes of an object to use Smooth Shading. The Smooth Shade tool controls how smoothly an

Chapter 5: Modeling Tools

object is rendered. Smooth Shade takes all faces in a model meeting at an angle greater than a specified degree and renders them as smooth blends instead of hard edges. This gives otherwise faceted surfaces a smooth look without inserting extra polygons (figure 5-378).

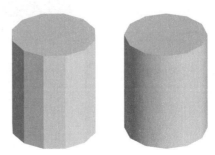

Fig. 5-378. A cylinder without and with Smooth Shading applied.

In the Smooth Shading Options dialog box (figure 5-379, you can specify whether an object is to use Smooth Shading and the criteria for the faces it should smooth. Smooth Shade can be set to affect all faces or only those that meet adjacent faces with an angle larger than the angle entered in the Edges With Angle Greater Than option.

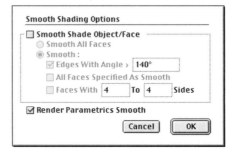

Fig. 5-379. Smooth Shading Options dialog box.

TIPS AND STUDY AND APPLICATION NOTES

1. Smooth Shade is available in RenderZone and Shaded Render only. The option for Smooth Shading must be turned on in the RenderZone and Shaded Render options.

2. The Smooth Shade tool only effects the faces of an object and not its edges (figure 5-378). Unfortunately, increasing the number of polygons is the only way to smooth out the edges of a model.

Tools 3: Render Attributes

LEARNING OBJECTIVE 5-72: Setting display characteristics for object rendering.

TOOL COVERED:

- Render Attributes

Fig. 5-380. Render Attributes tool.

The Render Attributes tool is used to determine the manner in which an object is displayed when rendered. The attributes determine such things as whether a model casts or receives shadows and whether objects render as shaded surfaces or wireframe.

The rendering attributes of an object can be set by selecting the Render Attributes tool and clicking on the object. The attributes assigned to that object are determined by settings in the Rendering Options dialog box, accessed by double clicking on the Rendering Attributes tool. Figure 5-381 shows the attributes that can be set with the Rendering Options dialog box.

Fig. 5-381. Rendering Options dialog box.

Figure 5-382 shows an assemblage of three objects, all with default rendering attributes for casting, receiving shadows, and shaded surfacing.

Fig. 5-382. Default rendering attributes of objects.

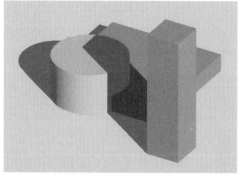

The tall object in the foreground has had its rendering attributes changed to Object Does not Cast Shadows (figure 5-383).

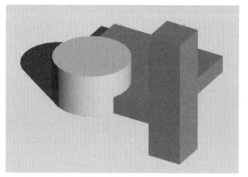

Fig. 5-383. Object Does Not Cast Shadows option applied to the object in the foreground.

The object in the foreground has been reset to cast shadows, but the middle object is set to no longer receive shadows. The shadow of the foreground object is cast across the ground and the cylinder, but does not cast across the middle object (figure 5-384).

Fig. 5-384. Object Does Not Receive Shadows option selected for center object.

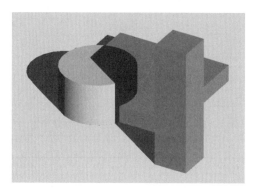

The foreground object has been set to Render As Wire Frame, but not to Shaded Surface. The cylinder has been set to render as both Wire Frame and Shaded Surface (figure 5-385).

Using the Tools

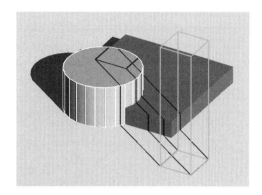

Fig. 5-385. Application of Render as Wire Frame option.

TIPS AND STUDY AND APPLICATION NOTES

1. Setting an object with the attributes to receive or cast shadow does not ensure your rendering will have shadows. For your rendering to have shadows, it is also necessary to ensure that Shadow Casting is specified as a lighting option, as well as making sure that the renderer is set to calculate shadows.

2. Certain rendering options such as Surface Render, Hidden Line, and Quick Paint will not recognize the Render As Wire Frame attribute. To have your object render as a wireframe it is necessary to use a higher-quality renderer such as RenderZone.

Tools 4: Texture Map

LEARNING OBJECTIVE 5-73: *Providing for exact positioning and repetition of textures.*

TOOL COVERED:

- Texture Map

Fig. 5-386. Texture Map tool.

Texture maps are patterns or images that can be "mapped" onto surfaces of objects to imitate materials and patterns. Textures can be applied to an object, individual faces, or a group of faces using Texture Map controls. The Texture Map tool allows for exact positioning and repetition of textures.

To create a texture map, first select Surface Style and double click it to open the Surface Style Parameters dialog box. Here, a predefined texture can be selected, or a texture can be imported through the Image Map option under the Color pull-down menu.

After the Surface Style containing the texture map has been assigned to an object using the Set Color tool, the texture can be manipulated with the Texture Map tool. Select the Texture Map tool and click on the object. This will bring up the Texture Map Controls dialog box (figure 5-387).

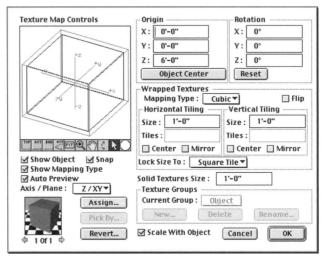

Fig. 5-387. Texture Map Controls dialog box.

The method by which a texture map is projected onto an object is determined by the Texture Mapping Type option. The Preview window in the Texture Map Controls dialog box shows the Texture Mapping Type indicated by the bounding shape enclosing the object. The manner in which the texture will be tiled and what the rendered object will look like can be previewed by selecting the Preview icon, located at the lower right corner of the Preview window (figure 5-388).

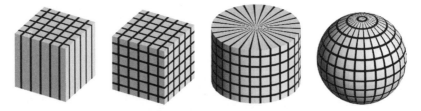

Fig. 5-388. Wrapped texture mapping types: Flat, Cubic, Cylindrical, and Spherical.

Parametric and UV Coordinates options in Mapping Type in the Texture Map Controls dialog box are used to map textures onto parametric and meshed objects that are often difficult to texture using other mapping types (figures 5-389 and 5-390).

Fig. 5-389. A Parametric Mapping Type was used for these toruses. This mapping type makes it possible to texture these objects without visible seams.

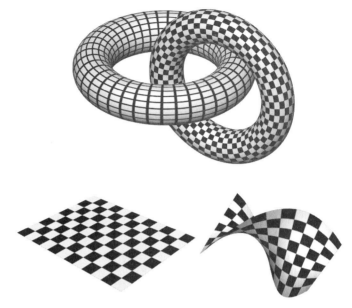

Fig. 5-390. If a UV Coordinates Mapping Type is applied to a surface or object, it will "stick" to the surfaces during deformations and transformations.

The Origin and Rotation of the texture can be altered directly using the control cursors in the Preview window. Alternatively, these can be numerically entered in the Origin and Rotation controls in the Texture Map Controls dialog box. Scaling textures is done by entering values in the Size or the Number of Tiles options in the Horizontal Tiling and Vertical Tiling dialog boxes. Figure 5-391 shows rotation and scaling of a simple texture.

Fig. 5-391. Rotating and scaling texture maps.

The Texture Groups option, located at the lower portion of the Texture Map Controls dialog box, is useful when mapping textures to individual faces or a group of faces. Select New to name the group. In the Preview window, select the faces that are to be in the group by clicking the edges that define the face. The selected faces will be highlighted with darker lines.

Any number of texture groups can be defined to suit the number of different textures needed for each face of an object. The first cylinder in figure 5-392 was mapped with a regular Cylindrical mapping type. In the second image, however, Texture Groups were used to group the top and bottom surfaces. Then, a Flat mapping type could be applied to this group while the Cylindrical mapping type was retained on the sides.

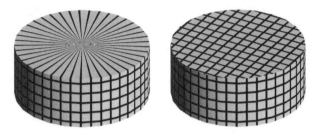

Fig. 5-392. Using Texture Groups to map textures to individual faces.

MacOS/PC NOTE:
Holding down Option (MacOS) or Ctrl + Shift (Windows) when selecting faces for Texture Groups will place the texture's coordinate system directly onto the surface of the selected face. This avoids having to relocate the coordinate system to fit the texture's orientation.

TIPS AND STUDY AND APPLICATION NOTES

1. The Texture Map Options dialog box, accessed by double clicking on the Texture Map tool, differs only slightly from the Texture Map Controls dialog box. This dialog box is used to preset most of the texture map parameters and apply those changes directly without opening the Texture Map Controls dialog box.

2. The Scale With Object option, located at the bottom of the Texture Map Controls dialog box, allows you to scale textures when the object onto which they are mapped is scaled.

3. See Chapter 11 for more information on texture maps and rendering.

Tools 5: Set Decal Attributes

LEARNING OBJECTIVE 5-74: Placing textures or surface styles onto previously assigned textures.

TOOL COVERED:

- Decal

Fig. 5-393. Decal tool.

The Decal tool allows you to place textures and decals onto surfaces of objects over previously assigned surface styles. The tool can be used effectively not only to place labels and pictures onto objects but to add details such as dirt, partial reflections, and texts over desired surfaces.

The Decal tool has many features in common with the Texture Map tool, such as controls for positioning textures. However, an additional

Fig. 5-394. Bottle with Decal label.

feature in the Decals dialog box is the group of options located in the lower right of the box that allows decals to be created and named.

When creating and placing a decal such as that shown in figure 5-394 onto a bottle, it is first necessary to create the texture and then assign it to a Surface Style. This is done in the Surface Style Parameters dialog box in the Color pull-down menu under Image Map. The Image Map options allow for precaptured textures to be loaded and edited.

With the Decal tool selected, click on the bottle. This brings up the Decals dialog box (figure 5-395). In this dialog box, click on New to assign the new texture to a decal. In the case of this bottle, the Mapping Type needs to be set to Cylindrical for the decal to match the shape of the bottle. The Horizontal and Vertical Size of the decal should also be adjusted to the appropriate size. The Preview window allows you to interactively move the location of the decal. A rendered preview can be viewed by clicking on the Preview icon located at the lower right corner of the Preview window.

Fig. 5-395. Decals dialog box.

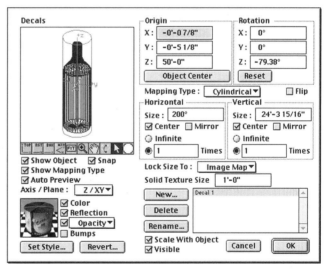

The Transparency Map option in the Surface Style Parameters dialog box under the Transparency pull-down menu can be used to place transparent decals on opaque surfaces or opaque decals on transparent surfaces. In figure 5-396, a Surface Style with a circular hole transparency map was placed as a decal using the Transparency decal option located at the lower left of the Decals dialog box.

Fig. 5-396. Placing transparent decals.

Similarly, in figure 5-397, a surface style consisting of a circle was created by transparency mapping away the periphery of the object the circle was mapped on. It was then placed as a decal on a transparent object with the Opacity decal option.

Fig. 5-397. Placing opaque decals.

TIPS AND STUDY AND APPLICATION NOTES

1. The Decal tool allows for up to 32 simultaneously overlapping decals.

Tools 6: Attributes

LEARNING OBJECTIVE 5-75: *Copying rendering attributes from object to object, and setting attributes for multiple objects.*

TOOLS COVERED:

- Set Attributes
- Get Attributes

Fig. 5-398. Tools covered.

The Set Attributes and Get Attributes tools are used to copy the rendering attributes of one object to another, and to set attributes for several objects simultaneously. The Set Attributes tool can be used to set multiple rendering attributes at once. The rendering attributes set with this tool could also be set using individual tools such as Color and Render Attributes, but this tool combines them conveniently into a single tool, making changing attributes easier and more efficient.

Using the Tools

The Set Attributes Options dialog box (figure 5-399) is where you define the attributes that will be set and those that will remain undefined. The default for this tool sets none of the attributes of an object.

Fig. 5-399. Set Attributes Options dialog box.

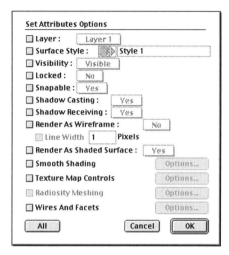

Figure 5-400 shows the Get Attributes Options dialog box. The Get Attributes tool is used to copy the attributes of an object and assign them to another object using the Set Attributes tool. The options in this dialog box determine which attributes will be copied from an object. The default is for all attributes to be copied.

Fig. 5-400. Get Attributes Options dialog box.

TIPS AND STUDY AND APPLICATION NOTES

1. The Set Attributes tool combines the functions of multiple attribute tools, making changing attributes easier and more efficient.

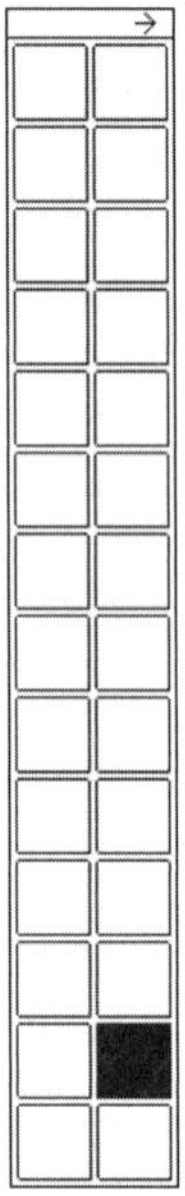

Fig. 5-401. Screen location of the Ghost and Layers tool palette.

2. The Get Attributes tool assigns the attributes of one object to another.

Ghost and Layers Tool Palette

The Ghost and Layers tool palette (screen location, figure 5-401; palette, figure 5-402) contain tools that help you manage the project by ghosting selected objects or by moving objects onto layers.

Fig. 5-402. Ghost and Layers tool palette.

Tools 1: Ghost and Unghost

LEARNING OBJECTIVE 5-76: *Graying out and reversing grayed-out objects.*

TOOLS COVERED:

- Ghost
- Unghost

Fig. 5-403. Tools covered.

A ghosted object is an object that has been made inactive. Ghosted objects are displayed in gray (default) in Wire Frame and cannot be selected or rendered. Ghosted objects usually result when an operation creates a new object, causing the original to become ghosted. Objects can also be intentionally ghosted so as to exclude them from an operation or to ease the viewing of complex models.

Figures 5-404 and 5-405 show a model before and after a Boolean Difference operation. In figure 5-405, the two operand objects are ghosted after the operation has been performed and the new object exists. The ghosted objects can be "unghosted" using the Unghost tool and used again in a different operation. Conversely, objects can be intentionally ghosted by using the Ghost tool.

Another method of ghosting and unghosting objects is to use the Objects Palette. The diamond symbol in the third column from the left of this dialog box shows the status of the object: a solid diamond represents unghosted, and a clear diamond represents ghosted.

Using the Tools

The ghosted/unghosted status of resultant objects from operations can also be controlled using thes Status Of Object section of the Tool Options Palette. The default is Ghost. Figure 5-406 shows the Edit menu, where you can select all ghosted objects by specifying the Select All Ghosted items. It is also possible to Hide Ghosted and Clear All Ghosted from the model.

Fig. 5-404. Object A and object B before Boolean Union operation.

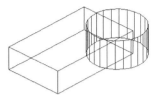

Fig. 5-405. Object B ghosted after Boolean operation.

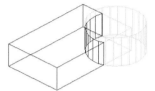

Fig. 5-406. Edit menu with additional Ghost/Unghost related operations.

TIPS AND STUDY AND APPLICATION NOTES

1. The color of ghosted objects can be changed in Project Colors, located under the Options menu.

2. Ghosted objects may be created as a result of operations that create new objects, leaving a residual trail of ghosted objects and making a model unnecessarily large and therefore slow to generate. It is often a good idea to Clear All Ghosted at regular intervals if you are sure you do not need the ghosted objects.

Tools 2: Set Layer

LEARNING OBJECTIVE 5-77: *Organizing and filtering information in a project, model, or drawing.*

Chapter 5: Modeling Tools

TOOL COVERED:

- Layers

Fig. 5-407. Layers tool.

The Layers tool and Layers palette work together to provide a powerful method of organizing and filtering the information in a project, model, or drawing. The Layers palette allows you to easily set up different layers to which objects can be assigned. Once a group of objects is on a layer, it can be easily selected, Ghosted, Hidden, Locked, or made unsnappable.

Use the Layer tool to assign an object to the currently active layer by clicking on it, or choose the layer you want it to be changed to by selecting that nonactive layer from the Layer Options dialog box (figure 5-408) and then selecting the object.

Fig. 5-408. Layer Options dialog box.

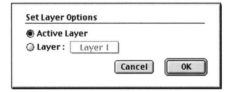

The Layers tool is one of several tools and palettes available that facilitate the organizing and filtering of information in a project. When the Layers palette (figure 5-409) is combined with the Symbol palette, the Objects palette, and tools such as Select By, it is possible to show, analyze, and manipulate a project in a variety of ways. These organizational tools are also critical in structuring a project so that it can be worked on by multiple users.

Using the Tools

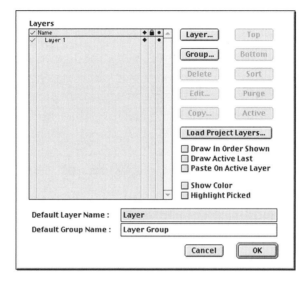

Fig. 5-409. Layers palette options.

The Layers palette controls hiding, ghosting, and locking of layers, as well as options for sorting, renaming, editing, grouping, copying, and purging layers. To make a new layer, click at the bottom of the list of Layer names and type in a new one. To control its attributes, you can toggle the various settings in the columns next to the name. The check mark indicates which layer is currently active.

The diamond symbol can be used to toggle between visible, hidden, and ghosted states. The bullet symbol determines whether the layer can be snapped to when using the Snap tools. The lock icon locks or unlocks a layer. When a layer is locked, it cannot be manipulated (figures 5-410 through 5-415).

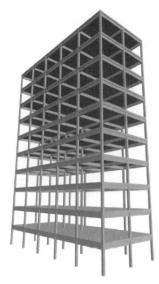
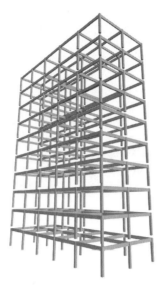
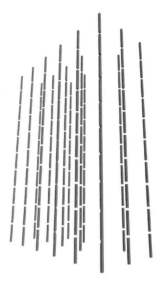

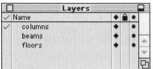
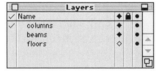
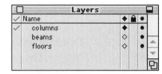

Figs. 5-410 and 5-411. The model with all the layers in the palette turned on and the corresponding palette.

Figs. 5-412 and 5-413. The model with the "floor" layer in the palette turned off and the corresponding palette.

Figs. 5-414 and 5-415. The model with the "beams" and "floor" layers in the palette turned off and the corresponding palette.

The previous illustrations are an example of how layers can be used to manage a large model. The model of the skyscraper is broken up into three separate layers: floors, columns, and beams. To view just a structural diagram of the project, all floors can be easily hidden by toggling visibility of the layer. To hide the beams and see just the columns, the beams layer can be hidden from view by toggling its visibility.

If each element (floor, column, and beam) is a separate symbol, the layers can be used to organize the building by floor. The two tools (Symbols and Layers) could be used in combination to filter information based on elements (floor, beam, column) or context (first floor, second floor, and roof).

Using the Tools

TIPS AND STUDY AND APPLICATION NOTES

1. When dealing with large models with extensive detail and multiple parts, it becomes a necessity to be able to separate objects and hide information in order to ease identification and manipulation of elements.
2. To increase the speed of rendering it is often best to hide elements that are too detailed, or elements that will not be seen in the final product.
3. The Layers palette is an effective tool for achieving both the goals of notes 1 and 2, as well as for maintaining a clear and manageable model.
4. Layers can be associated with a color to make viewing the organization of your model easier. See Chapter 4 for more discussion of the Layers palette.

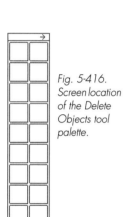

Fig. 5-416. Screen location of the Delete Objects tool palette.

Delete Objects Tool Palette

LEARNING OBJECTIVE 5-78: *Removing objects from a project.*

TOOL COVERED:

- Delete

Fig. 417. Delete tool.

The Delete tool (screen location, figure 5-416; palette, figure 5-417) is used to remove objects from a project. The delete tool will delete objects as well as groups of objects, depending on the modifier chosen in the Topological Level tool palette. To delete several objects at once, you can preselect several objects, select the Delete Tool, and click on any open space in the window.

The Delete tool will only delete entire objects and groups of objects. If the Topological Level is set to anything other than Object or Group, the delete tool will delete the object. If you wish to delete a point, segment, or other topological element other than objects, you will need to use the Delete Topology or Delete Geometry tool.

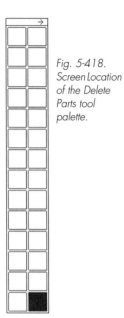

Fig. 5-418. Screen Location of the Delete Parts tool palette.

Delete Parts Tool Palette

Fig. 5-419. Delete Parts tool palette.

LEARNING OBJECTIVE 5-79: *Removing objects from a project.*

TOOLS COVERED:

- Delete Topology
- Delete Geometry

Fig. 5-420. Tools covered.

The Delete Parts tools (screen location, figure 5-418; palette, figure 5-419) are used to remove elements from an object to alter its geometry or topology. If Topological Level is set to Point, Segment, Face, Outline, or Hole, the element selected will be removed from the object it is part of, altering the object.

The Delete Topology and Delete Geometry tools work in an identical manner. The only difference in the two is how the geometry of the object is affected when an element is removed. When the geometry of an element is deleted using the Delete Geometry tool, the points associated with it are also deleted. When the topology of an element is deleted using the Delete Topology tool, the element is removed, but not its associated points.

Figure 5-421 illustrates the difference between deleting geometry and topology on a simple cube. On the left of the image is the original cube. In the middle of the figure is the cube with a face deleted using Delete Topology. Because the face is removed but not its four corner points, the cube shape remains but the side is now open. The right side of the figure shows the cube with a face deleted using Delete Geometry. Because the face is deleted as well as its four corner points, the cube collapses into a simple plane.

Using the Tools   331

Fig. 5-421. Deleting a face using Delete Topology and Delete Geometry.

Figure 5-422 shows how the geometry of an object can be changed by deleting topological elements using the Delete Geometry tool. At the top of the figure is the original object. The sequence shows the deletion of a point, segment, outline, hole, and face from the object. The deletion of the face can be made only after the hole has been deleted, to avoid creating an object with self-intersecting faces.

TIPS AND STUDY AND APPLICATION NOTES

1. It is important to note that because the Delete Topology tool removes topological elements but not their points, the resulting object is usually not well formed. A poorly formed object cannot be used in many operations, including Boolean Difference and Boolean Union.

2. Deleting topological elements using the Delete Topology tool will sometimes change a solid into a surface object. If topological elements are deleted using the Delete Geometry tool, the object will usually remain a solid.

3. Using the tools from the Delete Parts tool palette on parametric objects will cause their controls to be cleared.

4. form•Z will not allow you to delete points, segments, and faces from an object if it will result in an object with self-intersecting faces.

5. Multiple topological elements can be deleted at once by pre-picking them.

Fig. 5-422. Deleting a point, segment, outline, hole, and face with the Delete Geometry tool.

chapter 6

Window Tools

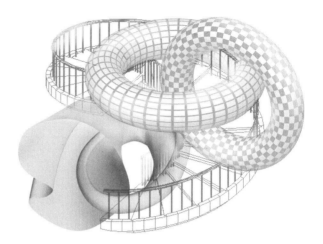

Introduction

The form•Z Window Tools are drawing and viewing aids that complement the Modeling and Drafting tools. Although they do not affect models directly, Window Tools facilitate the construction and manipulation of models by enhancing the environment within which the Modeling and Drafting tools operate.

The Interface

The Window Tools group is located in the lower left corner of the active window. The icons are either black or white, depending on whether they are currently active (black) or inactive (white). The icons show only the selected tools. However, when you click on an icon, the tool bar expands to reveal other tools within that category. Figure 6-1 shows the normal appearance of the Window Tools at the bottom left corner of the screen. Figure 6-2 shows the complete collection of expanded Window Tools.

Fig. 6-1. Window Tools palette, as displayed at the bottom left corner of the modeling window.

Fig. 6-2. Window Tools expanded.

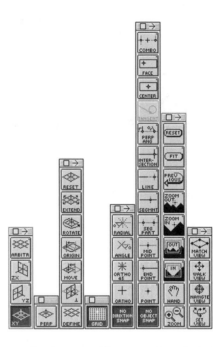

Like the Modeling Tools, these tool bars can be placed anywhere on the screen as a separate tool bar by clicking on the tool icon and dragging it out while keeping the mouse button depressed. As discussed in Chapter 4, the Window Tools can also be accessed through the Palettes pull-down menu. This will bring up a slightly larger strip of the same Window Tools icons. Because this Window Tools strip is larger and more legible, and represents exactly the same functions, you may find it convenient to view the Window Tools in this way.

The larger Window Tool icons contain a small dot at the upper right corner, which signals that there is a dialog box with additional options and numerical inputs available for that tool. The same dots are only visible from the regular Window Tools strip at the bottom of the screen if the tool bar has been expanded or dragged to another location on the screen. In either case, the options dialog box for a particular Window Tool is accessed by double clicking on the icon.

Fig. 6-3. Window Tools palette, as displayed via the pull-down Palettes menu.

Concepts and Terminology

There are several fundamental terms and concepts that must be understood in order to use Window Tools effectively. The following sections discuss these terms and concepts.

Reference Planes

form•Z uses Cartesian reference planes (i.e., X-Y, Y-Z, and Z-X) to orient models in space. Because the computer screen is a 2D viewing device, it is necessary to present these planes from which scale and location are referenced. In addition to the three Cartesian planes, form•Z includes an Arbitrary Plane option that facilitates certain operations within the model.

Snaps

Snaps are geometric constraints imposed on a modeling or drafting tool. There are three types of snaps: Grid, Directional, and Object. Grid Snap constrains the manipulation of entities within a model to a defined grid dimension. When this option is selected, all movements of the cursor and the model conform automatically to the grid. Directional Snap constrains the direction in which the entities within a model can be moved.

Object Snap constrains the movement of the cursor to topological elements (such as points, segments, and faces) within a model. When the cursor approaches the element designated by the current snap setting, the cursor will recognize the element and further operations will reference this element.

Zoom and Pan

The modeling window can be seen as a moveable window for viewing a model. In this sense, zooming allows for both closer inspection and a more general view of a model by moving the viewing window either toward or away from the model. Pan allows you to move the viewing window from side to side or up and down. The Walk View and Navigate View Window tools, new in version 3.0, combine Zoom and Pan with rotational controls, providing an approximation of a Virtual Reality display.

Window Tools

The sections that follow describe the various Window Tools, starting from the left end of the Window Tools palette. These include controls for reference planes, perpendicularity, grid snap, directional snap, object snap, pan and zoom, and view tools.

Reference Planes

Fig. 6-4. Active Reference Planes Window Tools palette.

LEARNING OBJECTIVE 6-1: *Orienting a model in space.*

TOOLS COVERED:

- XY Reference Plane Active
- YZ Reference Plane Active
- ZX Reference Plane Active
- Arbitrary Plane Active

SCREEN LOCATION: *First icon in the Window Tools palette in the bottom left-hand corner of the screen.*

The Active Reference Planes tools consists of the three Cartesian plane options XY, YZ, and ZX, as well as the Arbitrary Plane option, which can be custom defined. Each reference plane consists of a grid and x, y, and z axes. The grid gives an indication of scale and shows the active reference plane, which is your current drawing surface. The World Axes options, colored red, show the absolute orientation of the three Cartesian axes in space. The Plane Axes options, colored turquoise, show the relative axes of a custom-defined arbitrary plane.

Reference planes are used to orient a model in space. They affect how some modeling tools behave; in particular, those that generate shapes in relation to the active reference plane. Figures 6-5, 6-6, and 6-7 show a hexagonal pyramid generated with the X-Y, Y-Z, and Z-X reference plane options with the same height selected for each in the Heights menu. The height in each case is measured in reference to the active reference plane.

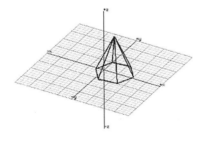 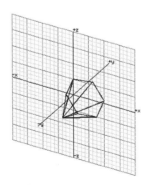 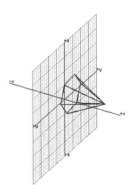

Figs. 6-5, 6-6, and 6-7. Identical objects created on the xy, xz, and yz reference planes

An arbitrary reference plane is a custom-defined plane that can be positioned according to specific modeling needs. Figure 6-8 shows an arbitrary reference plane defined to facilitate the creation of objects that lie at specific angles on the *xy* plane. The larger object was created on the *xy* plane, whereas all smaller objects were created on the arbitrary, slightly rotated, reference plane.

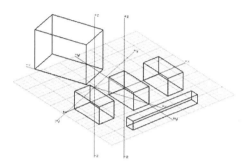

Fig. 6-8. Objects created on an arbitrary reference plane.

☛ **NOTE:** *For more information on defining arbitrary reference planes, see the third section of the Window Tools, which follows.*

The Reference Grid Options dialog box (figure 6-9), accessed by double clicking on the Reference Plane icon, allows you to adjust the scale of a grid by setting the increments in the *x, y,* and *z* directions, as well as the number of divisions between grid lines. There is also an option to turn the line grid into a grid of dots.

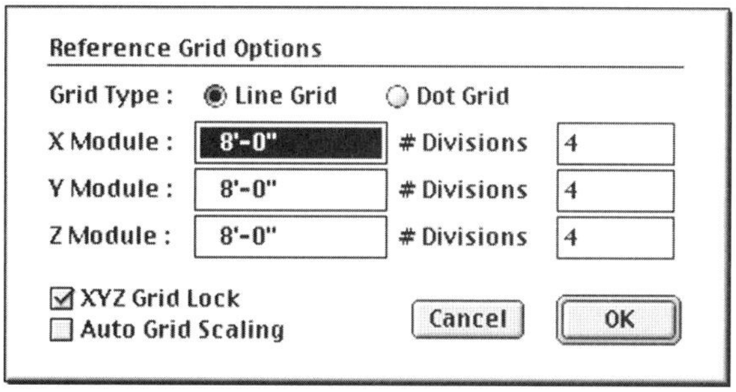

Fig. 6-9. Reference Grid Options dialog box.

TIPS AND STUDY AND APPLICATION NOTES

1. The following table lists the MacOS and PC (Windows) quick keys for the reference plane options.

Option	MacOS	Windows
X-Y Plane	Opt + X	Ctrl + Shift + X
Y-Z Plane	Opt + Y	Ctrl + Shift + Y
Z-X Plane	Opt + Z	Ctrl + Shift + Z
Arbitrary Plane	Opt + A	Ctrl + Shift + A

Perpendicular Switch

LEARNING OBJECTIVE 6-2: *Moving and drawing perpendicular to a reference plane.*

TOOL COVERED:

- Perpendicular Switch

Fig. 6-10. Perpendiicular Switch tool.

SCREEN LOCATION: *Second icon in the Window Tools palette in the bottom left-hand corner of the screen.*

The Perpendicular Switch limits the input of the cursor to motion in a direction perpendicular to the current reference plane. This important tool is for moving and drawing "up" or "down" from the reference plane. Normally, the drawing and transformation tools operate in relation to the active reference plane, allowing movement of the mouse in the two dimensions described by the plane. For example, if the *xy* plane is active, transformations and drawings can be done only on the *xy* plane.

The Perpendicular Switch is for working in the third dimension, the direction perpendicular to the active reference plane. If the *xy* plane is the active reference plane, the Perpendicular Switch will allow you to work in the *z* direction. The Perpendicular Switch is toggled on or off by clicking on it once. If the icon is black, the switch is on and all movement will be perpendicular to the reference plane. Figure 6-11 shows an object being moved up from the reference plane using the Perpendicular Switch.

Window Tools 339

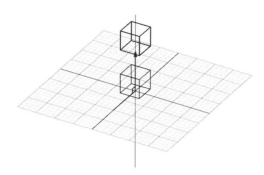

Fig. 6-11. A cube moved up from the reference plane using the Move tool and the Perpendicular Switch.

TIPS AND STUDY AND APPLICATION NOTES

1. When working in a 3D view, it is important to know whether the Perpendicular Switch is on or off. In certain 3D views, the action of moving things on the horizontal xy plane and moving things up, vertically, from the reference plane may appear the same. This confusion can cause drastic differences between what you think you are modeling and what you actually get.

Reference Planes

LEARNING OBJECTIVE 6-3: *Defining arbitrary and perpendicular planes; using the Move Plane, Move Plane Origin, and Rotate Plane options; extending a plane's grid; and resetting a plane.*

TOOLS COVERED:

- Define Arbitrary Plane
- Define Perpendicular Plane
- Move Plane
- Move Plane Origin
- Rotate Plane
- Extend Plane Grid
- Reset Plane

Fig. 6-12. Window Tools Reference Plane palette.

SCREEN LOCATION: *Third icon in the Window Tools palette in the bottom left-hand corner of the screen.*

Define Arbitrary Plane

Entirely new drawing planes can be custom-defined by using the Define Arbitrary Plane tool. An Arbitrary Plane can be defined by selecting one of the following definitions of a plane: three points, two segments, one outline, or one face. Figure 3-13 shows a roof plane selected as an Arbitrary Plane.

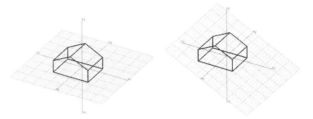

Fig. 6-13. Defining an Arbitrary Plane in reference to a roof.

Define Perpendicular Plane

The Define Perpendicular Plane option allows for a reference plane perpendicular to the current plane to be defined. This can be done by selecting the axis around which the plane needs to be rotated in order for it to be perpendicular (figure 6-14).

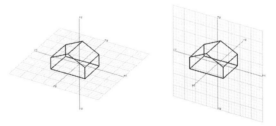

Fig. 6-14. Defining a Perpendicular Plane.

Move Plane, Move Plane Origin, and Rotate Plane

Move Plane, Move Plane Origin, and Rotate Plane can be used to reposition the current reference plane. The Move Plane command displays the translation of the entire reference plane grid; the Move Plane Origin keeps the edges of the plane in place and displays the movement of the Plane Axes; and Rotate Plane displays the plane spinning around the z axis.

> **NOTE:** *As you use these commands, keep in mind that, by default, the World Axes are shown in red and the Grid and the Plane Axes in blue. Reference plane adjustments are best demonstrated by the new position of the blue Plane Axes and the blue Grid in relation to the never-changing red World Axes.*

Extend Plane Grid

In the 3D views, the default grid is limited in size and sometimes there is a need to extend the grid to cover objects created beyond the grid. Extend Plane Grid allows the edge of the plane to be extended by clicking on the edge of the grid, moving it to the new location, and clicking again to finish the operation.

Fig. 6-15. Extending the reference plane grid.

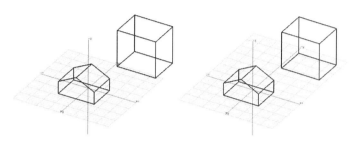

Reset Plane

Reset Plane resets the location and boundaries of the current reference plane to its default parameters, but applies only to orthogonal reference planes. If, for example, you have rotated the plane around the x axis, and then moved it, you will have to click on the XY Reference Plane Active tool to get back to the original plane.

➥ **NOTE:** *All defined planes can be stored, organized, and selected using the Planes Palette, as discussed in Chapter 4. For a step-by-step reference plane example, refer to the Portal exercise in Chapter 7.*

✓ **TIP:** *While Extend Plane Grid is selected, holding the Shift key down while clicking the mouse inside the window resizes and repositions the grid to cover all objects in the model.*

Grid Snap Switch

LEARNING OBJECTIVE 6-4: *Snapping to grid lines.*

TOOL COVERED:

- Grid Snap Switch

Fig. 6-16. Grid Snap Switch tool.

Chapter 6: Window Tools

SCREEN LOCATION: *Fourth icon in the Window Tools palette in the bottom left-hand corner of the window.*

If Grid Snap is active, the point you are inputting with the mouse will "jump" to the closest intersection of grid lines. This is true even if the grid lines are currently invisible. To toggle Grid Snap on and off, click on its icon in Window Tools. If the icon is black, it is currently on, and the mouse cursor will snap to the grid. In figure 6-17, a series of cones has been created with Grid Snap turned off. Notice how it is difficult to plant objects directly at the intersections of major grid lines, especially if modeling in a perspective window.

Fig. 6-17. Cones created with Grid Snap turned off.

The spacing for Grid Snap is set in the Grid Snap Options dialog box (figure 6-18). In figure 6-19, the Grid Snap Options are set to 8 feet. The XYZ Snap Lock option is activated so that when you change the size for one direction, all directions are updated. The Match Grid Module option adjusts the snap to the current settings of the reference grid, as defined in the Window Setup options.

Window Tools

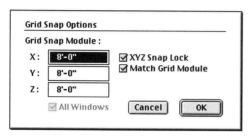

Fig. 6-18. Grid Snap Options dialog box.

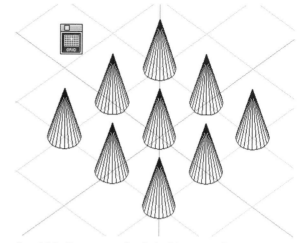

Fig. 6-19. Cones created with Grid Snap turned on.

Figure 6-19 shows a series of cones generated with Grid Snap set at the same spacing as the grid lines of the reference plane. Note that the cones are perfectly centered on the intersections of the grid lines.

Grid Snap's vertical (z axis) control is shown in figure 6-20. First, an 8-foot cubic frame module is created, as seen at left in the image. The Grid Snap interval is set at 8 feet. Continuous Copy mode is selected. Then copies of the module are stacked on top of one another with the Move command and the Perpendicular Switch.

Fig. 6-20. Snapping to the grid in the z direction.

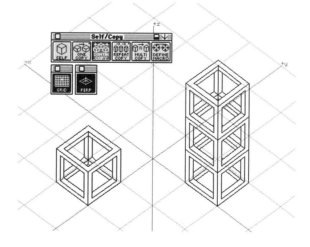

Chapter 6: Window Tools

Tips and Study and Application Notes

1. You can switch Grid Snap off or on in the middle of drawing commands. For example, the center of a circle could be snapped to the grid; the radius could then be defined without Grid Snap.

2. If you type a value directly into the Prompts window, rather than using the mouse, Grid Snap will be ignored.

Fig. 6-21. Window Tools Direction Snaps palette.

Direction Snaps

Learning Objective 6-5: Drawing straight and angled lines, and performing precise transformations.

Tools Covered:

- No Directional Snap
- Ortho Snap
- Ortho + Diagonal Snap
- Angle/Slope Snap
- Radial Snap

Screen Location: Fifth icon in the Window Tools palette, located in the bottom left corner of the screen.

The Directional Snaps options constrain the input of the cursor to a specific range of motion while using the drawing and transformation tools. These snap tools make it easy to draw straight and angled lines and to perform exact transformations. When a Directional Snap icon is black, it is active, or selected.

No Directional Snap

The No Directional Snap tool turns off directional snaps and allows you to move the cursor freely around when drawing and transforming objects.

Ortho Snap

Ortho Snap, short for Orthogonal Snap, limits movement of the mouse to directions parallel to the axes of the current reference plane. This tool produces objects with right angles only. This is a method of ensuring that everything in your model is square. Figure 6-22 shows a series of lines drawn with Ortho Snap active.

Fig. 6-22. Ortho Snap is used to produce right-angled lines.

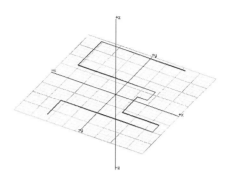

Ortho + Diagonal Snap

The Ortho 45 Snap tool constrains the movement of the mouse parallel to reference plane axes or at angles 45 degrees to axes. Figure 6-23 shows a line drawn using the Ortho 45 Snap tool.

Fig. 6-23. Ortho 45 Snap produces right angles and 45-degree lines.

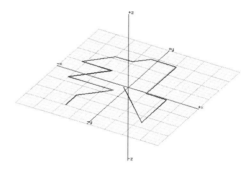

Angle/Slope Snap

The Angle/Slope Snap tool limits the movement of the mouse to directions defined by a specific angle measured from the reference plane axes. That angle is set in the Angle Snap Options dialog box (figure 6-24) by specifying an angle in degrees or desired slope. The resulting range of movement is limited to the direction of the angle or perpendicular to that angle.

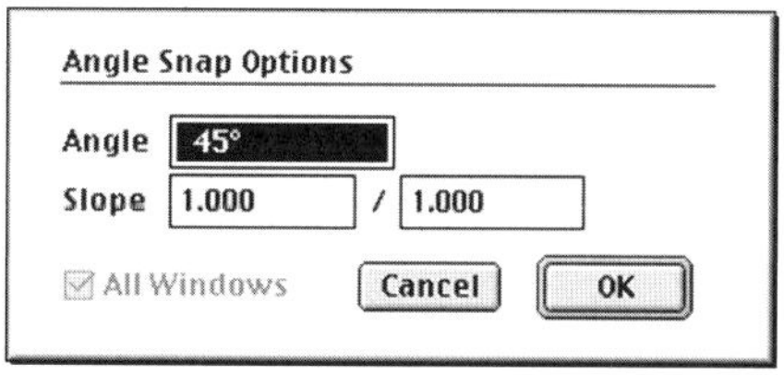

Fig. 6-24. Angle Snap Options dialog box.

Angle/Slope Snap can be used to create lines that give the impression that the entire reference plane had been rotated the specified angle. Figure 6-25 shows lines produced with Angle/Slope Snap.

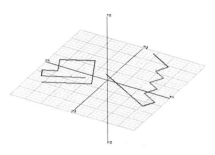

Fig. 6-25. Objects drawn at an angle to reference axes using Angle/Slope Snap.

Radial Snap

Radial Snap is similar to the Angle/Slope Snap tool. However, whereas Angle/Slope Snap constrains movement to an angle measured from reference plane axes, Radial Snap constrains movement to an angle measured relative to the last line drawn. Depending on the angle set in the Radial Snap Options dialog box (figure 6-26), this tool can be used to produce sharply pointed and toothed objects or gradually undulating and weaving lines.

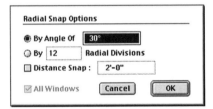

Fig. 6-26. Radial Snap Options dialog box.

This tool includes an option to limit the distance the mouse can move between clicks, which, for example, can be used to model curved objects one segment at a time. Figure 6-27 shows Radial Snap being used to draw sharp and smooth lines.

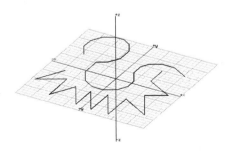

Fig. 6-27. A curved line and a zigzagged line produced with the Radial Snap tool.

Tips and Study and Application Notes

1. Directional snaps make it possible to draw various types of straight and angled lines, and to perform exact transformations when used in conjunction with the Move, Rotate, and Scale tools.

2. Ortho Snap is particularly useful when performing a Mirror operation, to ensure the perfect alignment of the mirror copy.

Object Snaps

Learning Objective 6-6: Constraining cursor behavior to part of an object's topology.

Fig. 6-28. Object Snaps Window Tools palette.

Tools Covered:

- No Object Snap
- Snap To Point
- Snap To Endpoint
- Snap To Midpoint
- Snap To Interval
- Snap To Segment
- Snap To Line
- Snap To Intersection
- Snap To Perpendicular/Angular
- Snap To Tangent
- Snap To Center
- Snap To Face
- Combination Snap

Screen Location: Sixth icon in the Window Tools palette at the bottom left-hand corner of the modeling window.

Object Snaps are drawing aids that constrain the cursor's behavior to a specified part of an object's topology. When a particular constraint (segment, midpoint, and so on) of an object snap is selected, the cursor will automatically jump to the corresponding part of the object. Object snaps are indispensable to the accurate alignment of entities with one another.

Fig. 6-29. No Object Snap tool.

Fig. 6-30. Snap To Poin

Figs. 6-31 and 6-32. A point in the triangle moved to a point in the rectangle using Snap To Point.

No Object Snap

The No Object Snap tool (figure 6-29) deactivates any object snapping currently active.

The Snap To Point tool (figure 6-30) will snap the cursor to the nearest point of an object. (See figures 6-31 and 6-32.) Note that the cursor will change to a double concentric square when it is in the vicinity of a snappable point.

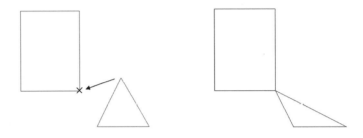

Snap To Endpoint

Fig. 6-33. Snap To Endpoint tool.

Figs. 6-34 and 6-35. A point in the triangle moved to the endpoint of the line using Snap To Endpoint.

The Snap To Endpoint tool (figure 6-33) will snap the cursor to the nearest endpoint of the closest line segment. (See figures 6-34 and 6-35.)

Fig. 6-36. Snap To Midpoint tool.

Snap To Midpoint

The Snap To Midpoint tool (figure 6-36) will snap the cursor to the point at the middle of the closest line segment. (See figures 6-37 and 6-38.)

Window Tools 349

Figs. 6-37 and 6-38. A point moved with the Snap To Midpoint snap tool.

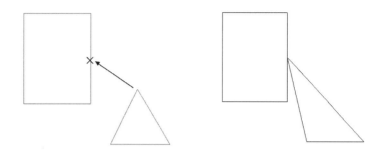

Fig. 6-39. Snap To Interval tool.

Snap To Interval

The Snap To Interval tool (figure 6-39) snaps the closest subdivision point on the closest line segment; the cursor will jump to the point designated in the Object Snap dialog box. There are two options: Interval Snap With n Divisions and Proportional Snap n% From Ends. The former snaps to the endpoints and points on a segment that are the result of dividing the segment into n number of divisions. The latter option snaps to the endpoints and the two points on the segment that are located at n% distance from the two ends of the segment. (See figures 6-40 and 6-41.)

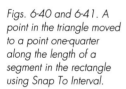

Figs. 6-40 and 6-41. A point in the triangle moved to a point one-quarter along the length of a segment in the rectangle using Snap To Interval.

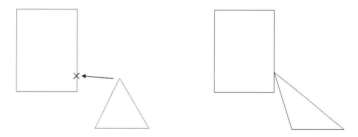

Fig. 6-42. Snap To Segment tool.

Snap To Segment

The Snap To Segment tool (figure 6-42) snaps the cursor to the closest point in the closest line segment (figures 6-43 and 6-44). It is affected by the settings in the Object Snaps dialog box.

Figs. 6-43 and 6-44. A point in the triangle moved to a segment in the rectangle using Snap To Segment.

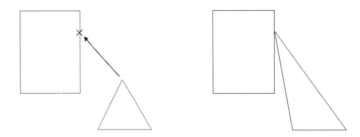

Fig. 6-45. Snap To Line tool.

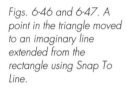

Snap To Line

The Snap To Line tool (figure 6-45) snaps the cursor to the points on the closest line segment, including points on the imaginary extension of that line segment (figures 6-46 and 6-47).

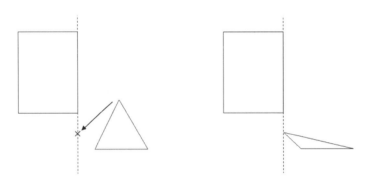

Figs. 6-46 and 6-47. A point in the triangle moved to an imaginary line extended from the rectangle using Snap To Line.

Fig. 6-48. Snap To Intersection tool.

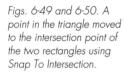

Snap To Intersection

The Snap To Intersection tool (figure 6-48) snaps the cursor to the point that is the intersection point of two segments (figures 6-49 and 6-50). It is affected by the settings in the Objects Snaps dialog box.

Figs. 6-49 and 6-50. A point in the triangle moved to the intersection point of the two rectangles using Snap To Intersection.

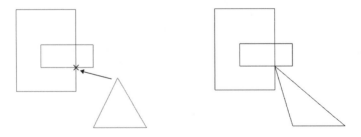

Window Tools

Fig. 6-51. Snap To Perpendicular/Angular tool.

Snap To Perpendicular/Angular

The Snap To Perpendicular/Angular tool (figure 6-51) can be set to Perpendicular Snap or Angular Snap At n Degrees in the Object Snaps dialog box. When creating a segment, the Perpendicular Snap tool forces a 90-degree angle while snapping the cursor to a point on a line. With the Angular Snap At n Degrees option, the angle between the segment and the object can be designated in the Object Snaps dialog box. (See figures 6-52 and 6-53.)

Figs. 6-52 and 6-53. A point in the line moved to be perpendicular with the rectangle using Snap To Perpendicular.

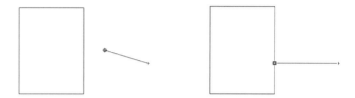

Fig. 6-54. Snap To Tangent tool.

Snap To Tangent

This option is available only in drafting mode (figure 6-54); in the modeling module, the icon appears dimmed. This tool can be used to draw lines tangent to circles and arcs. If tangency is crucial to your design, you should draw the necessary lines and arcs in the drafting mode and import them into the modeling module.

➥ **NOTE:** *See the discussion of Cut and Paste and Import in Chapter 3.*

Fig. 6-55. Snap To Center tool.

Snap To Center

The Snap To Center tool (figure 6-55) snaps the cursor to the point that is the centroid of the face closest to the cursor. (See figures 6-56 and 6-57.)

Figs. 6-56 and 6-57. A point in the triangle moved to the center of the rectangle using Snap To Center.

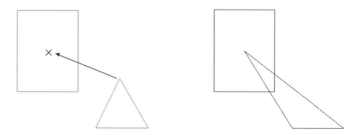

Chapter 6: Window Tools

Fig. 6-58. Snap To Face tool.

Snap To Face

The Snap To Face tool (figure 6-58) snaps the cursor to the points that are on the closest face of the object closest to the cursor. (See figures 6-59 and 6-60.)

Figs. 6-59 and 6-60. A point in the cone moved to the face of the cube using Snap To Face.

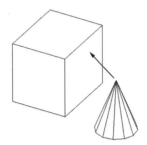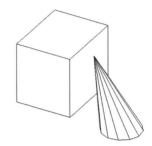

Fig. 6-61. Combination Snap tool.

Combination Snap

This tool will snap the cursor simultaneously to a combination of object parts selected in the Combination Snap dialog box (figure 6-62), which is accessed by double clicking on the Combination Snap icon. The dialog box displays eleven object snap icons. Some combinations may not make sense or produce any additional snapping. For example, Snap to Segment overrides Snap to Midpoint.

Figs. 6-62. Combination Snap dialog box.

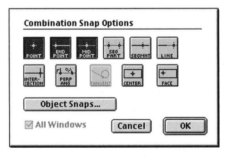

Figure 6-63 shows the Object Snaps dialog box. The Extended Intersections option makes the Snap To Intersections tool snap to points that are the imaginary intersections between segments extended infinitely. The Snap To Segment Part option controls how the Snap To Interval tool behaves. It designates the number of intervals along the length of a segment or the percentage from the endpoint of a segment to which the Snap To Interval snap will be constrained. In the Snap To Perpendicular/Angular options, you specify either the Perpendicular Snap or the angle to which the angular snap will be constrained.

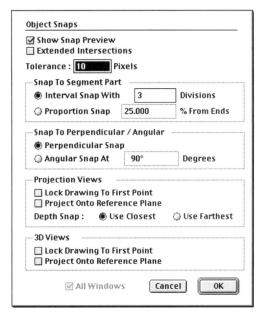

Fig. 6-63. Object Snaps dialog box.

Projection Views and 3D Views are options that help place points on the same plane. In cases where multiple points may overlap in the view of the model, it will force the object's snaps, regardless of the object's location, to be projected onto the designated plane of reference. You use the Projection Views option in an orthographic projection view when the view, points, and segments overlap and it is difficult to determine where the correct points to snap to are. Likewise, the 3D View option is used in 3D views when the model is too dense for points to be easily snapped to, or when all points need to be on the same plane.

The Lock Drawing To First Point option constrains all subsequent object snaps so that they will be projected onto the same plane as the first point selected. This allows you to locate a drawing on the same plane as its first point without having to worry about where subsequent object snaps will be snapped to. Project Onto Reference Plane does essentially the same thing; however, this option projects all object snaps onto the active reference plane.

In the Depth Snap options, the Use Closest option snaps to the points closest to the viewer when multiple points are overlapping in a view. Use Farthest does the opposite and selects the point farthest from the viewer.

Chapter 6: Window Tools

TIPS AND STUDY AND APPLICATION NOTES

1. The following table lists MacOS and PC (Windows) quick keys for Object Snap tools.

Tool	MacOS	Windows
No Object Snap	Shift + Opt + N	Ctrl + Alt + N
Snap To Point	Shift + Opt + P	Ctrl + Alt + P
Snap To Endpoint	Shift + Opt + E	Ctrl + Alt + E
Snap To Midpoint	Shift + Opt + M	Ctrl + Alt + M
Snap To Segment	Shift + Opt + S	Ctrl + Alt + S
Snap To Line	Shift + Opt + L	Ctrl + Alt + L
Snap To Intersection	Shift + Opt + I	Ctrl + Alt + I
Snap To Center	Shift + Opt + C	Ctrl + Alt + C
Snap To Face	Shift + Opt + F	Ctrl + Alt + F

Zoom and Pan

Fig. 6-64. Zoom and Pan Window Tools palette.

LEARNING OBJECTIVE 6-7: *Adjusting the view of a model.*

TOOLS COVERED:

- Zoom
- Hand
- Zoom In By Frame
- Zoom Out By Frame
- Zoom In Incrementally
- Zoom Out Incrementally
- Previous View
- Fit All
- Reset

SCREEN LOCATION: *Seventh icon in the Window Tools palette at the bottom left-hand corner of the modeling window.*

Zoom

Fig. 6-65. Zoom tool.

After selecting the Zoom tool (figure 6-65), click on the window to enlarge the displayed image by the factor specified in the Zoom Options dialog box (figure 6-66). The default setting for Zoom In/Out By Frame is Window Zoom By Center. This can be adjusted to Window Zoom By Corners so that the zoom frame can be defined as the corners of a rectangle.

Fig. 6-66. Zoom Options dialog box.

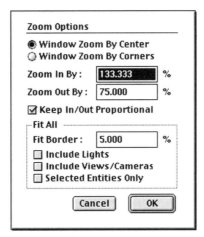

The location of the magnifying glass icon becomes the center for the zoom when the mouse is clicked. Holding down the option key (MacOS) or Control + Shift keys (Windows) will cause it to zoom out.

Hand

Fig. 6-67. Hand tool.

The Hand tool (figure 6-67) allows the view to be panned interactively. When you activate the Hand tool, a small hand icon appears on the screen. With this option active, you can click and drag the mouse to adjust the view according to the motion of the mouse (figures 6-68 and 6-69).

☞ **NOTE:** *Continuous zoom/hand can be turned off if counterintuitive to your work habits.*

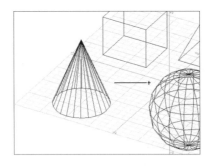
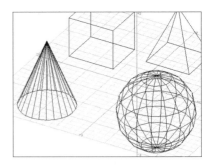

Figs. 6-68 and 6-69. Panning to the right with the Hand tool.

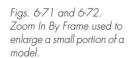

Fig. 6-70 and Zoom In/Out By Frame tools.

Zoom In/Out By Frame and Zoom In/Out Incrementally

Zoom In/Out By Frame (figure 6-70) draws a window around the portion of the model to be zoomed in or out (figures 6-71 and 6-72); this allows you to very quickly enlarge a small portion of the model. Zoom In/Out Incrementally (figure 6-73) zooms in or out at even steps designated in the Zoom Options dialog box (figures 6-74 and 6-75).

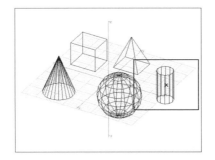
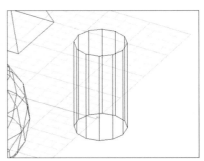

Figs. 6-71 and 6-72. Zoom In By Frame used to enlarge a small portion of a model.

Fig. 6-73. Zoom In/Out Incrementally tools.

Figs. 6-74 and 6-75. Zoom In Incrementally used to enlarge a model.

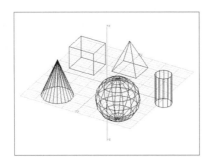
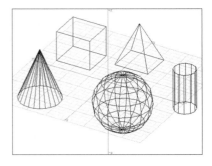

Window Tools

Previous View

Fig. 6-76. Previous View.

The Previous View tool (figure 6-76) returns the window to the view that was displayed before the most recent zoom or pan operation.

Fit All

Fig. 6-77. Fit All tool.

The Fit All tool (figure 6-77) adjusts the view so that all active objects are visible. (See figures 6-78 and 6-79.)

Figs. 6-78 and 6-79. Modeling window before and after use of the Fit All tool.

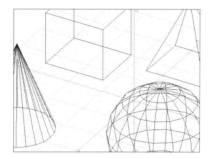 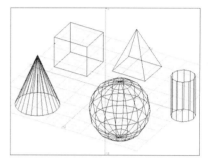

☛ **NOTE:** *Remember the Cmd + F key equivalent; you will use this tool often.*

The Fit All tool has two useful options, which are toggled on and off in the Zoom Options dialog box (figure 6-66). The first is Include Lights. The result can be seen in figure 6-80, where the view encompasses all of the objects, as well as all of the lights. This is useful if you are going to be working a lot with lighting effects in your model.

Fig. 6-80. Modeling window after use of Fit with Include Lights option.

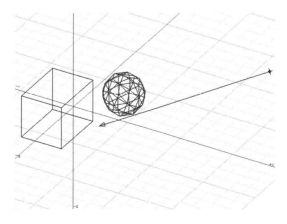

Chapter 6: Window Tools

The second option is Selected Entities Only, whose results are illustrated in figure 6-81. When using this option, the Fit tool will set the view to encompass only the selected items, not all items in the model. This is useful when you are having trouble finding an object on the screen. You can select in the Objects palette and then fit it to the window using this option and the Fit tool.

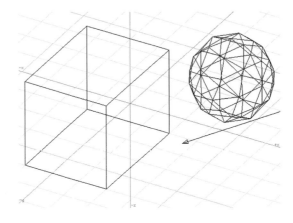

Fig. 6-81. Modeling window after use of Fit with Fit Selected Entities Only option.

Fig. 6-82. Reset tool.

Reset

The Reset tool (figure 6-82) returns you to the view in effect when the project window was first opened.

TIPS AND STUDY AND APPLICATION NOTES

1. It is useful to memorize some of the "quick" keys for the Zoom tools used regularly. The following table lists these key combinations for the Macintosh and PC (Windows) environments.

Tool	MacOS	Windows
Zoom In By Frame	Command + {	Ctrl + {
Zoom Out By Frame	Command + }	Ctrl + }(
Zoom In Incrementally	Command + [(	Ctrl + [
Zoom Out Incrementally	Command +](	Ctrl +]
Hand	Command + >(	Ctrl + >
Go To Previous	Command + <(	Ctrl + <
Fit All	Command + F(	Ctrl + F
Reset	Command + \(	Ctrl + \

Window Tools

View Tools

Fig. 6-83. Window Tools View palette.

LEARNING OBJECTIVE 6-8: *Changing the current view.*

TOOLS COVERED:

- Set View
- Navigate View
- Walkthrough
- Match View

SCREEN LOCATION: *Eighth set of tools in the Window Tools palette, located in the bottom left corner of the screen.*

Set View

Fig. 6-84. Set View tool.

View tools are for changing the current view in order to better see or present your model. The Set View tool (figure 6-84) allows you to interactively turn the view in two directions at once. The first mouse click defines a reference point; then moving the mouse horizontally rotates the view around the vertical axis. Moving the mouse vertically rotates the model around a horizontal axis. At the same time, pressing the Option key (MacOS) or Control + Shift Keys (Windows) zooms in and out.

Navigate View

Fig. 6-85. Navigate View tool.

The Navigate View tool (figure 6-85) displays a marquee, or interface device, that allows you to adjust many of the view parameters interactively. The marquee displays four arrows and four concentric circles, which form zones that respond differently to mouse input (figure 6-86).

Fig. 6-86. Dynamic Navigate View interface.

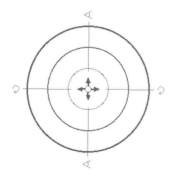

You control panning by moving the mouse in the inner zone between the small inner circle and the second circle. If you hold the cursor over the small inner circle, the cursor will change to a small black image of a camera, allowing you to zoom in and out along the line of sight by moving the mouse.

Place the cursor between the second and the outer circles, and then drag it, in order to rotate the view. Place the cursor over the spin icons at the left and right extents of the marquee, and then drag the mouse, in order to spin the view around the line of sight. Position the cursor over the zoom icons at the top and bottom of the marquee, and then drag them left or right, to zoom in or out of the view.

Walkthrough

Fig. 6-87. Walkthrough tool.

The Walkthrough tool (figure 6-87) allows you to move through a perspective view of the modeling scene, guided by the four arrows of its interactive marquee (figure 6-88). The further you move the mouse away from the marquee, the faster you will move through the model. Use the top and bottom arrows to move forward or back. Use the left and right arrows to turn.

Fig. 6-88. Walkthrough tool.

The Option key (MacOS) or Control + Shift keys (Windows) modifies the action of the top and bottom arrows for looking up and down. The Control key (MacOS) or Control + Alt keys (Windows) modifies the action of the top and bottom arrows for panning up and down, and the left and right arrows of the marquee for panning left and right. The Command key (MacOS) or Control key (Windows) modifies the action of the top and bottom arrows for rotating up and down, and the left and right arrows of the marquee for rotating left and right.

Match View

Fig. 6-89. Match View tool.

The Match View tool is for aligning perspective views to an underlay of a photograph or drawing so that you can "sketch" your model in perspective, or fit it into a preexisting perspective. If you load an underlay file while in perspective view, you can change the view angle to match that of the viewer or camera in the image.

↠ **NOTE:** *For more information on underlays, see Chapter 3.*

To match a view, click on the Match View tool (figure 6-89). The reference plane will be outlined in a black line with a dot at each corner. You can click on these dots and move them to line up the edges of the plane

Window Tools

with edges in the underlay to set the views to match. After each point is moved, the view will change. To move several points before the view changes, hold the Shift key down until all of the points are positioned; then release the Shift key and click again to set the view.

It is best to try to line the points up with a floor grid, wall, or piece of furniture you know is square. Once your view is matched, you can model in perspective and roughly sketch objects to match those in the underlay. The series of figures 6-90 through 6-92 shows this process. Figure 6-90 shows an underlay and the current view, which do not match. Figure 6-91 shows the control points that allow you to move the view and match the underlay. Figure 6-92 shows the new view and several walls that have been sketched in perspective.

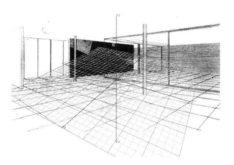

Fig. 6-90. The view is not in line with the underlay.

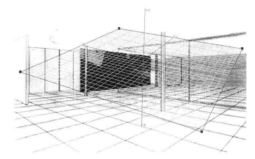

Fig. 6-91. Match View tool activated, with a point at each corner of the grid.

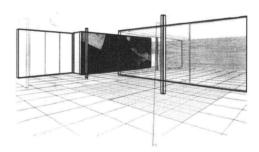

Fig. 6-92. View matched to underlay and elements drawn to match those in the view.

TIPS AND STUDY AND APPLICATION NOTES

1. It is often very useful for quick presentations or simply for realistic viewing to use the QuickDraw 3D renderer in conjunction with the Set View tool. With your model rendered with Quick-

Draw 3D, you can spin the view using the Set View tool and the object will remain rendered, allowing you to view it from all sides.

2. When using the Match View tool, you can prevent the view from changing until you have positioned all of the control points by holding down the Shift key.

3. When sketching and modeling in perspective, it is often necessary to turn on the Keep Vertical Lines Straight option in the Perspective options. This is because perspective drawings are not normally drawn with a third vanishing point (the vertical elements are not converging on a point high in the sky), but form•Z by default creates perspectives with this third vanishing point. If you are drawing on top of an underlay that does not use this vanishing point, and you do not turn on the Keep Vertical Lines Straight option, modeled objects will not match the underlay image.

4. Once you have set your view to match, save the view using the Views palette.

➭ **NOTE:** *For more information on the Views palette, see Chapter 4.*

MacOS/PC NOTE:
The Option key (MacOS) or the Control + Shift key combination (Windows) toggles the zoom options between zooming in and zooming out. The Control key (MacOS) or the Control + Alt keys (Windows) transforms the tool into the Zoom In/Out by Frame with Window Zoom by Corners option selected.

part 2

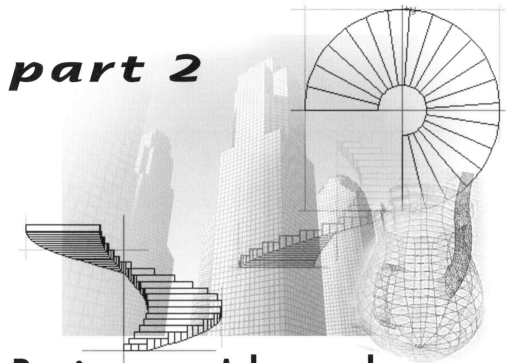

Beginner to Advanced Exercises

Plate 1. form•Z provides many tools for generating complex objects with a single mouse click. This illustration shows a bolt, screws, and a 3D logo, all of which are presented in Chapter 7 as five-minute step-by-step tutorials for the beginner.

Plate 2. Presented in Chapter 7 as a five-minute model, this spiral stair was generated with the Spiral Stair tool and rendered in RenderZone. By checking the Plain Object and Per Part options in the Spiral Stairs Edit dialog box, the stair will be created with each part generated as a separate object. This allows each element to be individually textured, as shown here.

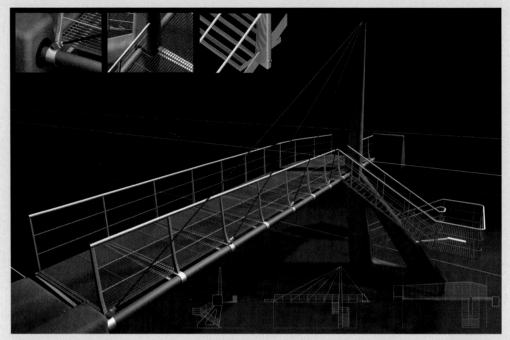

Plate 3. This illustration shows how form•Z's 2D drafting and 3D modeling modules can be used together to create sophisticated presentations. These images were rendered in RenderZone and then composited together with the 2D drawings in an image manipulation program. After a 3D model is built in form•Z, 2D images can be generated and imported into the drafting module, where dimensioning, labeling, and other 2D operations can be performed. *Project by Joseph Kosinski and Robert Quevedo of Columbia University's Graduate School of Architecture.*

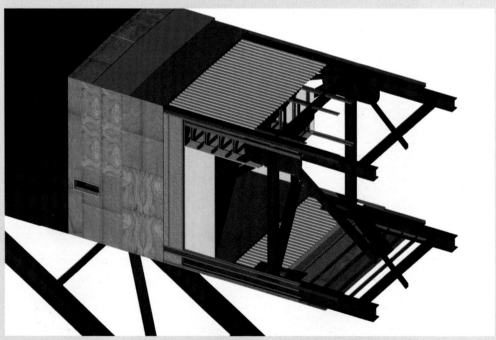

Plate 4. A RenderZone view of the construction system from the Chapter 9 exercises. *Project by M. Hoffman, C. Perry, and B. Yorker of Columbia University's Graduate School of Architecture.*

Plate 5. Texture mapped block.

This series of images (plates 5–15) from the Chapter 12 animation of an unbuilt Frank Lloyd Wright project, demonstrates form•Z's ability to handle complex architectural models. With models of this scope, rendering parameters should be kept as simple as possible; with over 400 lights in this model, only 6 cast shadows. In addition, all ornamentation is created through the use of bump maps rather than modeled geometry, and window mullions are simulated by using a gridded transparency map.
Project by Dean DiSimone and Joseph Kosinski (Faculty Advisor: Kenneth Frampton) of Columbia University's Graduate School of Architecture.

Plate 6. A high-resolution RenderZone rendering from the animation.

Plate 7. Frame 0 of 450.

Plate 8. Frame 45 of 450.

Plate 9. Frame 90 of 450.

Plate 10. Frame 135 of 450.

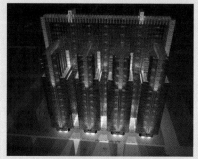
Plate 11. Frame 180 of 450.

Plate 12. Frame 225 of 450.

Plate 13. Frame 270 of 450.

Plate 14. Frame 315 of 450.

Plate 15. Frame 360 of 450.

Plate 16. The teapot and spoon (in the glass) from the Chapter 8 exercises are rendered with two cone lights, using RenderZone's Full Raytrace option. The tiled surface is modeled with rectangular objects with rounded edges, rather than with a tile texture, to produce a more realistic effect. To simulate realistic tumblers with ice, a water texture was applied to cubes and placed in a glass textured using the Predefined Glass surface style.

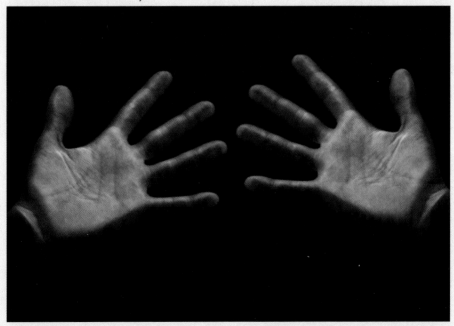

Plate 17. These hands were modeled without geometry, created solely from texture map techniques, as explained in the Chapter 9 Hands exercise.

Plate 18. This scene from Chapter 10 illustrates the realistic lighting conditions that can be achieved by performing a Full Raytrace rendering with a Radiosity solution. Because a radiosity solution calculates the iterative interaction of light with modeled surfaces, only two lights were required to illuminate this scene. Plates 19–22 show the same scene rendered with various form•Z options selected.

Plate 19. RenderZone rendering of the scene.

Plate 20. Scene rendered with Fog option selected.

Plate 21. Scene rendered with Accurate Glow option selected.

Plate 22. Scene rendered with Analysis option selected.

Plate 23. These sandals from Chapter 9 were modeled with form•Z patches and rendered in RenderZone.

Plate 24. This RenderZone view combines the umbrella and sandals models from Chapter 9. The umbrella was modeled with form•Z Nurbs surfaces.

This series of images (plates 25–32) illustrates the Fly example of Chapter 12, a step-by-step animation tutorial demonstrating the manipulation of Velocity Control Curves to produce frames simulating the strange and rapid flight of an insect in a room, starting in the upside-down position. This animation file is provided on the companion CD-ROM.

Plates 25 and 26. Frames 0–85: The fly starts upside down on the ceiling of the room. As it falls, it gains speed and turns right-side up.

Plates 27 and 28. Frames 86–161: The fly flies between the legs of the table and heads straight for the window and the view of the mountains.

Plates 29 and 30. Frames 162–243: Whack! The fly hits the window glass. It bounces off but, being an insect, tries again. Having no luck, it turns and heads for something to console itself: the fruit on the table.

Plates 31 and 32. Frames 244–299: The fly heads toward the table and the fruit bowl and finally lands on the fruit. All of this happens in 300 frames, or 10 seconds.

chapter 7

Exercises for Beginners

Introduction

One of the strengths of form•Z is that it includes powerful commands that allow the beginner to create relatively complex models in just a few operations. In this chapter, you will be lead through beginning exercises step by step. Most of these exercises, called Five-minute Models, can be completed very quickly.

The Five-minute Models vary from barns and geodesic domes to coins and goblets. These exercises are the best way for beginners to experience the satisfaction of creating meaningful 3D models in their first form•Z sessions. Even intermediate users might see a trick or two here they had not thought of before. The last eight exercises, starting with the paper clip, may take a bit longer to complete; you might call them "ten-minute models." The following are the Five-minute and other models presented in this chapter (and continued on the CD-ROM).

Five-minute Models

- Barn
- Mace
- Logo
- Coin
- Bolt
- Pear
- Goblet
- City
- Maze
- Landscape
- Geodesic Dome

Further Models for Beginners

- Paper Clip
- Monument Arch
- Portal on a Podium
- Ice Cream Cone
- Swiss Cheese
- Staircase

Exercises

Each of the exercises that follow assumes you have started a fresh form•Z project by invoking the New (Model) command in the File menu. Furthermore, it is assumed that Preferences and other user-adjusted parameters are at their default settings and that you are using English units.

Exercise 7-1: The Five-minute Barn

LEARNING OBJECTIVE 7-1: *Single-command modeling of blocks in 3D Enclosure mode.*

One of form•Z's best features is its ability to model buildings with a single command, such as Vector Line, while in 3D Enclosure mode. A simple building profile, drawn on a vertical plane, can generate a complete architectural shell with precisely specified wall, slab, and roof thicknesses. Windows and doors are easily cut into the walls while in Insert Opening mode. In the following exercise, you will draw a barn with windows and a doorway.

Exercise Steps

1. Turn Grid Snap on in Window Tools at the bottom of the screen (figure 7-1).
2. In the Heights menu at the top of the screen, define a custom setting of 80 feet. This dimension will actually define the length of the barn because you will draw the building profile on a vertical plane.
3. Select the vertical *zx* plane in Window Tools.
4. Double click on the 3D Enclosure mode icon in the Object Type tool palette (figure 7-2) to adjust the options in the 3D Enclosure Options dialog box.
5. In the dialog box, set the Wall Width, Top Thickness, and Base Thickness at 1'-0" and change the Top and Base settings to Closed. Click on OK to close the dialog box (figure 7-3).

Exercise 7-1: The Five-minute Barn

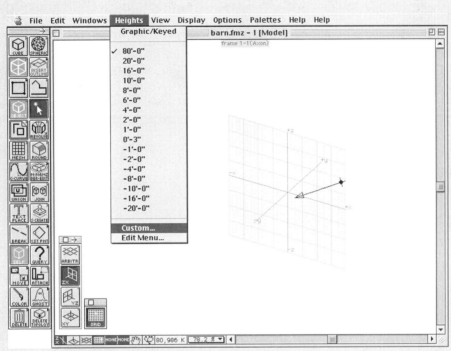

Fig. 7-1. form•Z screen with Window Tools and Heights menu selections.

Fig. 7-2. The Object Type tool palette with 3D Enclosure selected.

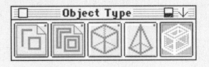

Fig. 7-3. 3D Enclosure Options dialog box.

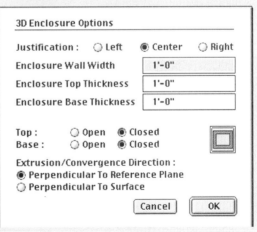

6. With the Vector Line tool, draw the profile of a barn as a single line (figure 7-4).

368 Chapter 7: Exercises for Beginners

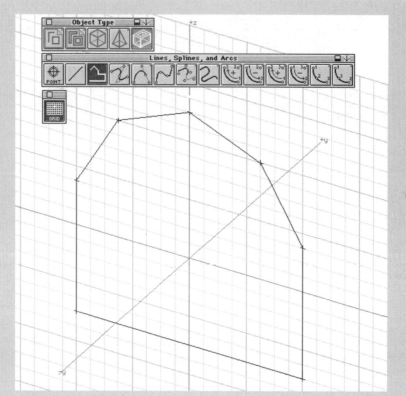

Fig. 7-4. Draw the barn profile on a vertical plane with Grid Snap active.

7. To close the shape, return to the starting point and double click. The length of the barn will reflect the current Heights setting (figure 7-5).

8. Set Topological Level to Face and use the Pick tool to select the surface of the barn wall by clicking on two of its edges (segments).

9. Select Insert Opening mode in the Insertions tool palette (second row, second column).

10. Use the Rectangle tool to draw the outlines of windows and doors. They will be cut out of the shell of the building as openings, as in figure 7-5.

11. Select the RenderZone display option from the Display menu to see the rendered barn (figure 7-6).

Exercise 7-1: The Five-minute Barn

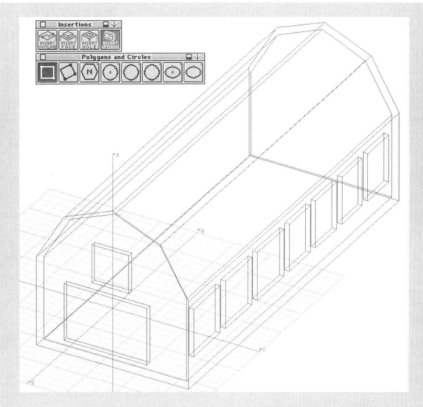

Fig. 7-5. Use Insert Opening mode to create windows and doors in the barn walls.

Fig. 7-6. View of the interior of the barn in RenderZone.

Tips and Study and Application Notes

1. Try selecting Perspective in the View menu, and adjusting the view using Edit Cone Of Vision (also in the View menu) to set the interior view so that it looks like that shown in figure 7-6.
2. Grid Snap is used throughout this exercise with an increment of 2 feet. This makes it easy to keep the roof symmetrical and the windows equally spaced (figure 7-7).

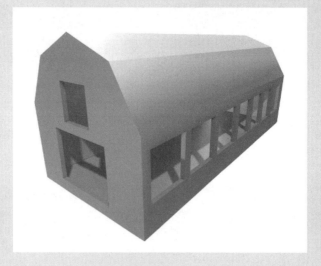

Fig. 7-7. View of the exterior of the Five-minute Barn.

Supporting and Further Reference

See discussions of the Insert Opening and 3D Enclosure tools in Chapter 5.

Exercise 7-2: The Five-minute Goblet

LEARNING OBJECTIVE 7-2: *Generating a complex object with a single curved line.*

This exercise makes a change in scale from a large object consisting of straight lines, like the barn, to a relatively small object drawn with curves. The goblet shows you how a curved line can generate a complex object. After you draw the profile of the goblet, you can turn it into a beautiful 3D model with two clicks of the mouse.

Exercise Steps

1. Select Front View from the View menu.
2. Activate the Grid Snap in Window Tools.
3. Select 2D Surface Object mode in the Object Type tool palette.
4. Use the Cubic B-Spline tool from the Lines, Splines, and Arcs tool palette to draw the profile of a goblet, starting on the z axis (point A in figure 7-8).
5. Turn off Grid Snap after inputting the first point. Draw the rest of the curve down to the base of the goblet (point B) without Grid Snap.
6. Turn Grid Snap on again, and complete the goblet base with the Vector Line tool, terminating at the z axis (point C) with a double click of the mouse. Note that the profile shape is not closed: it starts at point A and ends at point C.
7. Return to a 3D view. Select the Revolved Object tool in the Parametric Derivatives tool palette, and click on the profile line, then on the z axis. The final result will resemble figure 7-9 when displayed in Wire Frame, and figures 7-10 and 7-11 when rendered using the RenderZone display option from the Display menu.

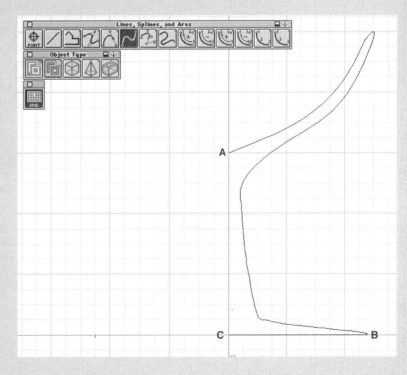

Fig. 7-8. Goblet profile drawn in side view.

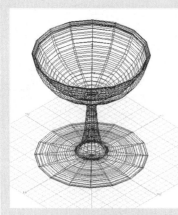

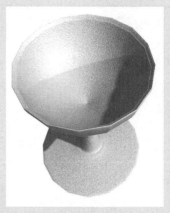

Fig. 7-9. Goblet in a 3D Wire Frame view.

Fig. 7-10. Goblet rendered in perspective view from the side.

Fig. 7-11. Goblet rendered in perspective view from above.

Tips and Study and Application Notes

1. The default # Of Points setting of the Revolved Object tool is 16. A higher setting, such as 48, would create a smoother curve and would eliminate the choppy faceting visible at the lip and base of the goblet. However, the model would be slower to render at the higher setting. At the other extreme, a setting of 4 would create a square goblet.

2. Keep in mind that for the resulting Goblet to be properly formed as a solid object, each end of the profile line must touch (but not cross) the z axis. If this is done, the resulting goblet is a solid object, and can be used in Boolean operations such as Union and Difference. If not, a surface will be generated, and less convenient Trim and Stitch operations would have to be used to modify it.

3. You can use the same technique for a wine glass, Coke bottle, or any other "lathed" object with a central axis of symmetry.

Supporting and Further Reference

See the Revolved Object (Revolve) tool in Chapter 5.

Exercise 7-3: The Five-minute Mace

LEARNING OBJECTIVE 7-3: *Using the Attach tool to attach repetitious elements to the surface of a 3D object, creating an odd-shaped object.*

The mace, a weapon commonly used in the Middle Ages, takes only a few quick steps to model in form•Z. To build this vicious-looking device, perform the following steps. Keep in mind that with the Attach tool, if you are not careful, you will end up creating a variety of strange and often unpredictable 3D objects.

> **NOTE:** *Be aware of how many faces are in your original object. Too many can cause problems with rendering. For more information on how to avoid this problem, see the Tool Options for the Spheric modeling tools in Chapter 5.*

Exercise Steps

1. Make sure you are in a 3D view such as Z=30° X=60°.
2. From the Balls tool palette, double click on the Spherical Object tool to open the Spherical Object Options dialog box. Select the Radius icon and Geodesic Sphere from the Shape pull-down menu. Click on OK (figure 7-12).

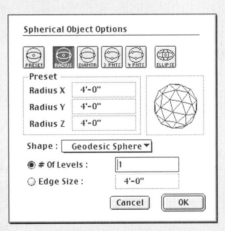

Fig. 7-12. Specifying a Geodesic Sphere in the Spherical Object Options dialog box.

3. Generate a sphere, as shown in figure 7-13. Click once to establish its center point, and again to define its radius.
4. Select 3D Converged mode from the Object Type tool palette. In this mode, every shape you create will be tapered into a pyramid.
5. Set the Heights menu to Graphic/Keyed.

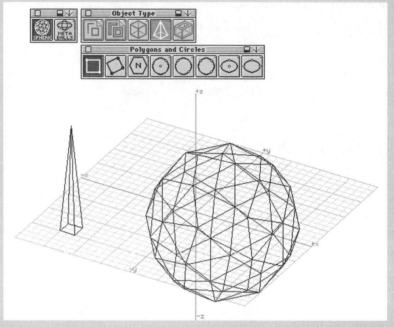

Fig. 7-13. Creating a sphere and a pyramid.

6. Use the Rectangle tool in the Polygons and Circles tool palette to draw a pyramid, as shown in figure 7-13. The first two clicks determine the shape of the base; the last click determines the height of the pyramid.

7. In the Relative Transformations tools palette, select the Attach tool. Click on the pyramid and then on the sphere. Copies of the pyramid will be automatically attached to each face of the sphere.

8. Use the Shaded Render display option from the Display menu to create a rendered view of the mace (figure 7-14).

Fig. 7-14. A mace formed of pyramids attached to a geodesic sphere.

Tips and Study and Application Notes

1. The center of the sphere is located on the *xy* plane at the point where you first click the mouse.

2. You can double click on the Attach tool to open the Attach Options dialog box or use the Tool Option palette. The options allow you to fine tune the location of the pyramid on the faces of the sphere with various scaling and positioning options.

3. The Attach tool can be used for more advanced applications such as attaching windows to the sloped sides of a building, or trees to meshed surface polygons of a terrain model.

Supporting and Further Reference

See discussion of the Spherical Solids and Relative Transformations tools in Chapter 5.

Exercise 7-4: The Five-minute City

LEARNING OBJECTIVE 7-4: *Creating city blocks with buildings using Parametric Cubes.*

The Cube tool under the Primitives tool palette lets you rapidly generate massing models of buildings sitting on the *xy* plane. If you vary the heights of the buildings, and arrange them in city blocks, you will have a city in just a few minutes.

Exercise Steps

1. Select a 3D axonometric view from the View menu such as that shown in figure 7-15.

2. Select the Cube tool in the Primitives tool palette. This tool will allow you to easily "sketch" buildings, varying their heights interactively.

3. Draw boxes of various heights. The first and second clicks define the base of the building; the third click determines the height.

   ✓ **TIP:** *Think of this as planting the two opposite corner stones of a building and then topping it off with a roof. Keep Grid Snap inactive to ensure a more random, realistic urban fabric.*

Fig. 7-15. Boxes generated using the Cube tool, sitting on the xy plane.

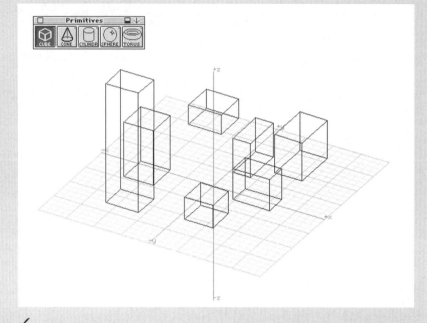

4. Repeat the previous step in various locations on the reference plane, but make every building slightly different in length, width, and height (figure 7-16).

   ◆◆ **NOTE:** *You can change colors as frequently as you want by selecting a different square in the Surface Styles palette.*

5. Try putting two boxes in the same space, with one slightly narrower and taller than the other. This is a technique for creating a building with a stepped profile.

6. View the final result using various View and Display options. Figure 7-17 shows a perspective view, displayed in RenderZone. Figure 7-18 shows the Front View with a sky texture added.

   ◆◆ **NOTE:** *See Chapter 10 for explanations of Display options.*

Exercise 7-4: The Five-minute City

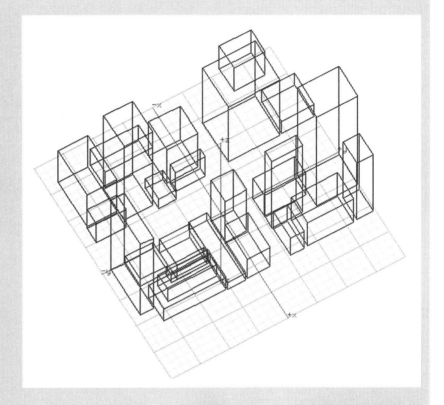

Fig. 7-16. Boxes of various heights organized around streets from blocks of a model city.

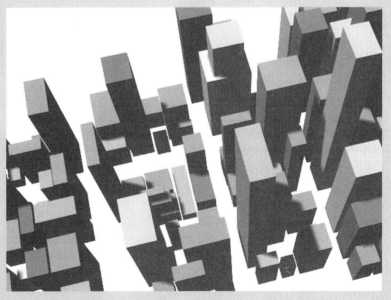

Fig. 7-17. Perspective View of the city rendered using the RenderZone display option.

Fig. 7-18. Elevation of the city showing sky.

Tips and Study and Application Notes

1. The sky can be created by selecting Sky in the Background pull-down menu in RenderZone Options in the Display menu. Click on the Options button next to the Background menu to add the setting sun.

2. The Cube tool is very efficient for creating boxes of various sizes. Similar results can be obtained using a multi-step process with the rectangle tool: (1) Heights (2) 3D Solid (3) Rectangle tool.

Supporting and Further Reference

See discussion of the Primitives tools in Chapter 5.

Exercise 7-5: The Five-minute Logo

LEARNING OBJECTIVE 7-5: *Creating a 3D logo.*

Take a break from designing buildings and cities to create something you could use as a graphic design element on your business card or Web page. One of the easiest things to do in form•Z is create a 3D model of your own name using the Text Place tool. You can easily define the font, size, style, and layout of the 3D text. Create your design with the following steps.

Exercise Steps

1. Select 3D Extrusion mode in the Object Type tool palette.
2. Select a 3D view.
3. Click on the Text Place tool in the Text tool palette and then click on the reference plane to specify the location of the text. The 3D Text Editor dialog box will appear.
4. Enter the desired text in the text box, replacing the word *Text* (figure 7-19).

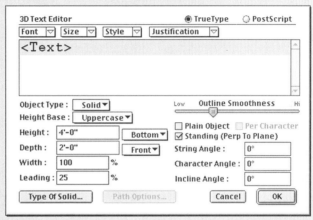

Fig. 7-19. Using the Text Place command. Note the option to have the letters stand perpendicular to the reference plane.

5. Select a font, style, and justification from the pull-down menu items at the top of the 3D Text Editor dialog box.
6. Set a Height for the letters, and a Depth setting for their 3D extension.

> **NOTE:** *Width is expressed as a percentage of the normal width of the letter. Leading is the vertical spacing between rows of text, and is measured as a percentage of the font Height.*

7. Select the Standing (Perp to Plane) setting. The result is shown in figure 7-20.

Fig. 7-20. Wire Frame view of text generated as 3D objects perpendicular to the reference plane.

8. Click on OK and view the result in RenderZone. You can now use the Text Edit tool to adjust the text font and size to achieve the desired effect. Figure 7-21 shows a perspective view of the 3D logo in RenderZone. Figure 7-22 shows a perspective view of the logo with a high Depth setting.

Fig. 7-21. Perspective view of a 3D logo rendered with shadows in RenderZone. Note that the word Architect is parallel to the reference plane.

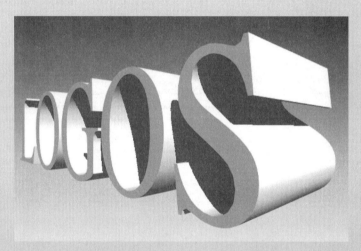

Fig. 7-22. Perspective view of 3D text with a high Depth setting.

Tips and Study and Application Notes

1. The Outline Smoothness slider bar in the 3D Text Editor dialog box allows you to generate very smooth letters, but there is a cost; the smoother the 3D letters, the more surfaces, and therefore the longer the rendering time.

2. The Leading value controls the vertical spacing between text lines. Use the Depth control to achieve dramatic 3D extensions on the letters.

3. The Text Place tool creates 3D solids that can be used in a variety of operations, such as Scale and Deformation, as well as Boolean operations such as Union and Difference.

Supporting and Further Reference

See discussions of the Text Place and Text Edit tools in Chapter 5.

Exercise 7-6: The Five-minute Maze

LEARNING OBJECTIVE 7-6: *Creating walls in 3D Enclosure mode.*

There are many tools in form•Z that facilitate modeling architectural elements. Walls are one of the easiest elements to create in form•Z. In 3D Enclosure mode, any line you draw can automatically be extruded into a 3D partition. In a few minutes you can create corridors and rooms, or even build a complex labyrinth.

Exercise Steps

1. Double click on the 3D Enclosure icon to bring up the 3D Enclosure Options dialog box, shown in figure 7-23.

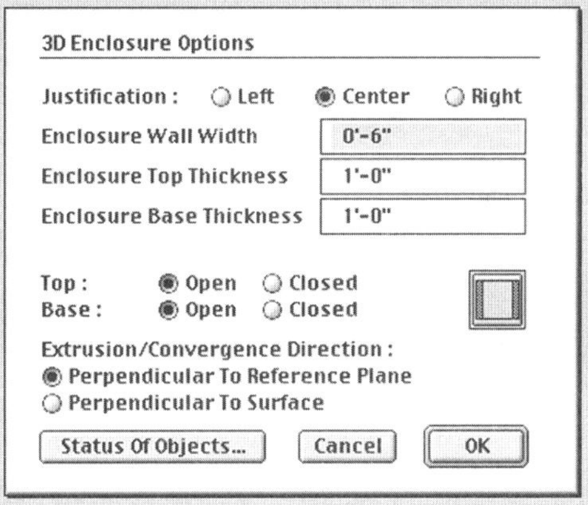

Fig. 7-23. 3D Enclosure Options dialog box.

2. Set Enclosure Wall Width of 6 inches and set the Top and Bottom options to Open. Click on OK.

3. Double click on Grid Snap in Window Tools to bring up its options. Set Grid Snap Module to 8'-0" and click on OK. Make sure the Grid Snap icon is highlighted.

4. Select 8 feet in the Heights menu.

5. Use the Vector Line tool to define the centerline of the wall. The first click locates the beginning of the wall. Subsequent single clicks define corners. A double click locates the end of a wall. Try to match the wall shown in figure 7-24.

Fig. 7-24. One wall system has been built; the second one is being traced on the plane.

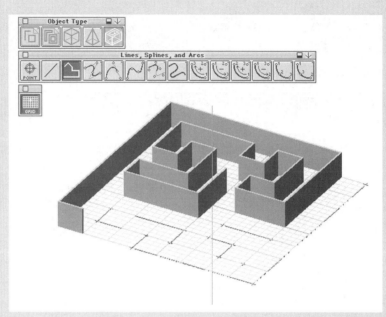

6. Draw another line to define a second wall system to complete the maze. The final result should resemble figures 7-25 and 7-26.

Fig. 7-25. Perspective view of the 3D maze, generated with shadows in RenderZone.

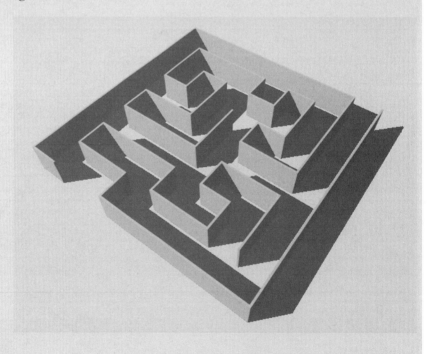

Fig. 7-26. Overhead perspective view of the 3D maze.

Tips and Study and Application Notes

1. Grid Snap is essential to this exercise; the wall sections are kept parallel and perpendicular to each other, as well as equally spaced.
2. The Vector Line tool is one of the most fundamental building tools in form•Z. Remember that the first click defines the starting point. Subsequent clicks determine corners of the wall. A double click defines the end of a wall.

> ☛ **WARNING:** *When using the Vector Line tool, a triple click of the mouse will simultaneously return the line to its starting point and close the wall system, which of course is not desirable in the case of an open maze.*

Supporting and Further Reference

See discussions of the 3D Enclosure mode and the Vector Line tool in Chapter 5.

Exercise 7-7: The Five-minute Coin

LEARNING OBJECTIVE 7-7: *Tactical use of texture maps.*

Sometimes the best models do not depend on the detail and quality of the modeled form but primarily on the tactical use of texture maps. This five-minute exercise will show you how easy it is to use textures to make extremely realistic coins and similarly textured objects.

Exercise Steps

1. Under the Windows menu, select Window Setup. Change the reference grid module to 1", and the # Divisions to 2. Then use the Zoom In Incrementally tool in the Window Tools until the grid is clearly visible.
2. In the Heights menu, set a Custom height of 1/16".
3. With the 3D Extrusion mode selected in the Object Type tool palette, use the Circle by Center and Radius tool to create a Quarter coin-sized cylinder with a radius of 1/2" (figure 7-27). You may find it helpful to set your Grid Snap by double clicking on the Grid Snap icon and selecting the Match Grid Module.

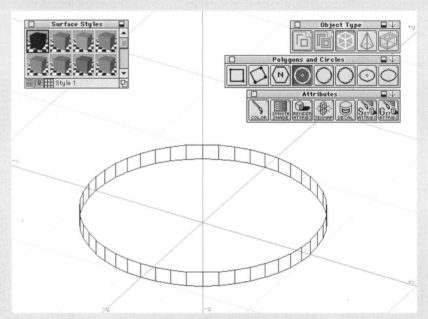

Fig. 7-27. A simple model of a Quarter created with the Circle by Center and Radius tool.

4. Double click on the icon of the active Surface Style to open the Surface Style Parameters dialog box.
5. In the pull-down menu under Color, select Image Map (figure 7-28).
6. Click on Options to open the Image Map Options dialog box; then click on Load to bring up the standard Open File dialog box. Locate the image file of the head of the Quarter (named *quarter.tif* on the companion CD-ROM) and click on Open.
7. In the Image Map Options dialog box, specify 1 for both Horizontal and Vertical Repetitions and toggle the option to center the image map for both fields. Click on OK (figure 7-29).

Exercise 7-7: The Five-minute Coin

Fig. 7-28. Select Image Map in the Surface Style Parameters dialog box and click on Options.

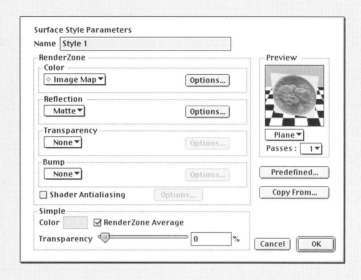

Fig. 7-29. Image Map Options dialog box with Repetitions set to 1.

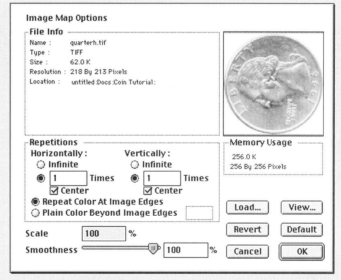

8. Return to the Surface Style Parameters dialog box. Select Plane as the Texture Mapping Type from the pull-down menu located under the texture preview window on the right. Click on OK to return to the modeling window.

9. Select the Texture Map tool in the Attributes tool palette and click on the coin. This will open the Texture Map Controls dialog box. To the right of the preview window, select Flat in the Mapping Type pull-down menu. For both Horizontal and Vertical Tiling, enter 1 Tiles and check Center (figure 7-30).

Fig. 7-30. Texture Map Controls dialog box.

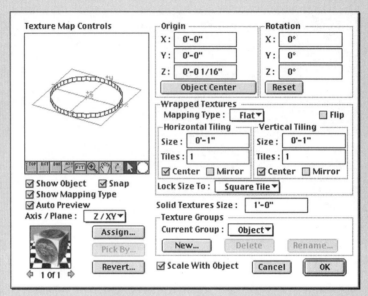

10. Change to Perspective view under the Views menu and use RenderZone to render your coin (figure 7-31).

Fig. 7-31. Rendered coin.

> **NOTE:** *You can make some virtual cash by duplicating the coin, using the Copy tool (figure 7-32).*

Fig. 7-32. A virtual mint!

Tips and Study and Application Notes

1. Using the One Copy modifier with the Move and Scale tools, you can make copies of your original coin. Duplicate the coin and use a combination of Move and Rotate tools until you have a pile of money (figure 7-33).

2. Generally, it is much more efficient to use texture maps to achieve detail rather than actual geometry. For instance, you could have modeled Washington's head on the surface of the coin. A good rule of thumb is to never model details that can be achieved more effectively through the use of texture maps. The result is often much more realistic as well.

3. In addition to using the image file named *quarterb.tif* as an Image Map, you can also load it as a Bump map to further enhance the realism of your image.

Supporting and Further Reference

See discussion of the Texture Map tool in Chapter 5 and the discussion of RenderZone in Chapter 10.

Exercise 7-8: The Five-minute Landscape

LEARNING OBJECTIVE 7-8: *Using the Terrain Model tool to turn 2D contours into a 3D landscape.*

Make yourself some virtual real estate! Once you have drawn or traced some 2D contour lines, you can almost instantly create a 3D landscape using the Terrain Model tool. By adjusting the Terrain Model options you can obtain a variety of surface treatments, from smooth meshes to abruptly stepped terraces.

Exercise Steps

1. You must draw the 2D contour lines and define a boundary outline that intersects the contour lines before you can use the Terrain Model tool.
2. In Top view, select 2D Surface Object from the Object Type tool palette.
3. Use the Rectangle tool to draw a boundary for your terrain (figure 7-33).

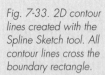

Fig. 7-33. 2D contour lines created with the Spline Sketch tool. All contour lines cross the boundary rectangle.

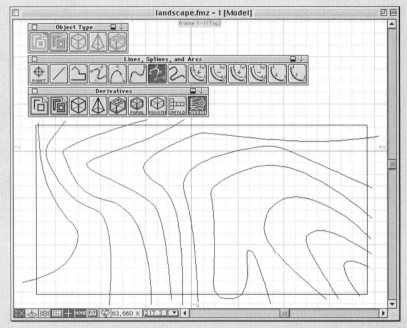

4. Use the Spline Sketch tool to draw a series of contour lines. Both ends of these contour lines must be drawn to cross the boundary line. The lines should not overlap or intersect. Draw the minimum number of contour lines required to define the terrain adequately. Too dense a pattern of lines will increase your file size and may cause problems with rendering.

5. Use the Pick tool to select the contour lines in order, from lowest to highest in elevation, but do not select the boundary rectangle.

6. Double click on the Terrain Model icon to bring up the Terrain Model Options dialog box (figure 7-34). Select Adjust Heights At Intervals Of, and specify 2 units. Select Site (Starting) Height, and specify 1 unit. This sets the elevation of the first contour line. Click on OK.

7. With the contour lines selected in the correct order, select the Terrain Model tool. You will be prompted to click on the boundary rectangle to generate the 3D model (figure 7-35).

Exercise 7-8: The Five-minute Landscape

Fig. 7-34. Terrain Model Options dialog box.

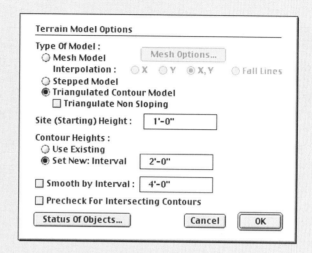

Fig. 7-35. Using the Terrain Model tool to create a triangulated-mesh landscape.

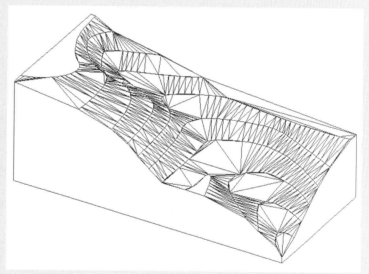

8. You can also try creating a different type of Terrain Model. Undo the previous operation. Adjust Terrain Model Options to mesh model to generate a square mesh, and use the Terrain Model tool to generate a different style of landscape (figure 7-36).

> **NOTE:** Use of the Spline Sketch tool produces satisfyingly rounded contours with all styles of terrain models. The Vector Line tool could be used instead with less line segments, but the results in this particular case would be unpleasantly ragged with the stepped and triangulated-mesh terrain styles.

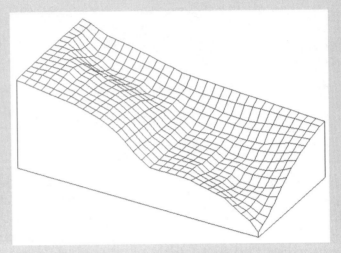

Fig. 7-36. An alternative surface treatment: a square mesh surface.

9. Undo again, modify the Terrain Model options, and regenerate the terrain to achieve the effects shown in figure 7-37.

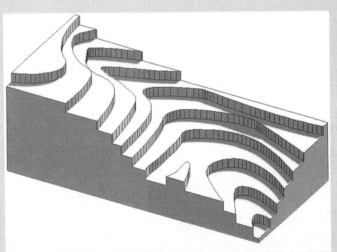

Fig. 7-37. An alternative terrain treatment: a layered topographical model.

Tips and Study and Application Notes

1. The Undo command in the Edit menu makes it easy to sample a variety of terrain styles. After generating a terrain, Undo, adjust the Terrain Model options, and try another style.
2. It is important to be conscious of the scale at which you are working. If you specify a mesh size too small for the scale of the model, it will unduly slow down the rendering process.

3. Be aware of the Smooth Interval setting of the Spline Sketch tool as you draw contours. The smaller the Step Size, the more points will be in the line; consequently, your terrain model will be more difficult to manipulate and render.
4. Note that if you were modeling a real terrain, you would scan the topographical map of the area and use it as an underlay.

Supporting and Further Reference

See discussion of the Terrain Model tool in Chapter 5, and discussion of the Underlay tool in Chapter 3.

Further Exercises

The following illustrations show the exercise models contained on the companion CD-ROM. For instructions on using the CD-ROM, see the Introduction.

The Five-Minute Bolt

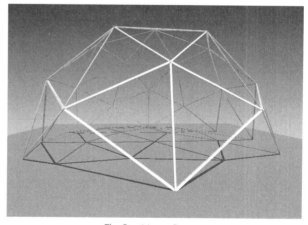

The Five-Minute Dome

Chapter 7: Exercises for Beginners

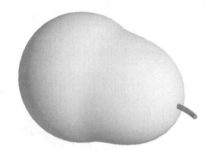

The Five-Minute Pear

Paper Clip

Ice Cream Cone

Monumental Arch

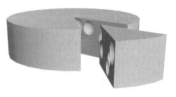

Swiss Cheese

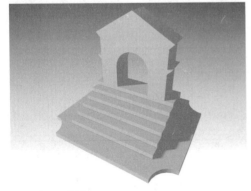

Portal on a Podium

Spiral Staircase

chapter 8

Intermediate Exercises

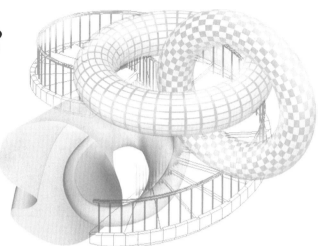

Introduction

The exercises in Chapter 7 introduced the basic concepts and tools for modeling, as well as a few built-in features for creating complex models in one step. This chapter builds on the basic tools you learned in the previous chapter, and shows you how to combine them in order to produce more involved and complicated models.

These intermediate exercises teach more than just how to use tools; they show you how to approach projects and how to choose which tools to use. In the exercises that follow, many of the tools from the previous chapter are revisited in a more in-depth manner. Along with revisiting tools previously covered, some of the more complicated functions of the program are addressed, such as modeling curvilinear forms, texturing, and rendering.

Like the earlier examples, the exercises range in scale from a soda can to an urban landscape. The exercises become progressively more difficult, the assumption being that your working knowledge of the software will increase. When you finish this chapter, you should have almost a complete understanding of the tools. Chapter 9 will carry this approach even further, applying the tools to even more complicated projects and forms. The following are the exercise models covered in this chapter and on the companion CD-ROM.

- Skyscraper Towers
- Picture Frame
- Screwdriver
- Tape Dispenser
- Magnifying Glass
- Chair

- Knife
- Spoon
- Hammer
- Fork
- Soda Can
- Teapot

Exercise 8-1: Modeling Skyscraper Towers

LEARNING OBJECTIVE 8-1: *Modeling and duplicating buildings to create a cityscape.*

An urban landscape is a much-modeled, computer-generated environment, and form•Z allows you to model and render a city of skyscrapers in a very short time. In the segments of this exercise, you will model four tower types that could be used in a computer-generated cityscape.

The buildings will be grouped on a single city block. The entire block can then be duplicated several times to create the effect of a dense downtown area. The goal is not to achieve realistic detail with street-level views but to generate effective aerial views of the virtual skyline with the simple 3D models. Figure 8-1 shows a view of the towers to be modeled.

Fig. 8-1. Overall view of the towers.

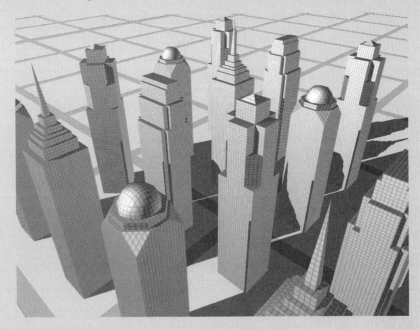

Analysis and Method

Four generic skyscraper shapes will be created to demonstrate the tools needed to fill a virtual skyline. The buildings begin with simple rectangular bases. The buildings are then differentiated by the geometry of their towers. The first tower has a series of setbacks defined in a side view, the second has a ziggurat-like tower, the third has more complex setbacks, and the fourth is capped with a dome. Simple massing models will be created for these four types. By keeping the complexity and number of polygons in the model low, you will be able to duplicate the massing models several times, which would not be practical with a very detailed model. By applying texture maps to approximate the window patterns, images of several city blocks can be easily generated.

For simplicity, the buildings will be modeled at a small scale; all bases are 16 feet wide. Once the towers are complete, it is easy to scale them to any size. Of course, if you were modeling an actual building for which you had complete dimensions, you would use the true dimensions from the beginning.

Skyscraper Modeling Process for Four Skyscraper Designs

Setback Design

Many famous vintage skyscrapers are defined by a distinctive setback profile that allows light and air to reach the streets below. It is a simple matter to model such buildings by drawing the profile in a side view. Keep in mind that skyscrapers and most other architectural subjects should be modeled with Grid Snap activated and set at an appropriate module so that it will be easy to keep dimensions consistent. For this setback design, perform the following steps.

1. Select the Right Side view from the View menu. Set the Heights menu to 16 feet. Because you will be drawing in a side view, this will be the width of the building.

2. Set the Grid Snap Module to 1 foot. Do this by double clicking on the Grid Snap icon and adjusting the Grid Snap Options.

3. Use the Vector Line tool in the Lines, Splines, and Arcs tool palette with the 3D Extrusion modifier in the Object Type tool palette to draw the profile of a building with setbacks, as shown in figure 8-2. Double click at the last point to finish drawing the profile. The setbacks decrease in size as the building rises. Notice that the last setback near the top of the building is only 1 foot deep.

Fig. 8-2. Drawing the profile of the building with setbacks.

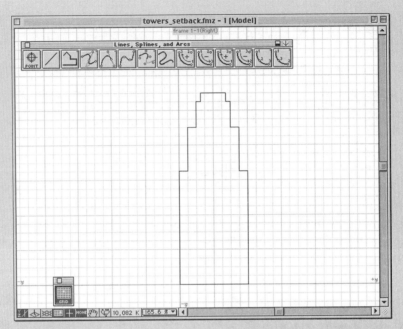

4. Set the Heights menu to 20 feet for the second portion of the building, which is wider.
5. Use the Vector Line tool with the 3D Extrusion modifier to draw the second profile, which is entirely within the first outline, as shown in figure 8-3.

Fig. 8-3. Drawing the second profile.

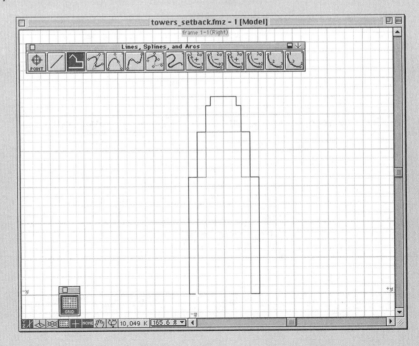

Exercise 8-1: Modeling Skyscraper Towers

6. In Top view, use the Move tool in the Geometric Transformations tool palette to center the second, wider portion of the building, as shown in figure 8-4.

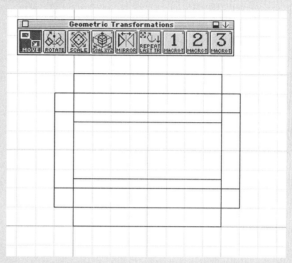

Fig. 8-4. Centering the second profile in Top view.

7. Use the Union tool in the Booleans and Intersections tool palette to join the two parts of the building. Figure 8-5 shows the resulting model with the Hidden Line display option. (Figure 8-24 shows the model with texture maps combined with three other tower models.)

Fig. 8-5. Building after Boolean Union operation rendered with Hidden Line display option.

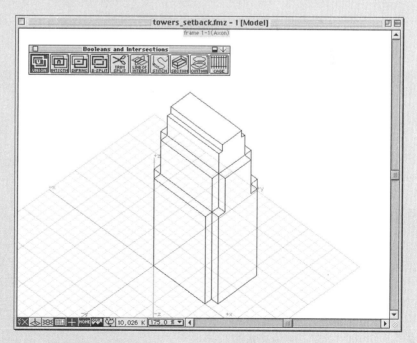

Ziggurat Design

A familiar architectural device for capping skyscrapers is the stepped roof, reminiscent of the Babylonian ziggurat. This type of stepping is easily accomplished in form•Z with the Insert Face/Volume tool.

1. Set the Grid Snap Module to 1 foot. Select a 3D view to work in.
2. Select the Cube tool in the Primitives tool palette. Make a 16-foot square base with the first two clicks. Then set the height of the building at 32 feet by typing *32'* in the Prompts box. This will create the main mass of the building.
3. Set Topological Level to Face, and use the Pick tool to select the top face of the building by clicking on any two of its segments.
4. Select the Insert Face/Volume tool in the Insertions tool palette.
5. Set the Heights menu to 4 feet.
6. Use the Rectangle tool to add the first step to the roof, as shown in figure 8-6. When you do this operation, you get the warning box shown in figure 8-7. This is because the cube has parametric properties that will be lost when you perform the Insert Face/Volume operation on it. Click on OK to continue.

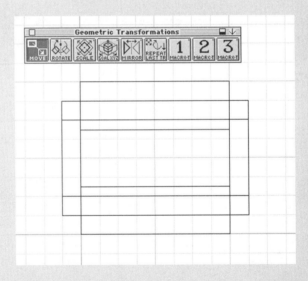

Fig. 8-6. Adding the first step of the roof.

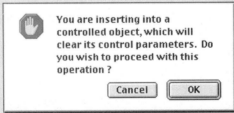

Fig. 8-7. Warning box that appears when inserting a face into a parametric cube.

Exercise 8-1: Modeling Skyscraper Towers

7. The top of the new step should be highlighted. This will allow you to continue to add steps. Set the Heights menu to 2 feet.

8. In Top View, continue to add steps, as shown in figure 8-8.

Fig. 8-8. Adding steps.

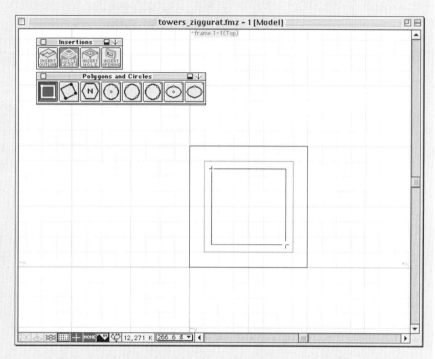

9. Check the results each time in a 3D view as you continue to add steps. Vary the Heights setting, as shown in figure 8-9.

10. To complete the tower with a spire, switch to the 3D Converged modifier in the Object Type tool palette. Set the Heights menu to 16 feet.

11. With the Snap to Point tool in the Window tools active, use the Rectangle tool to snap the spire on the highest step of the building, as shown in figure 8-9.

Fig. 8-9. Setting variable heights for steps.

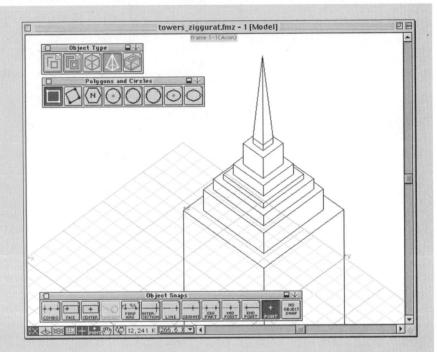

Faceted Setback Tower

Skyscraper towers that include faceted setbacks with 45-degree angles are easily modeled, as the following procedure demonstrates. To model this type of tower, perform the following steps.

1. Set the Heights menu to 16 feet.
2. Set the Grid Snap module to 2 feet.
3. In a 3D view, use the Rectangle tool with the 3D Extrusion modifier to draw the first tier of the building by drawing a 16-foot square, as shown in figure 8-10.
4. Set the Heights menu to a Custom Height of 28 feet.
5. Use the Vector Line tool to draw the outline of the second tier of the building, with the corners cut at 45-degree angles, as shown in figure 8-11.
6. Set the Heights menu to 40 feet.
7. Use the Vector Line tool to draw the cruciform outline of the third tier of the building, with the corners notched, as shown in figure 8-12.

Exercise 8-1: Modeling Skyscraper Towers

Fig. 8-10. Drawing the first tier of the building.

Fig. 8-11. Drawing outline of second tier.

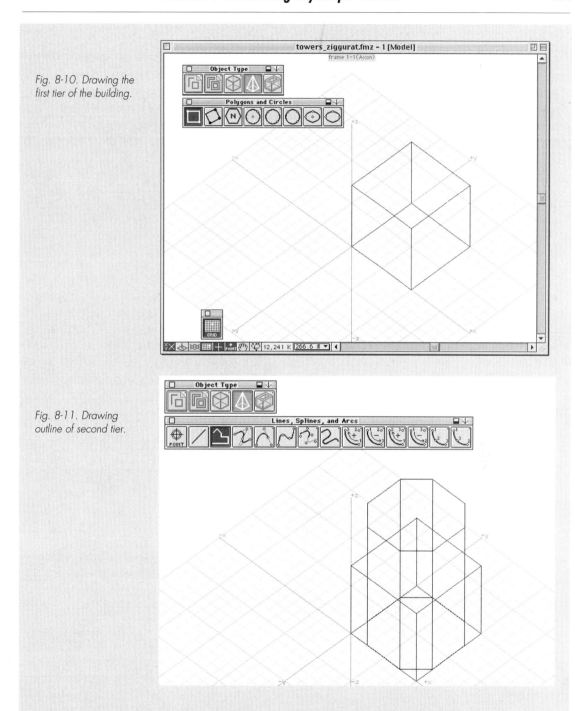

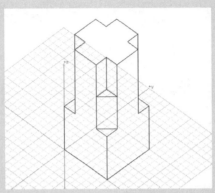

Fig. 8-12. The three tiers in Hidden Line display.

8. Draw a final tier that is 48 feet in height and has an 8-foot square base at the center of the building.
9. Use the Boolean Union tool to join into one mass all four parts of the building.
10. To plant a pyramid on the roof, select Segment in the Topological Levels tool palette, and use the Pick tool to select the two segments of the roof, so that the face is highlighted. Click on the Define Arbitrary Plane tool located in the Reference Planes tool palette at the bottom of the modeling window. Then click anywhere to define the roof as the new active reference plane, as shown in figure 8-13. You can now create the pyramid directly on the roof, saving you the trouble of creating it on the *xy* reference plane and moving it later to the top of the roof.
11. While in 3D Converged mode, use the Rectangle tool to add a pyramid (with a height of 2 feet) to the roof. The result is shown in the Hidden Line display in figure 8-14.

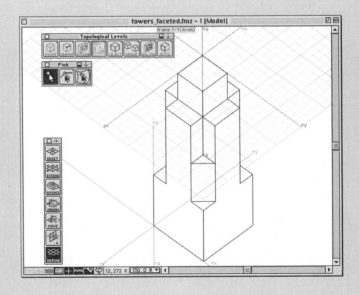

Fig. 8-13. Using Define Arbitrary Plane to make the roof the new drawing surface.

Exercise 8-1: Modeling Skyscraper Towers

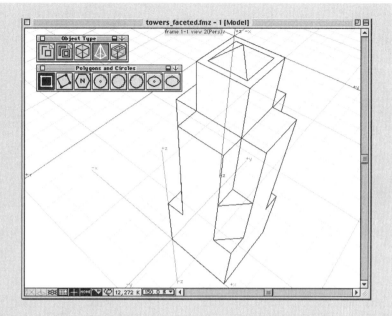

Fig. 8-14. Adding a pyramid to the top of the tower.

Domed Tower

Modern skyscraper design and ancient architecture share the traditional puzzle of how to make the transition between a square tower and a circular dome. form•Z provides beveling tools to ease the transition between the shapes, as well as spherical shapes that can be used to model domes. To perform such a transition, perform the following steps.

1. In a 3D view, use the Cube tool to draw the main block of a building 16 feet square at the base and 32 feet high.

2. Double click on the Round tool (next to the Meshes and Deform tool palette) to open the Rounding Options dialog box, shown in figure 8-15. Use the Controlled Rounding option. Click on OK.

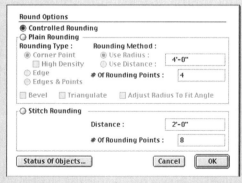

Fig. 8-15. Round Options dialog box.

3. With the Round tool still selected, click on the model. This opens the Rounding Edit dialog box (figure 8-16). Select the Use Radius and Bevel options from the Object Charac-

teristics box. Then select the Round option with 5'-0" as the Radius. In the preview window, select the four corners of the roof, and click on the Preview button. Click on OK if you are satisfied with the way the corners are rounded.

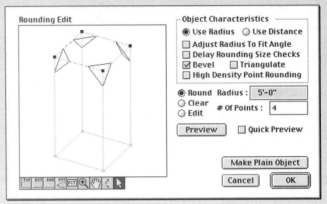

Fig. 8-16. Rounding Edit dialog box.

4. Double click on the Polygon tool in the Polygons and Circles tool palette to bring up the Polygon Options dialog box (figure 8-17). Select the octagon.

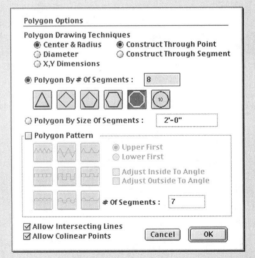

Fig. 8-17. Polygon Options dialog box.

5. Set the Heights menu to 2 feet. Draw an octagon 14 feet wide. In the Front view, move the octagon to the height of the roof. The result should resemble figure 8-18.

6. Select the Spherical Objects tool and draw a sphere with a radius of 6 feet, as shown in figure 8-19. In the Right Side view, move the sphere to the roof level, as shown in figure 8-20.

Exercise 8-1: Modeling Skyscraper Towers   405

Fig. 8-18. Aligning the octagon shape with the roof.

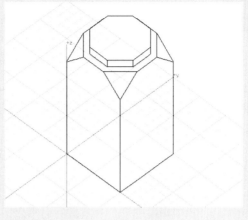

Fig. 8-19. Drawing a sphere.

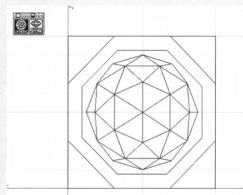

Fig. 8-20. Moving the sphere to the roof level location.

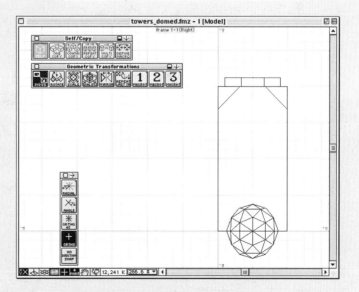

Chapter 8: Intermediate Exercises

7. Use the Boolean Union tool to join all three building parts. A 3D view of the final building is shown in figure 8-21.

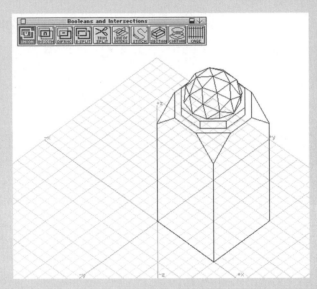

Fig. 8-21. Sphere and building after Boolean Union operation.

Adding Windows and Mullions as a Texture Map

Using texture maps to simulate windows and mullions is a quick and efficient way of making your skyscrapers look more realistic without having to model all windows individually, which could be extremely time consuming.

1. Click in an empty area in the Surface Styles palette to create a new Surface Style. In the Surface Style Parameters dialog box, select Grid under the Color pull-down menu (figure 8-22).

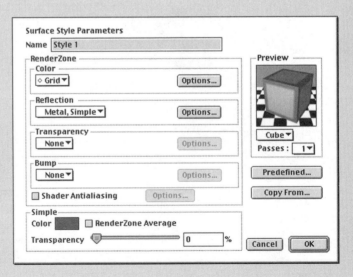

Fig. 8-22. Selecting Grid and Metal, Simple options in the Surface Styles Parameters dialog box.

2. Click on Options to view the settings. Select a gray as the Grid Color for the mullions and white as the Background for the windows (figure 8-23). Click on OK

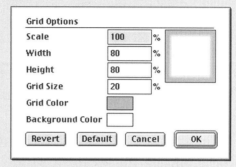

Fig. 8-23. Choosing the colors for the windows and mullions in the Grid Options dialog box.

3. In the Reflection pull-down menu, select Metal, Simple to give the windows a shine. Click on OK.

4. Using the Color tool in the Attributes tool palette, click on your skyscraper to assign the Surface Style you have just created.

Four Skyscrapers Together

Figure 8-24 shows the four skyscraper models assembled and duplicated in block formations. A few of the buildings have been rotated. The heights of some of the buildings have been increased. This is done by selecting the four points at the base of a building and then dragging them downward with the Move command. The model was then repositioned back to the ground plane. Rectangles have been duplicated on the *xy* plane to represent city blocks.

Half of the image is shown in Wire Frame; the other is rendered with a texture map of grids, using RenderZone. This split-screen effect is easy to achieve: render first in Wire Frame, then select Set Image Size in the RenderZone Options dialog box and drag the mouse around half the area of the screen to render only the selected area in RenderZone.

Chapter 8: Intermediate Exercises

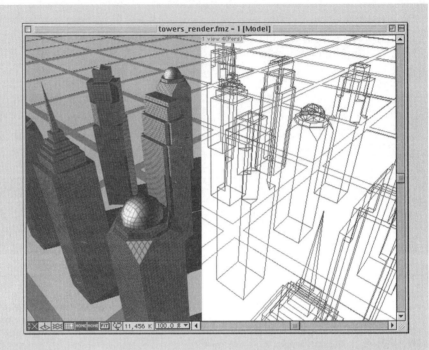

Fig. 8-24. Partial rendering of cityscape.

Exercise 8-2: Modeling a Tape Dispenser

LEARNING OBJECTIVE 8-2: *Modeling in 2D using the Line Editing tools and deriving a 3D object.*

Many models and objects are best made by first creating a 2D outline of the object and then extruding it to create a 3D model. This technique is useful when the 2D outline is complicated and requires the Line Editing tool, available only in 2D. The following tape dispenser model (figure 8-25) is an exercise in modeling and editing 2D surface objects to be extruded into solid form.

Analysis and Method

An object with a complicated shape in plan or section, but a simple profile from all other views, is often easier to model in 2D using line tools instead of in 3D using more complex tools. Such is the case with this tape dispenser. You will create a 2D profile of the tape dispenser using tools from the Line Editing tool palette and the Lines, Splines, and Arcs tool palette. Once you have the object modeled in 2D, you will use the 3D Extrusion and 3D Enclosure tools from the Derivatives tool palette to create a 3D model.

Exercise 8-2: Modeling a Tape Dispenser

Fig. 8-25. Tape dispenser to be modeled.

Tape Dispenser Modeling Process

The following exercise takes you through three basic phases in producing a tape dispenser. The first phase involves creating a series of separate 2D lines and arcs as the basis for generating the profile of the tape dispenser. The second phase involves using the tools from the Line Editing palette to join these 2D lines into a single profile defining the outline of the tape dispenser. The final phase involves extruding this profile into a 3D tape dispenser.

Drawing the Elements of the Tape Dispenser

To draw the elements of the Tape Dispenser you will use tools from the Lines, Splines, and Arcs tool palette. The elements will be created as separate pieces so that they can be manipulated with the Line Edit tools in the following phase.

1. Specify a finer resolution for the Numeric Accuracy setting in the Project Working Units dialog box under the Options menu (figure 8-26). Numeric Accuracy should be set to a size appropriate for the scale of the tape dispenser, such as 0'-1/128".

Fig. 8-26. Setting the Numeric Accuracy in the Project Working Units dialog box.

2. You should also change the Reference Grid Module in the Window Setup option under the Windows menu to a 1" grid with four subdivisions. Grid Snap Module should be set to 1/16".

3. Select the 2D Surface modifier in the Object Type tool palette. This will remain toggled for most of the exercise. Switch to the Front View.

4. Select the Segment tool from the Lines, Splines, and Arcs tool palette and draw a line approximately 1-1/2" in length, this will be the bottom edge of the tape dispenser. Be sure to use the Grid Snap. Your first line should resemble figure 8-27.

Fig. 8-27. Create a line using the 2D Surface modifier and the Segment tool.

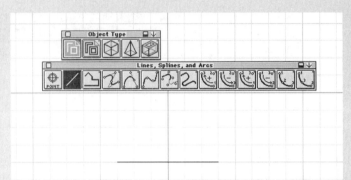

• **NOTE:** *This example does not attempt to precisely replicate the shape of an actual tape dispenser, which would be done using the Underlay command with a scanned image. Rather, a similar outline, consisting of arcs and lines, is sketched on the reference plane. You can construct the same outline by comparing the lines and arcs to the background grid as you work through the exercise.*

5. Switch to the Arc, Counterclockwise, End-point Last tool. Select By Max Size Of Segments in Tool Options palette, setting the value for this option to 0'-0 1/8". This will increase the resolution and detail of the arcs you create in this exercise. With the Grid Snap on, snap to the end of the line from the previous step to begin creating your arc, click again to set the center of the arc's sweep, and a double click to end the arc. The arc should resemble that shown in figure 8-28, part of the bottom edge of the tape dispenser.

6. Switch to the Vector Line tool and toggle off the Grid Snap. Using Point Snap, begin drawing a line at the end of the arc from step 5. With the Vector Line tool still active, turn off the Point Snap and toggle the Grid Snap back on. Continue to draw a crooked line like that in figure 8-29. This is the tearing edge of the tape dispenser.

7. Using a combination of the Arc tools and Vector Line, draw the outline of the top of the tape dispenser. Be sure to keep the Grid Snap on and begin the outline at the end of the line drawn in the previous step. Your model should resemble the model shown in figure 8-30.

Exercise 8-2: Modeling a Tape Dispenser   411

Fig. 8-28. Drawing the arc at the bottom of the tape dispenser.

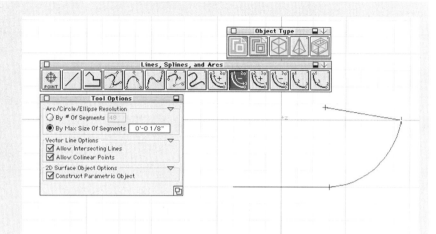

Fig. 8-29. Using Vector Line to create the tearing edge of the tape dispenser.

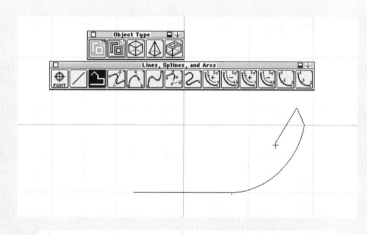

Fig. 8-30. Use the Vector Line and Arc tools to model the top outline.

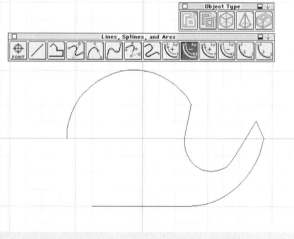

8. The final step in this phase of the modeling process is to use the Segment tool to create the back edge of the tape dispenser. Toggle off the Grid Snap and use the Point Snap tool to ensure the line begins at the end of the arc from the previous step. Use the Orthogonal Snap to ensure the line is straight. Do not attempt to close the shape yet; when you have completed this step your model should resemble figure 8-31.

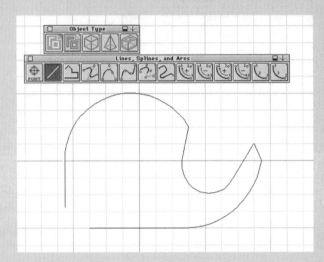

Fig. 8-31. Use the Segment tool to create the back edge of the tape dispenser.

Editing and Joining the 2D Elements

The next phase in the modeling process is to use the Line Editing tools to edit and join the 2D elements from the previous phase to create a single outline.

1. Select the Fillet/Bevel Lines tool from the Line Editing palette. Select Fit Fillet from the Tool Options palette. Set the # Of Edges to 4 and the Fillet Size to 0'-0 1/16". Click on the crooked line that was modeled as the profile of the tearing edge in the previous phase (marked with an arrow in figure 8-32). The corner of this line will be rounded with a 1/16" radius.

2. Select the Trim Lines tool from the Line Editing palette. In the Tool Options palette, select Trim Pairs Of Segments and Fit Fillet. In the Fit Fillet section of the options, set the # Of Edges to 8 and the Fillet Size to 0'-0 1/2". Click on the line that was created as the bottom edge (labeled A in figure 8-33) in the previous phase of the exercise and then on the line that was created as the back edge (labeled B in figure 8-33). The two lines will be joined by a radius.

Exercise 8-2: Modeling a Tape Dispenser

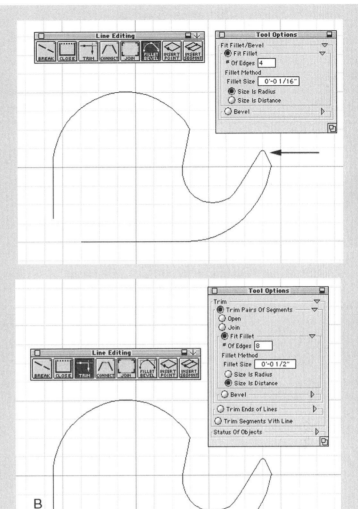

Fig. 8-32. Using Fillet/Bevel Lines tool to round the corner of the tearing edge.

Fig. 8-33. Join two edges with a radius using the Trim Lines tool.

3. To join all of the 2D elements to create a continuous outline and a single surface, select the Pick tool and drag a rectangle to select all of the lines. Select the Close Line tool from the Line Editing palette and select Connect from the Tool Options palette. Click in an empty space in the modeling window to join and close the lines (figure 8-34).

Fig. 8-34. Using the Close tool to join all of the lines into a single, continuous outline.

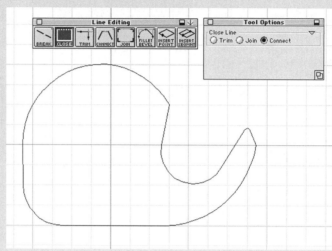

4. Select the Circle, Center and Radius tool and create an outline for the hole in the center of the tape dispenser (figure 8-35).

Fig. 8-35. Create the hole using the Circle/Center and Radius tool.

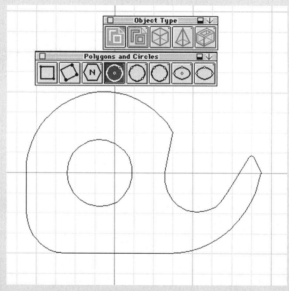

5. Use the Difference tool from the Booleans and Intersections tool palette to subtract the center "hole" from the outline of the tape dispenser. When using the Boolean tool, select Keep as the Tool Options for Operand Status; you will need to keep the original outline and hole for a later step. Figure 8-36 shows a rendered version of the tape dispenser's outline before and after subtracting the hole.

Exercise 8-2: Modeling a Tape Dispenser

Fig. 8-36. The outline before (top) and after (bottom) subtracting the hole.

➣ **NOTE:** *When using a Boolean tool on 2D surface objects, form•Z is sensitive to how the 2D objects were generated and the direction of their lines. You may receive a warning message when using the Boolean Difference. If you receive a warning, use the Reverse Direction tool from the Topologies tool palette (figure 8-37) to reverse the direction of one of the pieces. See Chapter 5 for more about Booleans and Reverse Direction.*

Fig. 8-37. Reverse Direction tool.

Extruding into 3D

The final phase is to use the 3D Extrusion tool and the 3D Enclosure tool to derive a 3D solid from the surface object.

1. Select the 3D Extrusion tool from the Derivatives tool palette and select Graphic Keyed from the Heights pull-down menu. Click on the outline shape to extrude it; this will be the side plate of the tape dispenser. Be sure to change the Operand Status back to Ghost before using the tool (figure 8-38).

➣ **NOTE:** *The Operand Status set to Ghost will preserve the original objects for later use.*

Fig. 8-38. Extrude the outline into a solid form using 3D Extrusion.

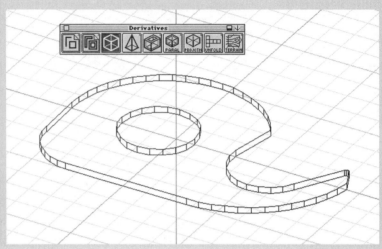

2. To avoid accidentally selecting the 3D solid generated in the previous step, use the Ghost tool to hide it, or place it on a new layer and then hide that layer. Select the 3D Enclosure tool from the Derivatives palette. Set the Justification to Left in the Tool Options palette, and set the Enclosure Wall Width to 0'-0 1/16". Click on the circle that was used as the center hole and extrude it to produce a thin, solid wall (figure 8-39).

Fig. 8-39. Extrude the "hole" to produce a thin wall.

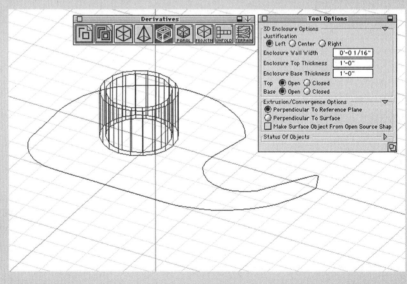

3. Before extruding the remaining outline to produce the outer shell of the tape dispenser, you will need to break the outline to leave a gap through which the tape will be dispensed. Select the Break Line tool from the Line Editing palette and select the Intersection Snap from the Window tools. Click on the outline in the locations marked A and B in figure 8-40 to break the line.

Exercise 8-2: Modeling a Tape Dispenser

Fig. 8-40. Break the outline with the Break Line tool.

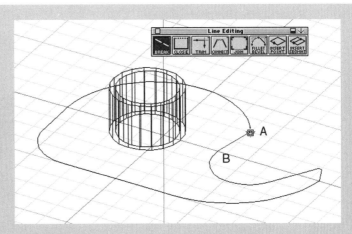

4. Select the 3D Enclosure tool and set Justification to Right in the Tool Options palette. Click on the outline and extrude it to create the outer shell of the tape dispenser. The wall should resemble figure 8-41.

Fig. 8-41. Extrude the outline to produce the outer shell of the dispenser.

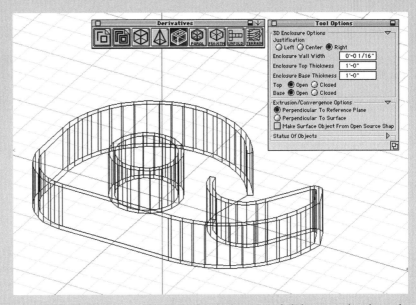

5. Turn on the layer or use the Unghost tool to bring back the 3D solid that is to be the side panel of the tape dispenser. Select the Union tool from the Booleans and Intersections tool palette to join all of the pieces into a single solid object (figure 8-42).

Fig. 8-42. The completed solid tape dispenser.

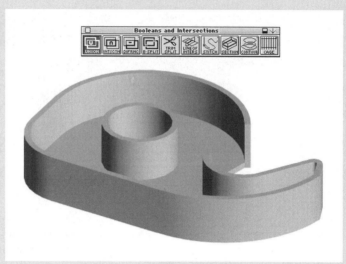

6. For a realistic rendering, create a surface style with transparency and high reflectivity and apply it to the tape dispenser. To model a roll of tape, create a solid cylinder and remove the center using the Difference tool.

Supporting and Further Reference

See the discussion of the tools in the Line Editing palette in Chapter 5. For more about using the Boolean Difference tool, see the Swiss Cheese example in Chapter 7.

Exercise 8-3: Modeling a Picture Frame

LEARNING OBJECTIVE 8-3: *Employing and manipulating image and texture maps.*

This exercise shows you how to model a simple picture frame for a photograph. You will place an image on an object using the Image Map option in the Surface Style palette. You will then use the Texture Map tool to position and scale the image map correctly on the model.

Analysis and Method

In this exercise, a photograph is imaged onto a rectangular object. The frame and mat are simple source shapes swept along a rectangular path around the picture using the Sweep tool. Figure 8-43 shows examples of rendered picture frames.

Fig. 8-43. Rendered picture frames in an interior space.

Picture Frame Modeling Process

The following picture frame requires just three quick phases. The first two phases involve modeling a picture plane and mapping an image onto the plane using Surface Styles. The final phase shows you how to frame your picture using the Sweep tool.

Modeling a Picture Plane

You will first create a picture plane on which the photograph is placed. Because the picture frame is vertical, it is easier to work in the *zx* reference plane. It is important to model your picture plane to the same proportions as your image to prevent stretching and distortions. In this example, a 600-pixel by 400-pixel image is used that exactly matches the proportions of a 36" by 24" picture plane (3:2 ratio).

1. Set the Grid Module to 1'-0" with 12 divisions. This makes it easy to work in both feet and inches simultaneously. Set the grid snap to 1" or Match Grid Module.

2. Select ZX Reference Plane from the Reference Plane window tool located at the lower left of the modeling window.

3. Select a Custom Height (thickness) of 1" for the picture plane and draw a rectangle on the *zx* plane that is 36" wide by 24" tall (figure 8-44).

Chapter 8: Intermediate Exercises

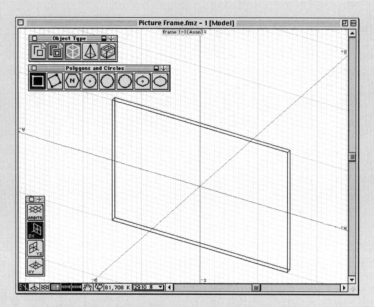

Fig. 8-44. Modeling the picture plane on the zx reference plane.

Mapping an Image onto the Plane Using Surface Style

To map an image onto the plane, perform the following steps.

1. Open the Surface Style Parameters dialog box of the surface style onto which the image is to be assigned by double clicking on the corresponding icon in the Surface Styles palette.

2. In the Surface Style Parameters dialog box, select Image Map in the Color pull-down menu (figure 8-45).

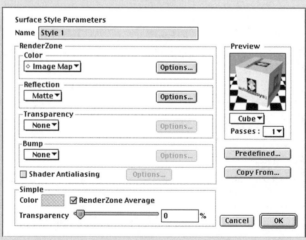

Fig. 8-45. Image map selected in the Surface Style Parameters dialog box.

3. Import the image named *picture.tif* provided on the companion CD-ROM by clicking on Options to open the Image Map Options dialog box; and then clicking on Load to bring up the standard Open File dialog box. Locate the image file and click on Open.

4. Make sure that Horizontal Repetition and Vertical Repetition are both set to 1. This ensures that the image does not repeat itself across the object. If the Infinite option is selected, the image will be tiled to fill the entire object or area. Select the Center option. This will ensure that the image is centered on the face of the picture plane (figure 8-46).

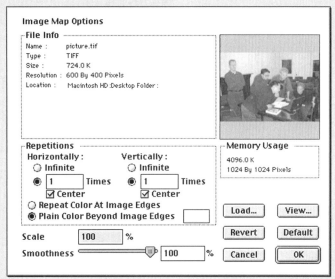

Fig. 8-46. Image Map Options dialog box.

5. In the modeling window, select the Texture Map tool located in the Attributes tool palette and click on the object. This will bring up the Texture Map Controls dialog box. Select Cubic for the Mapping Type, and select Image Map under the Lock Size To option so that the proportions for the original image are preserved. Enter 3'-0" (the dimensions of the picture plane) under Horizontal Tiling, and 2'-0" under Vertical Tiling, in the Size fields (figure 8-47). You can view a rendered preview by selecting the Rendered option when you click on the circular icon located at the lower right corner of the preview window. Click on OK.

Fig. 8-47. Texture Map Controls dialog box.

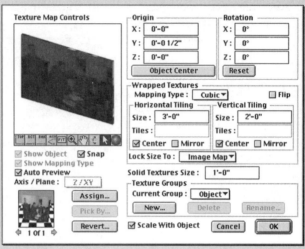

6. Use RenderZone to view the picture plane with the applied image map. Your results should resemble figure 8-48.

Fig. 8-48. Picture plane showing the applied image map rendered in RenderZone.

Using Sweep to Create the Frame and Mat

To create the frame and mat, perform the following steps.

1. Using the 2D Surface modifier and the Rectangle tool, draw paths for both the framing mat (the white cardboard around the picture) and the frame (figure 8-49). These rectangles should be 42" by 30" and 50" by 38", respectively. These paths will be used to sweep the source shapes for the mat and frame using the Sweep tool.

Exercise 8-3: Modeling a Picture Frame

Fig. 8-49. Creating paths for the framing mat and picture frame.

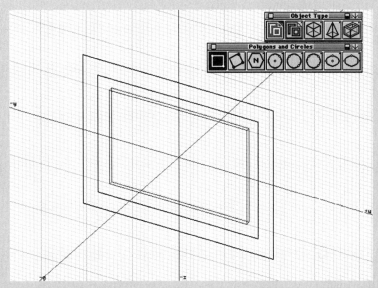

2. Draw source shapes for the mat and frame using the Vector Line tool with the 2D Surface modifier. You may find it easier to draw these items in Front View (figure 8-50). Be sure to toggle on Grid Snap and Ortho Snap as needed.

Fig. 8-50. Drawing source shapes for the framing mat and picture frame in Front View.

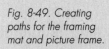

3. Select the Sweep tool. Click on the source shape and then the path for the framing mat. This automatically opens the Sweep Edit dialog box (figure 8-51). Click on OK to continue.

Fig. 8-51. Sweep Edit dialog box.

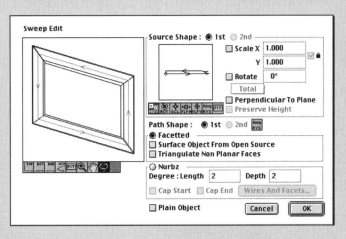

4. Repeat step 3 for the picture frame (figure 8-52).

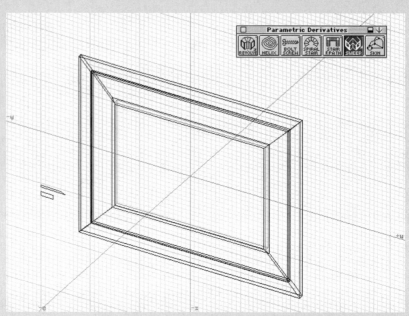

Fig. 8-52. Sweeping the source shapes to create the framing mat and picture frame.

5. Select a white color for the mat and a suitable color for the frame in the Surface Styles palette. Use the Color tool in the Attributes tool palette to apply the selected surface style to the mat and frame (figure 8-53).

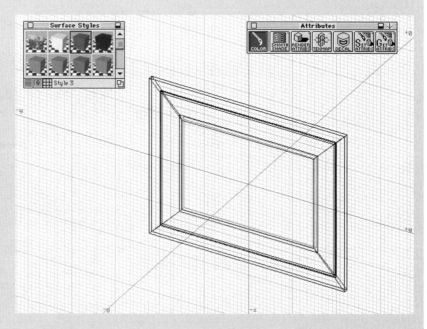

Fig. 8-53. Setting colors for the framing mat and picture frame using the Color tool.

6. Use RenderZone in the Display menu to display the final object with its image maps (figure 8-54).

Fig. 8-54. Rendered picture frame.

Exercise 8-4: Modeling a Magnifying Glass

LEARNING OBJECTIVE 8-4: *Creating a model with special rendering and material properties.*

The RenderZone renderer in form•Z has the ability to accurately simulate material properties such as reflectivity and refraction. Because of this ability to mimic real behavior, it is possible, with the correct settings, to model a magnifying glass (figure 8-55) that appears to function like a real one. A simple revolved object with the appropriate shape and material properties can be used to simulate the optical properties of a magnifying glass, warping the rendered view of objects seen through the lens.

Fig. 8-55. Magnifying glass to be modeled.

Analysis and Method

The magnifying glass has a round lens, a frame, and a handle. All of these components, because of their uniform shape, are easily generated using the Revolve tool. To create this model, you will make three profiles (one for each component) and revolve them to produce the parts. Before you begin, set your Reference Grid to 1", with four divisions; set Grid Snap to 1/16"; and set Numeric Accuracy to 1/32".

Magnifying Glass Modeling Process

The magnifying glass represents a three-phase modeling exercise. The first two phases, creating the lens and frame and modeling the handle, use the Revolved Object (Revolve) tool. The final phase, adding materials, uses surface styles to make the magnifying glass look realistic.

Creating the Lens and Frame

To generate the lens and the thin metal frame that surrounds it (referred to in this exercise as the frame), you will draw 2D profiles and use the Revolved Object tool to generate the 3D form. To do this, perform the following steps.

1. Using the Spline, Quadratic Bezier tool in conjunction with the 2D Surface Object modifier, draw a profile of one-quarter of a lens in the front view. Before you draw the curve, set the smooth interval for the Spline/Quadratic Bezier tool to 1/8". Because this curve will be used in a Revolve operation, make sure it ends exactly on the z axis (figure 8-56).

Exercise 8-4: Modeling a Magnifying Glass   427

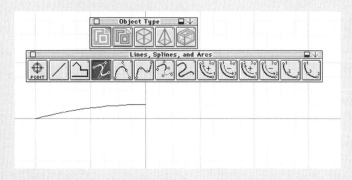

Fig. 8-56. Using the Spline/Quadratic Bezier tool to model a lens outline.

2. Using the Mirror tool, and with the One Copy modifier toggled on, select the Spline and mirror it across the *xy* plane. To ensure that the new line remains touching the *z* axis, it is best to perform the Mirror operation with Orthographic Snap or Grid Snap on (figure 8-57).

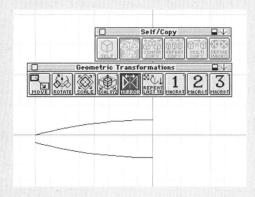

Fig. 8-57. Mirroring the lens profile.

3. Switch to the Connect Lines and click on the end of one curve, and then on the other. A straight segment will be generated, joining the two profiles (figure 8-58).

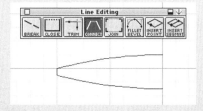

Fig. 8-58. Connecting the two profiles to create one.

4. To draw the profile of the frame, change Grid Snap to 1/32". Using the Vector Line tool and the Arc tools, draw a thin profile with rounded corners at the edge of the lens (figure 8-59).

Fig. 8-59. Drawing the profile of the frame.

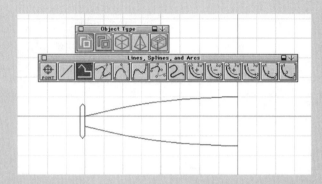

5. To finish modeling the lens and its frame, select the Revolve tool from the Parametric Derivatives tool palette, select the Facetted option, and enter *32* for # Of Steps with the Total option selected. Select the profiles and then the *z* axis to create a revolved shape from the lens profile and the frame profile (figure 8-60).

Fig. 8-60. Revolving the frame and lens profiles.

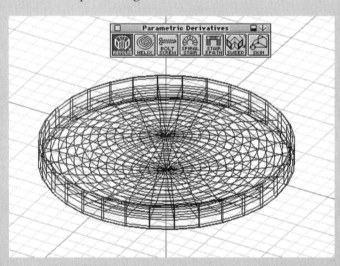

Modeling the Handle

The next phase in making your magnifying glass is to model the handle. Like the lens and the frame, this rounded shape is easiest if done with the Revolve tool.

1. Switch to the Right side view. Using the Vector Line tool and the Arc tools, draw a profile for the handle. Both ends of the profile should touch the *x* axis to guarantee that a solid object is generated when it is revolved (figure 8-61).

2. Select the Revolve tool from the Parametric Derivatives tool palette.

3. Click on the profile and then on the *x* axis to produce a handle (figure 8-62).

Exercise 8-4: Modeling a Magnifying Glass

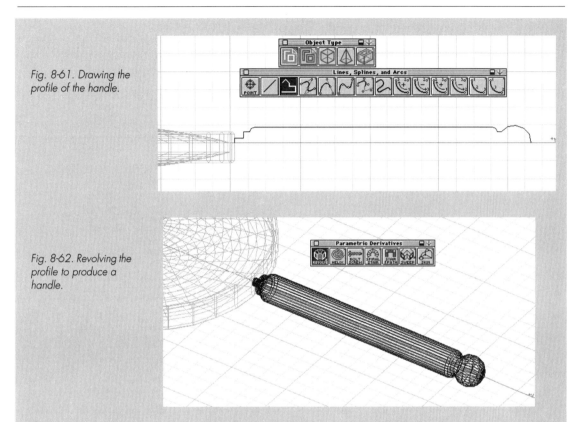

Fig. 8-61. Drawing the profile of the handle.

Fig. 8-62. Revolving the profile to produce a handle.

4. You will have to switch to a side view and use the Move tool to properly position the handle relative to the frame and lens (figure 8-63).

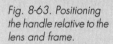

Fig. 8-63. Positioning the handle relative to the lens and frame.

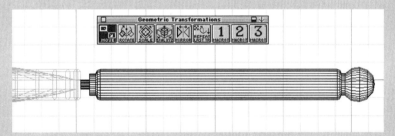

Adding Materials to the Magnifying Glass

With the proper material qualities added, the magnifying glass model can be made to look very convincing. With the right parameter settings for the glass, the model can even appear to operate like a real magnifying glass.

1. Click on a blank space in the Surface Styles palette to create a new surface style. This will bring up the Surface Style Parameters dialog box (figure 8-64). Click on the Color button and set the color to a gray. This will be the base color of the metal. Select Metal, Accurate from the pull-down menu under Reflection. Rename the Surface Style as Chrome in the Name field (figure 8-64).

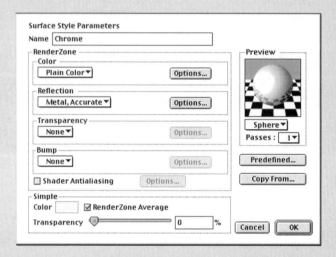

Fig. 8-64. Surface Style Parameters dialog box.

2. With the new Chrome surface style active in the Surface Styles palette, select the Color tool from the Attributes tool palette and click on the lens frame to assign the Chrome surface style (figure 8-65).

3. Create a new Surface Style named Handle and set its color to be black. Do not change any other parameter. Assign this surface style to the handle using the Color tool again.

4. Switch Topological Level to Face and select the faces at the two ends of the handle by drawing a rectangle around each end with the pick tool. Use the Color tool to assign these the Chrome surface style. This will make the ends look metallic, whereas the middle portion of the handle will remain black (figure 8-66).

5. Create a third new surface style and name this one Glass. Click on the Predefined button in the Surface Style Parameters dialog box. In the Predefined Material dialog box, select *synthetic.zmt* from the category pull-down menu and select Glass as a predefined material. Click on the Apply Selected Material button to return to the Surface Style Parameters dialog box. Click on OK to return to the model (figure 8-67).

Exercise 8-4: Modeling a Magnifying Glass

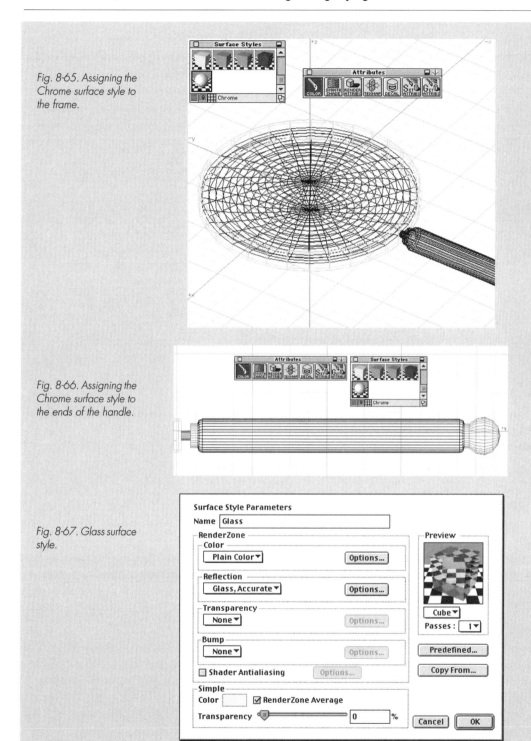

Fig. 8-65. Assigning the Chrome surface style to the frame.

Fig. 8-66. Assigning the Chrome surface style to the ends of the handle.

Fig. 8-67. Glass surface style.

6. Use the Color tool to apply this new surface style to the lens of the magnifying glass.7. Because you do not want the lens to cast a shadow on what it is magnifying, you need to turn off its ability to cast shadows. Select the Render Attributes tool from the Attributes tool palette. Select the Object Does Not Cast Shadows option from the Tool Options palette, and click on the lens to apply this attribute (figure 8-68).

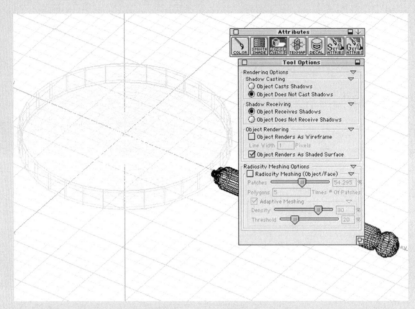

Fig. 8-68. Setting the Render Attributes tool to Object Does Not Cast Shadows.

7. Now that you have set all of the rendering parameters, you can view your model using RenderZone. Under the RenderZone options, make sure that the Reflections, Transparencies, and Shadows options are turned on.

> ✓ **TIP:** *This may take a long time to render if you do not have a lot of RAM in your computer. If it will not render, try turning Shadows or Reflections off in the RenderZone options and rerendering the model.*

8. For added effect, try putting a distorted surface with text under the magnifying glass. Figure 8-69 shows the rendered magnifying glass.

Fig. 8-69. Final rendering of the magnifying glass.

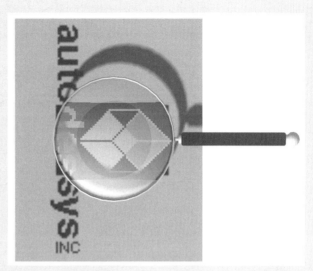

→ **NOTE:** *For more about texture mapping, see the Picture Frame exercise in this chapter.*

Exercise 8-5: Modeling a Screwdriver

LEARNING OBJECTIVE 8-5: *Employing controlled-mesh and Boolean operations to create a complex form.*

A screwdriver presents several interesting 3D modeling challenges. One of these is the way the screwdriver blade in this exercise is formed by shearing off the end of a widening metal cylinder. The other is the familiar hexagonal handle. Figure 8-70 shows the screwdriver to be modeled.

Analysis and Method

You will build the flared screwdriver blade with a controlled-mesh solid. You will then complete it with a series of Boolean operations, almost like slicing it off with a knife. The handle will be made with the Revolve tool, and the grooves will be carved away with a Difference operation. You will use actual dimensions, which, because they are small compared to form•Z's default settings, will necessitate zooming in.

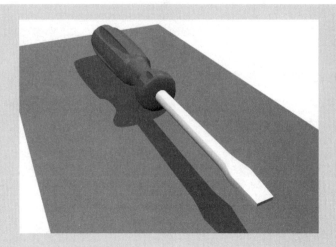

Fig. 8-70. Rendered screwdriver.

Screwdriver Modeling Process

Modeling a screwdriver is a three-phase process. The following exercise takes you through the phases of forming the shaft, creating the blade, and cutting the grooves in the handle.

Forming the Screwdriver Shaft

Begin by setting up form•Z for modeling at a small scale. Because the object is relatively small, you will have to zoom in closer than the default camera setting, and you will have to set Grid Snap to a more convenient module.

1. Set the Grid Snap module to 1/16" (figure 8-71).

Fig. 8-71. Selecting a finer grid.

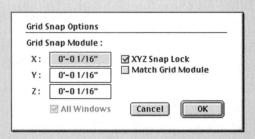

2. Select *yz* as the active reference plane, and select the 2D Surface Object from the Object Type palette.
3. Adjust the Window Setup options, located in the Windows pull-down menu, so that the grid lines are at 1-inch increments with four subdivisions.
4. With Grid Snap active, draw three circles (figure 8-72) with radii of 1/8, 3/16, and 1/4 inch, respectively.

Exercise 8-5: Modeling a Screwdriver 435

Fig. 8-72. Three concentric circles.

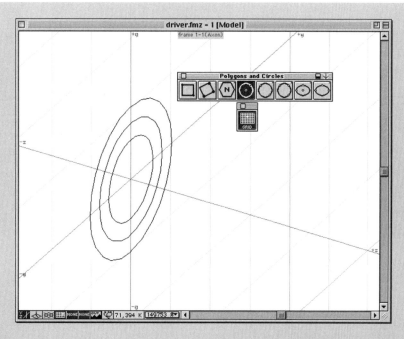

5. Activate the Perpendicular switch, and use the Move command to position the circles along the *x* axis, as shown in figure 8-73. Note that a copy of the smallest circle is moved, not the circle itself. This is done with the Copy tool from the Self/Copy tool palette (figure 8-73).

Fig. 8-73. Moving the circles to their appropriate locations.

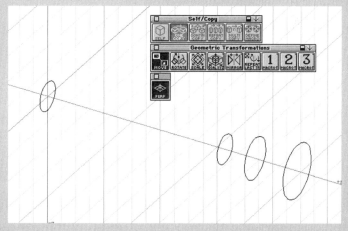

6. Use the Pick tool to select the four circles, starting from one end of the shaft.
7. Select the Controlled Mesh tool. Set the Mesh type to All Closed in the Type Of Controlled Mesh Object dialog box (figure 8-74). With the circles still selected, click any-

where in the modeling window. The flared shaft of the screwdriver will be generated, as shown in figure 8-75.

Fig. 8-74. Setting the Type of Controlled Mesh to All Closed.

Fig. 8-75. Using the C-Mesh tool to create the screwdriver shaft.

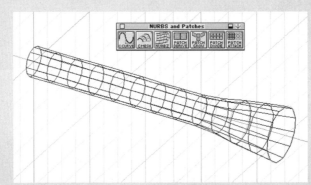

Creating the Screwdriver Blade

The next operation is a bit like whittling a stick with a sharp knife. You will use Boolean Difference commands to slice off portions of the screwdriver shaft to form a blade.

1. Select 3D Object mode. In Front view, draw the 3D double wedge shape shown in figure 8-76 and figure 8-77. Make sure the Heights setting is greater than the diameter of the shaft.

Fig. 8-76. Two wedge shapes.

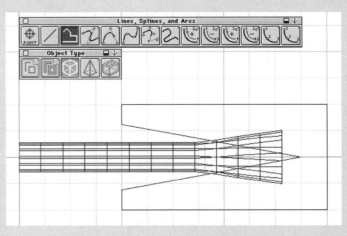

Exercise 8-5: Modeling a Screwdriver

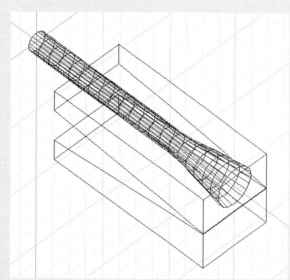

Fig. 8-77. Cutting the blade.

2. Use the Boolean Difference command to subtract the wedge shape from the shaft. The result is shown in figure 8-78.

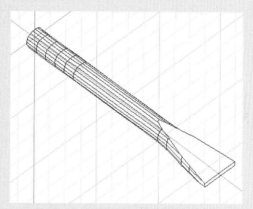

Fig. 8-78. The blade sliced once.

3. Switch to Top view to model another 3D wedge shape, as shown in figure 8-79. Take care to position the lines as shown.

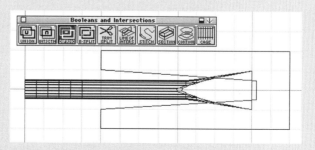

Fig. 8-79. Cutting the blade again.

4. Use another Difference operation to subtract the wedge from the shaft. The final screwdriver blade should resemble figure 8-80.

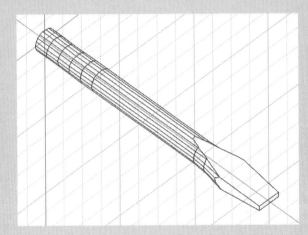

Fig. 8-80. The completed blade.

The Screwdriver Handle

The main form of the screwdriver handle is best created with a Revolve (Revolved Object) command. Take the time to get the profile curve right, or the result will not look like the standard screwdriver handle you expect to see.

1. Select the 2D Surface modifier from the Object Type tool palette.

2. With Grid Snap active for the first and last points so that they touch the x axis, draw the profile line shown in figure 8-81. It is essential that the line begin and end exactly on the x axis.

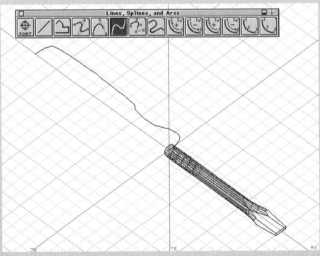

Fig. 8-81. The handle profile.

3. Select the Revolved Object tool. To ensure a smoother handle, select the Nurbz option in the Revolved Object dialog box (figure 8-82).

Exercise 8-5: Modeling a Screwdriver

Fig. 8-82. Revolved Object dialog box.

4. Revolve the profile curve around the *x* axis. The result is shown in figure 8-83.

Fig. 8-83. The revolved handle.

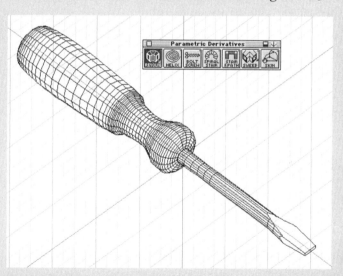

Cutting Grooves in the Handle

To cut grooves in the screwdriver handle, perform the following steps.

1. Set the Heights menu to be greater than the length of the handle.

2. Working in the Right Side view, and in 3D Extrusion mode draw one of six hexagons using the Polygon tool in the Polygons and Circles tool palette, as shown in figure 8-84. Check that the hexagonal tube is positioned to intersect with the handle in two places.

Fig. 8-84. Six hexagon tubes.

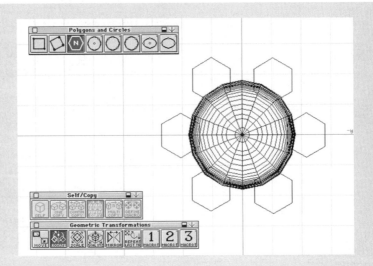

3. Select Repeat Copy, and Rotate a copy of the hexagon around the point 0,0 by 60 degrees.

4. Continue to rotate copies until you have six hexagons around the handle, as shown in figure 8-84.

5. Use the Boolean Difference command to subtract the hexagonal tubes from the handle, as shown in figure 8-85. The finished screwdriver is shown in figure 8-86.

Fig. 8-85. Carving grooves.

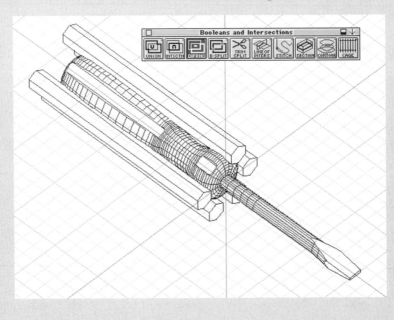

Further Exercises   441

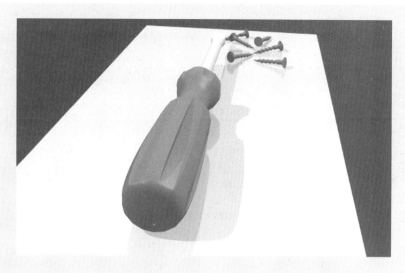

Fig. 8-86. Final rendering of screwdriver with screws.

6. Apply appropriate Surface Styles to make the screwdriver more realistic.

Further Exercises

The following illustrations show the exercise models contained on the companion CD-ROM. For instructions on using the CD-ROM, see the Introduction.

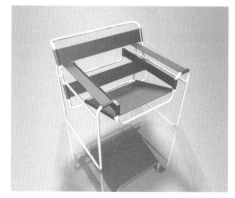

Chair

Knife

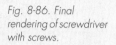

Fork

Chapter 8: Intermediate Exercises

Spoon

Soda Cans

Hammer

Teapot

chapter 9

Advanced Exercises

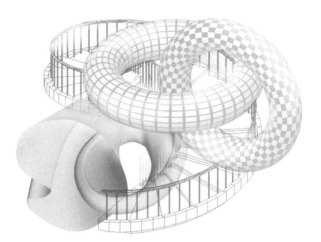

Introduction

After working through the beginner and intermediate exercises in the previous two chapters, a form•Z modeler should have a firm grasp of the modeling tools available. This chapter deals less with instruction in specific modeling methods and places an emphasis on technique and application. The following exercises and on the companion CD-ROM range from the scale of the hand to that of an office, each of them focusing on the formal possibilities and potentials of various form•Z tools.

It is assumed that you have a basic understanding of the modeling tools and concepts in form•Z. This chapter includes discussion of how collections of these tools can be applied to create specific effects. Where the previous chapters dealt with the "how" of modeling tools, these exercises present the "when" and "why."

The Exercises

The exercises in this chapter get incrementally more difficult. As you progress through the exercises, your proficiency with the modeling tools should build. By the time you reach the final few exercises, there will no longer be a discussion of the specific methods required to create the model, but a discussion of what effects are trying to be created and what approach you would use to orchestrate these effects. The following exercises are presented in this chapter and on the companion CD-ROM.

- Hand
- Telephone
- Umbrella
- Flashlight
- Fruit Bowl

- Row Boat
- Toothpaste Tube
- Sandals
- Space Frame

- Construction System
- Model Analysis—The Villa Savoye
- Lighting an Interior

Exercise 9-1: The Hand—Modeling Using Images

LEARNING OBJECTIVE 9-1: *Creating complicated and convincing models using only textures.*

With form•Z RenderZone and form•Z RadioZity comes the ability to apply images directly onto models using the Surface Styles palette. The use of these images, usually referred to as image maps or texture maps, can give your model a level of detail that would otherwise be difficult to achieve. Texture maps can save you a lot of time, taking the place of time-consuming modeling of minute details.

Photographic textures in conjunction with image files used as "displacement maps" and "transparency maps" can cut corners in modeling and shorten modeling time. Using multiple image files as texture maps, displacement maps, and transparency maps, a realistic model of a hand can be made from a simple square plane. The outline of the hand—and the hand's thickness, color, and texture—can all be created by applying images using the Surface Style palette and the Displacement tool. Figure 9-1 shows the hand to be modeled.

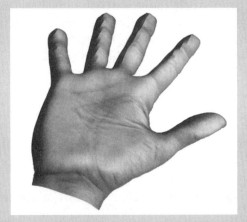

Fig. 9-1. A hand modeled using only textures.

Analysis and Method

To create this model you will be using three image files: one to mimic the color of the hand, one to create the outline, and the last to create the thickness and solidity of the hand. The image files will need to be mapped onto an abstract object that will become the hand through the subsequent layering of these textures. Because all of the modeling will be done with images, the entire model can be generated from a meshed 2D plane.

Figures 9-2 through 9-4 show the images that will be used in this exercise. The first image will be applied to the simple object as the texture map (figure 9-2) to provide the photorealistic detail. The second image will be the transparency map (figure 9-3) used to make the areas around the hand transparent. The last image will be the displacement map (figure 9-4), used to deform the object with its black and white values as a reference, so that a simple object will be transformed to the shape of a hand.

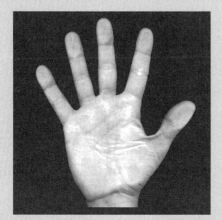 

Fig. 9-2. Texture map.   Fig. 9-3. Transparency map.

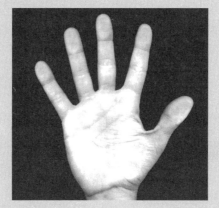

Fig. 9-4. Displacement map.

Hand Modeling Process

The hand will be created using a three-phase modeling process. The first phase will be to create the 2D surface to accept the image files. The second phase will be to apply the color image map and transparency map. The final phase will be to apply an image as a displacement map.

Creating the Surface

To generate the surface to accept the texture maps, you will need to model the surface using the 2D Surface modifier and the Rectangle tool.

1. Set up your modeling environment by setting the Reference Grid to 1'-0" with 4 Subdivisions.

Chapter 9: Advanced Exercises

2. Change the Grid Snap to 1" and zoom in until the mesh of the grid becomes visible.
3. With the Grid Snap on, select the 2D Surface modifier from the Object Type tool palette and the Rectangle tool from the Polygons and Circles tool palette. Draw a surface that is 9" square (figure 9-5).

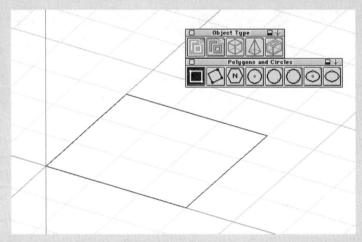

Fig. 9-5. Using the 2D Surface modifier and the Rectangle tool to create a surface.

4. Select the Mesh tool from the Meshes and Deform tool palette and set the Mesh increment to 1/8" in the Tool Options palette. Click on the 9" square to subdivide it into a 1/8" mesh (figure 9-6).

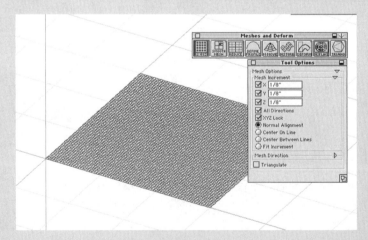

Fig. 9-6. Using the Mesh tool to subdivide the 2D surface.

Applying the Color Image Map

The next phase in the process of modeling the hand is to apply a photograph of a hand as a Color Image Map and a Transparency Map.

1. Double click on the current active Surface Style in the Surface Styles palette to open the Surface Style Parameters dialog box. Select Image Map from the Color pull-down menu and click on Options to open the Image Map Options box.

2. In the Image Map Options box, click on Load and locate the file *color.tif* (available on the companion CD-ROM). The color image of a hand should appear in the upper right (figure 9-7). Click on OK.

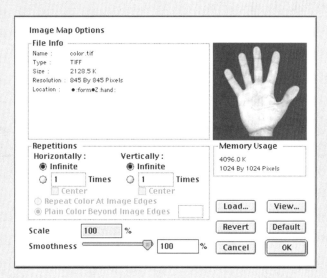

Fig. 9-7. Image Map Options.

3. Select Transparency Map from the Transparency pull-down menu in the Surface Style Parameters dialog box and click on the corresponding Options button to open the Transparency Map Options.

4. In the Transparency Map Options box, click on the Load button and locate the file *transpar.tif* (available on the companion CD-ROM). A black-and-white image of the outline of a hand should appear in the upper right of the Transparency Map Options box (figure 9-8). A transparency map renders an object transparent based on a black-and-white image file. Wherever the image file is black, the object to which the transparency map is applied will appear transparent; where the image is white, the object will appear opaque. In the case of the hand, the transparency map of a white hand on a black background will create the outline of a hand. Click on OK when you are done.

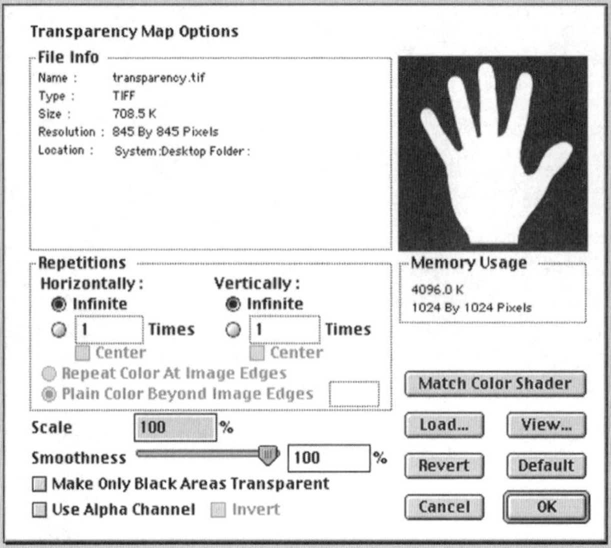

Fig. 9-8. Loading the Transparency Map.

5. When you have completed loading the Transparency Map and Image Map, change the name of the Surface Style to Hand in the Name box at the top of the Surface Style Parameters box. The Surface Style Parameters should appear as it does in figure 9-9.

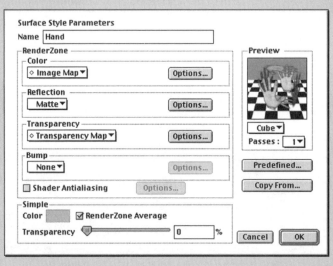

Fig 9-9. The Hand Surface Style.

6. Once you have created the Hand Surface Style, you need to apply it to the 2D surface. Select the Texture Map tool from the Attributes palette and click on the 2D surface. This will open the Texture Map Controls window (figure 9-10), allowing you to adjust the way a Surface Style is applied to an object.

Fig 9-10. Texture Map Controls window.

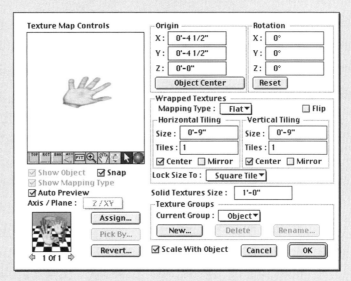

7. Select a Mapping Type of Flat from the Wrapped Textures section of the Texture Map Controls. This sets the type of geometry used in applying the texture to the surface (see the Texture Map tool in Chapter 5 for further explanation of Mapping Type). Check the Center option under Horizontal and Vertical Tiling and enter *1* for Tiles under Vertical Tiling; the number of Tiles under Horizontal Tiling will change automatically. To preview the texture map, click on the icon of a sphere at the bottom right of the preview box and select Rendered. The texture image should appear as it does in figure 9-7. Click on OK.

Displacing the Hand

Now that the Hand Surface Style has been created, you can apply it to the 2D surface and set the proper mapping parameters using the Texture Map tool.

1. Select the Displacement tool from the Meshes and Deform palette and click on the 2D surface object. The Displacement Map Edit dialog box will appear (figure 9-11).

2. Click on the icon at the bottom left of the window to locate and load your displacement map file. In the file browser, select the file *displace.tif* (available on the companion CD-ROM). When you click on OK, the image of the hand should appear at the bottom left of the window.

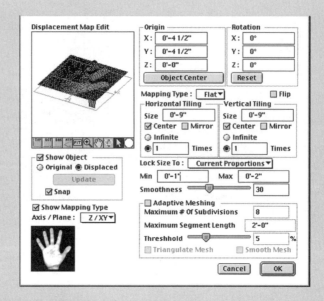

Fig. 9-11. Displacement Map Edit dialog box.

3. Select Flat from the pull-down menu for Mapping Type, and select Center under Horizontal and Vertical tiling. Set the Min to 1" and the Max to 2", and adjust the Smoothness slider to 30. If you want a preview of your displacement map, click on the radio button next to Displaced in the Show Object box on the left side of the controls. When you click on OK, your 2D surface should resemble figure 9-12.

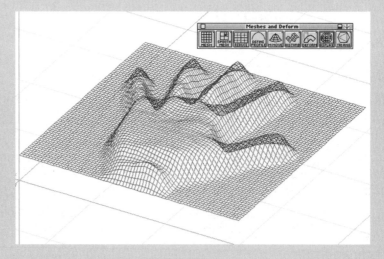

Fig. 9-12. Displaced surface.

4. Once the surface has been displaced, use RenderZone to render the Hand. For best effect, select Shader Anti-aliasing and Super Sampling of High in the RenderZone Options. The final rendering is shown in figure 9-13.

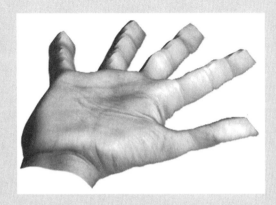

Fig. 9-13. A final rendering of the hand.

TIPS AND STUDY AND APPLICATION NOTES

1. Effective use of Image Maps, Displacement Maps, and Transparency Maps can save a lot of time in modeling. The hand is only one example of how you can save yourself the time-consuming task of modeling details. Other things that can be modeled easily using textures instead of polygons are objects such as leaves, wire mesh, picket fences, windows, and trees.

2. Using textures to create modeling effects instead of actually modeling all polygons not only saves on modeling time but lowers the number of polygons in your model, making rendering time much shorter.

3. When rendering objects with transparency maps, you should use the Soft shadows (Mapped) option for the lights casting shadows of your object. This option will only cast a shadow of the visible portion of the model. The Hard shadows (Raytraced) option will inappropriately cast a shadow of the entire mesh.

Exercise 9-2: Modeling a Telephone

LEARNING OBJECTIVE 9-2: *Applying advanced form•Z functions.*

Modeling an industrially designed object, such as a telephone, presents a fun challenge to the advanced form•Z modeler. The subtle curves and smooth, sleek lines require practiced control of the Parametric Derivative tools: Sweep, Skin, and Revolved Object. The subtle bends and bulges of these types of objects require an understanding and application of the Mesh tools: C-Curve, Mesh, and Deform.

The following exercise of modeling a telephone will apply many of the advanced functions of form•Z, such as Skin and Deform. It will also test your skills with the more general tools, such as Attach and Difference. Figure 9-14 shows the telephone to be modeled.

Fig. 9-14. Telephone to be modeled.

Analysis and Method

The telephone will be modeled in two pieces: body and handset. The primary form of the body will be made using the Sweep tool and then deformed with the Deform tool to add the subtle bends and bumps. The buttons and speaker will be made using the 3D Extrusion and Decal tools. The handset, because of its complicated curvilinear shape, will be made using the C-Curve and Skin tools.

Modeling the Body

In the following steps you will be using the Sweep and Deform tools to create the body of the telephone.

1. Set up your modeling environment by changing the Grid Snap module to 1/4" and setting Window Grid to 1", with four subdivisions. Change Numeric Accuracy to 1/32".

2. Using the Vector Line and Arcs tools in conjunction with the 2D Surface modifier, draw a 2D outline of the telephone (figure 9-15). This will be used as a path for a Boundary Sweep; therefore, be sure to triple click the mouse at the last point to close the line.

3. Using the Spline, Quadratic Bezier tool and the 2D Surface modifier, draw a small profile of the edge of the telephone (figure 9-16). This will be the source shape for the Boundary Sweep; therefore, this profile should be left open (double click on the last point only).

Exercise 9-2: Modeling a Telephone   453

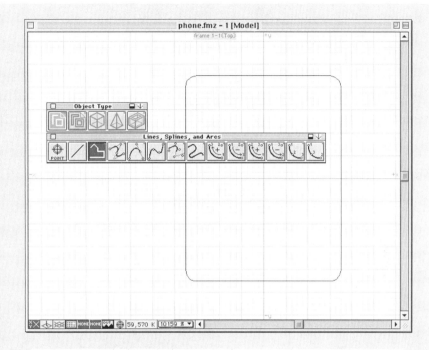

Fig. 9-15. Outline of the telephone.

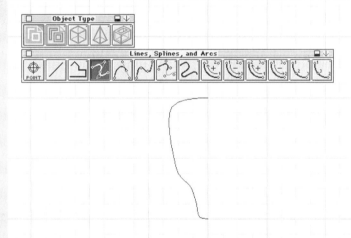

Fig. 9-16. Edge profile of the telephone.

4. To generate the body of the telephone, use the Sweep tool in the Parametric Derivatives tool palette. Open the Sweep dialog box and select the Boundary Sweep option. With the Sweep tool active, click on the edge profile, and then on the closed outline of the telephone. You will be presented with the Sweep Edit dialog box (figure 9-17). Preview your object to ensure it will be generated properly. If the object does not appear to be correct, you can adjust the parameters for the Sweep tool directly in the dialog box. Click on OK to generate the telephone body (figure 9-18).

Fig. 9-17. Sweep Edit dialog box.

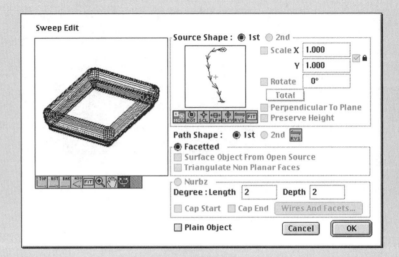

Fig. 9-18. Telephone body generated with Boundary Sweep.

5. To create a more realistic telephone, you need to add a tilt to the body using the Deform tool. Open the Deform Options dialog box (figure 9-19) and select Radial Bend. Also make sure that Base Reference Plane is set to Active Plane. Switch to the 30°-60° view and make sure the reference plane is set to XY. With the Deform tool active, click on the telephone body. A control box will appear. Grab an edge and bend the body to apply a tilt (figure 9-20).

6. The telephone should also have a slight taper when looking down at it. Switch to Top View and with Topological Level set to Face, pick all of the faces on the front of the phone. Using the Independent Scale tool, scale these faces around the origin in the X dimension only. This will have the effect of tapering the object (figure 9-21).

Exercise 9-2: Modeling a Telephone

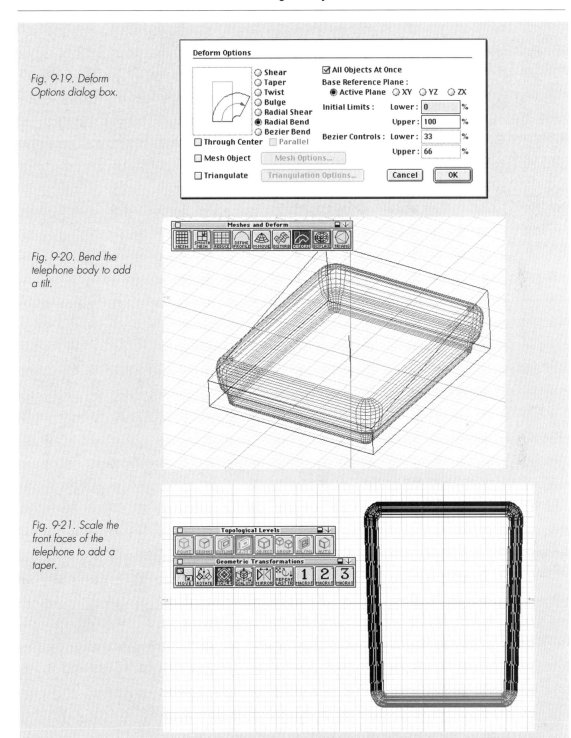

Fig. 9-19. Deform Options dialog box.

Fig. 9-20. Bend the telephone body to add a tilt.

Fig. 9-21. Scale the front faces of the telephone to add a taper.

Creating the Handset

Leaving the body of the telephone for now, the next phase involves modeling the handset. Because of its complex curvilinear form, the handset will be made using the C-Curve and Skin tools.

1. Switch to the Right side view and draw three rough lines outlining the top, bottom, and middle of the handset. This should be done using the Vector Line tool and the 2D Surface modifier (figure 9-22).

Fig. 9-22. A rough outline of the handset using the Vector Line tool.

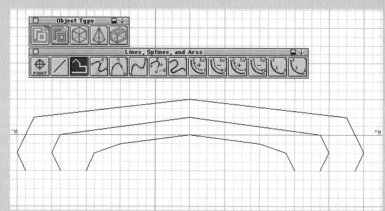

2. Select the C-Curve tool, with the default Quadratic Quick Curves option, in the Controlled Curve Options dialog box to smooth out these lines and adjust their shape (figure 9-23). When using the C-Curve tool, set the Controlled Curve Options to have a Smooth Interval of 1/4".

Fig. 9-23. Smooth the curves using the C-Curve tool.

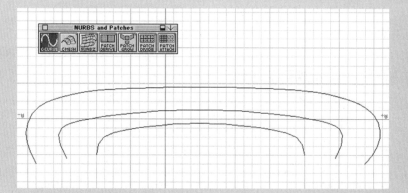

3. Switch to the front view and, using the Move tool and the Rotate tool, move these profiles to the right side of the axis and tilt each of them (figure 9-24).
4. Use the Mirror tool and the One Copy modifier to mirror the three curves across the yz axis and create a second set of curves (figure 9-25). These six curves will be the paths for a Skin operation.

Fig. 9-24. Tilt the curves using the Rotate tool.

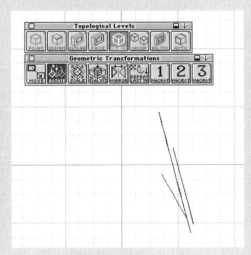

Fig. 9-25. Mirror the curves across the yz axis to create a second set.

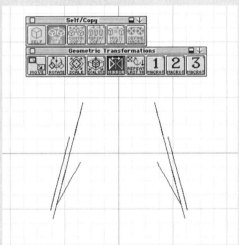

5. With the paths created, the next step is to model the source objects to be used in the Skin operation. Using the Vector Line tool with the 2D Surface modifier and Point Snap, draw a closed hexagonal shape by connecting the end points of the curves (figure 9-26).

458　Chapter 9: Advanced Exercises

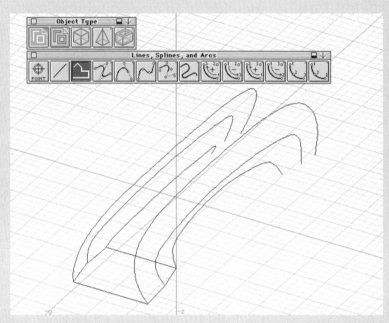

Fig. 9-26. Draw a hexagonal shape using the Vector Line tool and Point Snap.

6. In the Display menu, open the Wire Frame Options dialog box (figure 9-27). Toggle on the Show Direction option. This operation will show the line's direction when you click on OK.

Fig. 9-27. Wire Frame Options dialog box.

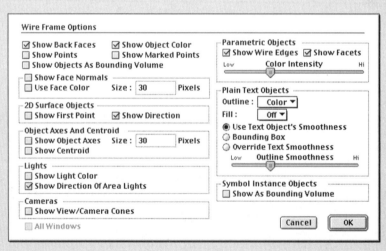

7. Check the line directions and reverse them if necessary with the Reverse Direction tool. The result should resemble figure 9-28.

Exercise 9-2: Modeling a Telephone

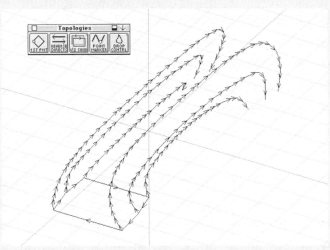

Fig. 9-28. Reversing line direction.

8. The final step is to use the six paths and one source to create a skin. Open the Skin Options dialog box and select the Skinning Along Paths option. Within the Skinning Along Paths option, type in *1* for the # Of Sources and *6* for the # Of Paths, and check off Close At Ends of Open Paths and Boundary Based. This last option will generate a solid object from the skin operation. With the Skin tool active, select the source shape and then the six paths to generate the handset (figure 9-29).

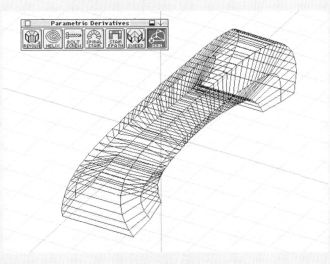

Fig. 9-29. Using the Skin tool to generate the handset.

Creating the Hang-up

With the handset completed, you can return to the body to add detail. The next step is to create a recess for the handset.

1. The first step is to "hang up the phone" by positioning the handset in the proper place on the body. Use the Rotate tool and Move tool to locate the handset parallel to the face of the telephone. You may need to scale the handset in the *x, y,* and *z* directions to make its proportions fit the scale of the telephone body (figure 9-30).

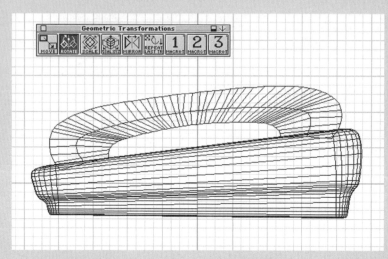

Fig. 9-30. Positioning the handset using the Move and Rotate tools.

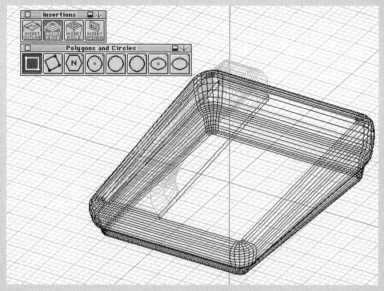

Fig. 9-31. Creating a recess to accept the handset.

2. Use the Arbitrary Reference Plane tool to create a reference plane on the top face of the telephone body. Select a Topological Level of Segment or Face. With the Arbitrary Reference Plane tool active, select two edges of the top face. A new reference plane will be created. Place the handset on a separate layer and ghost this layer so that the handset will not interfere with the modeling process.

3. Using the Insert Face/Volume tool and the Rectangle tool, create a recess in the telephone body to accept the handset (figure 9-31). Use the Graphic/Keyed setting from the Heights menu to interactively set the depth of the recess.

Creating Buttons

In this section of the exercise, you will add detail to the telephone body by creating a raised area for number buttons and a speaker button.

1. To create a raised area for the number buttons, you will use the Deform tool and deform the top face of the phone body. To deform the face, you must first subdivide it using the Mesh tool. Set the Mesh options to a scale of 1/4" in the x, y, and z directions. With the Mesh tool active and the Face Topological Level selected, apply a mesh to the top face of the telephone (figure 9-32).

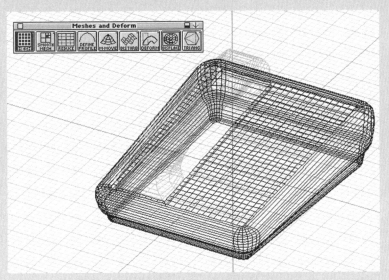

Fig. 9-32. Meshing the top face of the telephone body.

2. When you are selecting the edges of the face to mesh, select the edge toward the bottom of the telephone body first. The reason for this is that the Mesh tool creates a mesh grid parallel to the first segment selected, and for the later steps it is best to have a grid parallel to the top and bottom edges rather than parallel to one side.

3. Another necessary step before applying the Deform tool to the telephone body is to create a custom profile. The surface of the body will be deformed using a profile, but none of the profiles already in the Profiles palette have the proper shape. To create the profile, use the Smooth Lines tool to draw a 2D line similar to that shown in figure 9-33. Select the Define Profile tool in the Meshes and Deform palette and click on the curve to define a profile. A new profile icon will be added to the Profiles palette. Once you have done this, you can delete the original curve.

➥ **NOTE:** *See Chapter 4 for a discussion of the Profiles palette.*

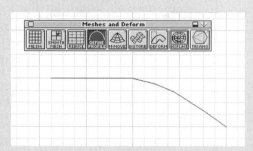

Fig. 9-33. Define a profile using the Define Profile tool.

4. Select Move Mesh and open its options dialog box. Select the Radial movement shape. Set # Of Sides to 4. With the Move Mesh tool selected and the newly created Profile active in the Profiles palette, click on the top face of the telephone body and create a raised area similar to the one shown in figure 9-34. When using the Move Mesh tool, be sure to toggle the Perpendicular Switch on so that the deformation will be performed perpendicular to the reference plane.

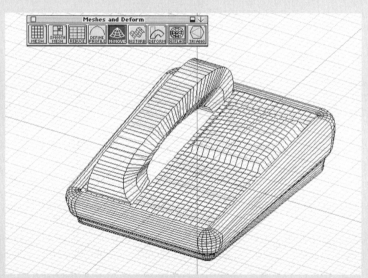

Fig. 9-34. Deform the face of the telephone body to create a raised area for the buttons.

5. Now it is time to create the telephone buttons. Use the Vector Line and Arcs tools and the 3D Extrusion modifier to create a rectangular button (figure 9-35).

6. When you have made one button, you can use the Attach tool to place it on the face of the telephone body. Set Attach Type to Entire Object in the Attach Options. Set the Adjust To options for Face to Center Of Face. With Topological Level set to Face and the Attach tool selected, select the bottom face of the button and then a rectangle of the mesh where you wish the button to be placed. The button will be moved and rotated into place on the face of the telephone (figure 9-36).

Exercise 9-2: Modeling a Telephone

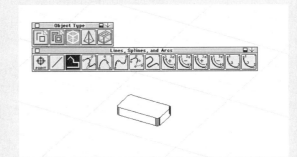

Fig. 9-35. A single button.

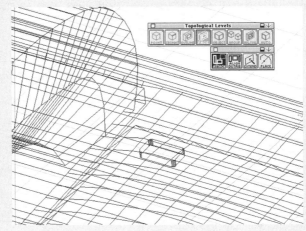

Fig. 9-36. Use the Attach tool to place the button on the face of the telephone body.

7. Once the first button is placed, you can use the Move tool and the Copy modifier to create the remaining eleven buttons in the number keypad. If you use the Arbitrary Reference plane you defined earlier in the exercise, you can use the Grid Snap tool to ensure the buttons are evenly spaced. Now use the same process to create a speaker button and some small, triangular volume buttons. Place them toward the bottom of the telephone (figure 9-37).

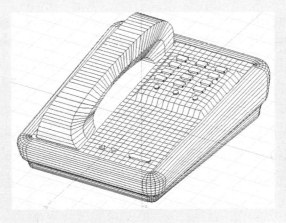

Fig. 9-37. Create copies of the first number button to finish the keypad.

Finishing Touches

The telephone is almost complete. The last few steps are some finishing touches you can add to make your model especially convincing.

1. Use the Difference tool from the Boolean palette to subtract the handset from the body of the telephone. This will create two small indentations that would be used to hang up the phone. Once you have created the indentations, take the handset off the hook and place it on its side.

2. Using the Vector Line and C-Curve tools, create a curve between the telephone body and the handset (figure 9-38). This will be used to generate a coiled phone cord.

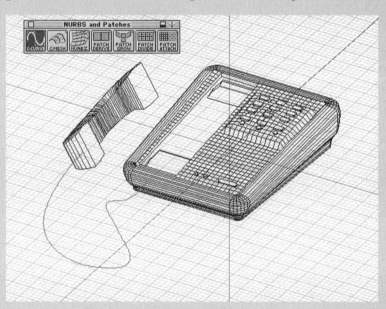

Fig. 9-38. Draw a curve between the telephone body and handset.

3. Select the Helix tool in the Parametric Derivatives tool palette. Open its options and select Wire Helix Along Path. Then set # Of Cycles to 30, Wire Helix Radius to 1/4", and # Of Steps to 12. Next, select Per Cycle in the field adjacent to # Of Steps. These options will produce a spiral helix with 30 spirals of 1/4" radius along a path. Close the options, and with the Helix tool active, click on the curve. A spiral will be created.

4. Draw a very small 2D circle using the Circle tool and the 2D Surface object modifier. Select the Sweep tool and select the Axial Sweep option from its options dialog box. Sweep the circle along the spiral to create a phone cord.

5. Apply decals of numbers to each of the buttons on the keypad, and a decal saying Speaker to the speaker button. To create a speaker, use a decal of a grid of black dots with a transparency map. For best effect, the speaker decal should be placed using the Opacity option in the Decals dialog box.

> **NOTE:** For more information on how to apply decals, see the Soda Can exercise in Chapter 8 and the Decal tool in Chapter 5. Image files for the decals are provided on the companion CD-ROM.

6. Use RenderZone to render your model. To make your decals as effective as possible, select the Full Raytrace rendering type in the RenderZone options. Figure 9-39 shows the rendered telephone.

Fig. 9-39. Final rendering of the telephone.

Exercise 9-3: An Umbrella—Modeling Surfaces Using Nurbz

LEARNING OBJECTIVE 9-3: *Creating a curved surface model with Nurbz surfaces.*

A key strength of form•Z 3.0 is its ability to generate and manipulate parametric curves and meshes. One type of parametric object available is NURBS (Non-Uniform Rational B-Spline) based objects, called Nurbz. Nurbz objects have editable control parameters, making it easy to adjust the shape of curved surfaces and lines. This example uses Nurbz to modify a revolved surface to produce the familiar shape of an umbrella (figure 9-40). From this example you will learn the basic generation and manipulation techniques for Nurbz surfaces.

466 **Chapter 9: Advanced Exercises**

Fig. 9-40. A Nurbz umbrella.

Analysis and Method

The umbrella is an introductory exercise in using the Nurbz tool and manipulating Nurbz surfaces. You will build a section of the umbrella by creating a surface between several curves. Although this could be done with the C-Mesh or Skin tool, a Nurbz surface allows you to edit the shape of an object after its generation and gives you more specific and direct control over its form. Nurbz surfaces have a network of control points that can be individually moved, rotated, and scaled to finely sculpt the surface of an object.

Umbrella Modeling Process

The first phase is to model a simple panel of the umbrella canopy by creating curves and generating a surface using the Nurbz tool. The second phase is to shape the surface using the Edit Controls tool. The third phase is to use the Copy tool to complete the canopy of the umbrella, and the Sweep tool to generate the support struts. The final phase is to make a handle and render the umbrella.

Creating the Surface

To generate the umbrella's canopy, you will model a single panel as a Nurbz surface using the 2D lines and the Nurbz tool.

1. Using the 2D Surface modifier from the Object Type tool palette and the Cubic Bezier Spine tool from the Lines, Splines, and Arcs palette, draw a curve in the Right View that is approximately 2-1/2 feet wide and 1 foot high. This curve is a sectional profile of your umbrella (figure 9-41).

Exercise 9-3: An Umbrella—Modeling Surfaces Using Nurbz 467

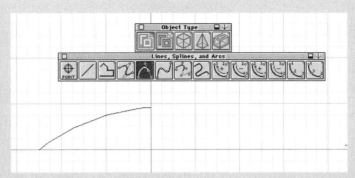

Fig. 9-41. Using the 2D Surface modifier and the Cubic Bezier tool to create a curve.

2. Select the Repeat Copy modifier from the Self/Copy palette and the Rotate tool from the Geometric Transformations palette. Rotate the curve 22.5 degrees around the origin, then click again and the Repeat Copy modifier will make a second copy. You should end up with three curves, as in figure 9-42.

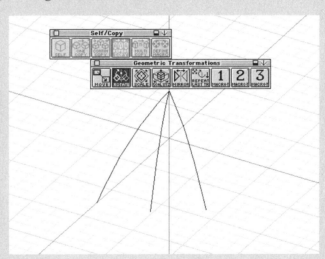

Fig. 9-42. The umbrella's 2D profiles ready to turn into a surface.

3. Select the Nurbz tool from the NURBS and Patches tool palette and set the Tool Options to a Length and Depth degree of 2 (figure 9-43). Using the Nurbz tool, click on the three curves in order from left to right. The surface will be generated automatically on the last click.

468 Chapter 9: Advanced Exercises

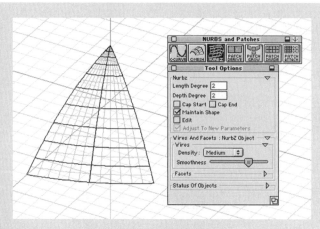

Fig. 9-43. Generate the Nurbz surface using the Nurbz tool.

Editing the Nurbz Surface

To shape the section of umbrella canopy to more closely resemble stretched fabric, you will need to use the Edit Controls tool.

1. Select the Edit Controls tool from the Pick tool palette. With the Edit Controls tool active, click once on the surface. It will change color and a series of control points will appear.

2. With the Edit Controls tool still active, hold down the Option key (MacOS) or Control + Shift keys (PC) and click in an empty space on the window. A small contextual pop-up menu will appear. Select Uniform Scale from the pull-down menu.

3. With the Edit Controls still active, hold down the Shift key and click on one of the lines in the center row of control points. They will all turn the active color to show they are selected, as in figure 9-44.

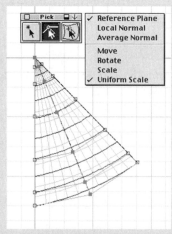

Fig. 9-44. Change the name of the Surface Style.

4. Click on one of the active points on the center row to finish the process of selecting the points. Click on the last point in the row (the one closest to the origin) to set the center for the scaling operation. Click again at the other end of the row, to start the scaling

operation and move the mouse to scale the control points around center. The scaling operation will move the center row of points toward the center and bend the surface panel inward, as in figure 9-45. Click a final time to end the operation.

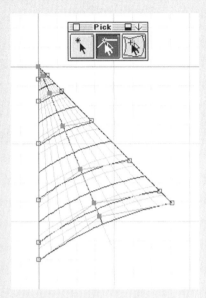

Fig. 9-45. Bend the surface inward using the Edit Controls.

Completing the Umbrella Canopy

Now that you have created a single panel of the umbrella canopy, you will need to copy it to complete the canopy. You also need to create the struts to support it.

1. With the Repeat Copy tool still active in the Self/Copy palette, select the Rotate tool from the Geometric Transformations palette. Click on the surface and rotate it around the origin 45 degrees. After rotating once, click in an open area of the window to rotate additional copies until the umbrella canopy is complete (figure 9-46).

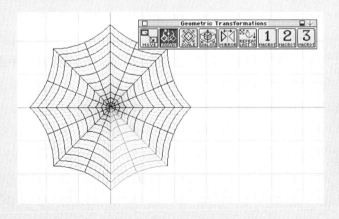

Fig 9-46. Copy the section seven times to complete the umbrella canopy.

Chapter 9: Advanced Exercises

2. Place the umbrella canopy on another layer and hide it to avoid selecting it by accident. Use the Unghost tool to unghost one of the original curves used to create the Nurbz surface. You will use this as a path for creating the struts.

3. Use the Vector Line tool to draw a small triangle on the *xy* reference plane. This will be the source shape for the sweep operation to create the struts (figure 9-47).

4. Select the Sweep tool from the Parametric Derivatives tool palette. Select Axial Sweep as the sweep type from the Tool Options and select the Nurbz option. In the Nurbz option section, enter *1* for the Length Degree and *2* for the Depth Degree (figure 9-47). The degree of the Nurbz surface determines how faceted it will appear. The choice of 2 in the Depth will make it smooth along the strut so that its curvature will match that of the canopy. The choice of 1 degree in the Length will make the strut faceted in section.

5. With the Sweep tool active, click on the triangle and then on the curve. The Sweep Edit dialog box will open. Adjust the position and orientation of the Source Shape so that the preview resembles that in figure 9-48. Click on OK and your strut should be complete.

6. Use the Repeat/Copy tool and Rotate to copy this strut to complete the umbrella's support structure (figure 9-49).

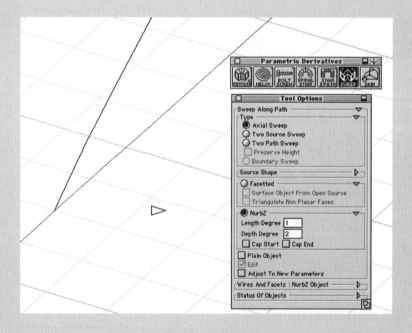

Fig. 9-47.
Sweep tool options and generation shapes of the Sweep operation.

Exercise 9-3: An Umbrella—Modeling Surfaces Using Nurbz

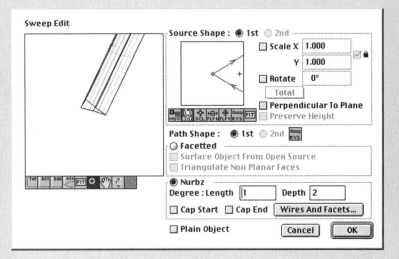

Fig. 9-48. Sweep Edit dialog box.

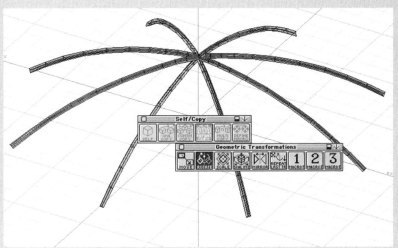

Fig. 9-49. Use Repeat/Copy and Rotate to create the remaining struts.

Making a Handle

The final phase in modeling your umbrella is to create a shaft and handle.

1. Switch to the Right side view; select the Segment tool from the Lines, Splines, and Arcs tools; and select the 2D Surface modifier from the Object Type tool palette. Draw a single line from slightly above the top of the umbrella to a point about 2-1/2 feet below the bottom edge of the canopy (figure 9-50). This will define the path of the shaft.

Fig. 9-50. Draw the path for the shaft using the Segment tool and the 2D Surface modifier.

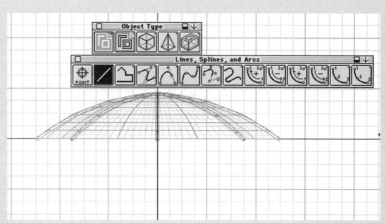

2. With the 2D Surface modifier still active, switch to the Arc, Clockwise, End-Point Last tool. Set the Tool Options to have a Max Size of Segments of 1". Beginning at the end of the segment you created in the previous step, draw a semicircular arc (figure 9-51). This will be the hooked handle of the umbrella.

Fig. 9-51. Draw the path for the handle using the Arc tools.

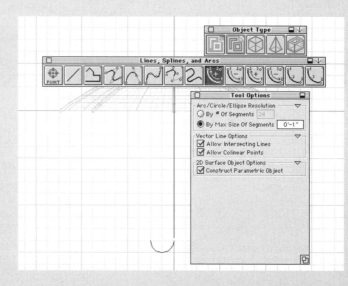

3. Using the Circle by Center and Radius tool, draw a small circle and a larger circle on the *xy* reference plane. When creating the circles, set the Max Size of Segments to 1/2" in the Tool Options (figure 9-52). These are the source shapes of the handle and shaft.

Exercise 9-3: An Umbrella—Modeling Surfaces Using Nurbz 473

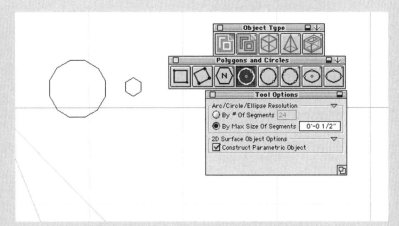

Fig. 9-52. Draw the source shapes for the handle and shaft.

4. To complete the handle and shaft, use the Sweep tool from the Parametric Derivatives palette. Select the small source shape and then the path for the shaft to create the shaft. Repeat the process with the source and path for the handle (figure 9-53).

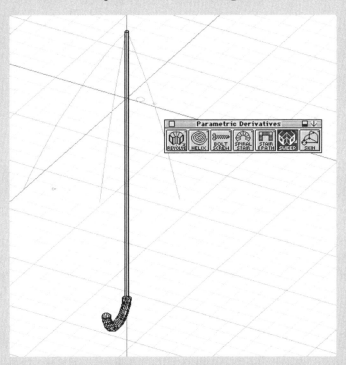

Fig. 9-53. Umbrella handle and shaft created with the Sweep tool.

5. To finish the umbrella, assign appropriate colors to the canopy, shaft, and handle, and use RenderZone to render your model (figure 9-54). For added effect, try turning on the Depth Effects in the RenderZone options and select Snow from the pull-down menu (figure 9-40).

474 **Chapter 9: Advanced Exercises**

Fig. 9-54. Completed umbrella rendered in RenderZone.

TIPS AND STUDY AND APPLICATION NOTES

1. The umbrella's surface could be created in a number of ways. You could, for example, do this with the C-Mesh or Skin tools, explained in Chapter 5.

Exercise 9-4: Modeling a Flashlight

LEARNING OBJECTIVE 9-4: *Using rendering effects such as lighting and texturing.*

Lighting, texturing, and other rendering effects can help create convincing modeling results in form•Z. A little time spent with these details can replace many needless hours of modeling. In the following flashlight exercise, the modeling technique is relatively simple, using mostly the line tools and Revolved Object. However, the final effect is rather complex. The realistic effect of a flashlight is created with four lights and a variety of reflectivity and transparency textures. Figure 9-55 shows the flashlight to be modeled.

Analysis and Method

The flashlight is a cylindrical object, ideally suited for the Revolved Object tool. The body of the flashlight, the reflective lamp, and the light bulb can all be derived from 2D lines and curves using the Revolved Object tool. The thumb switch will be derived from the surface of the flashlight body. It will be created with the Line of Intersection tool and then shaped to fit the thumb using the Deform tool.

Exercise 9-4: Modeling a Flashlight

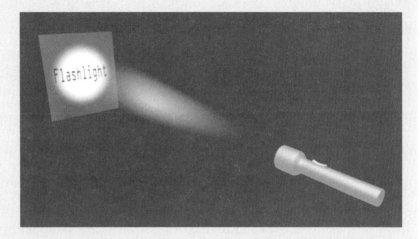

Fig. 9-55. Glowing flashlight to be modeled.

The lighting effects will be created with a series of lights: a Directional light for ambient effect, a Cone light to mimic the light of the flashlight, and a Point light to represent the actual light bulb. The final effect of the glowing cone emanating from the flashlight is created with a Neon Transparency surface style on a solid cone.

Creating the Body and Lamp

In the next few steps you will create the body, lamp, and light bulb of the flashlight using 2D lines and the Revolved Object tool.

1. Set up your modeling environment by changing the Grid Snap module to 1/16" and setting the Window Grid to 1" with four subdivisions. Set the Numeric Accuracy to 1/32".

2. Draw a 2D profile of the flashlight body using the Vector Lines and Arcs tools. Because the profile will be used in a Revolve operation, it should touch the x axis on one end (figure 9-56).

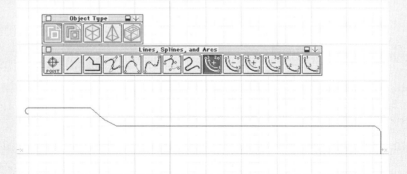

Fig. 9-56. Profile of the flashlight body.

3. Using the same tools, draw the profiles of the light bulb and lamp. The profile of the light bulb should touch the x axis on both ends so that it will create a solid form when revolved (figure 9-57).

Chapter 9: Advanced Exercises

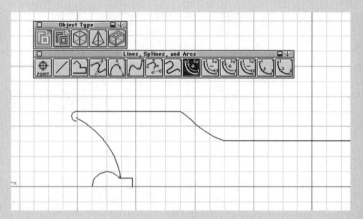

Fig. 9-57. Profiles of the light bulb and lamp.

4. Use the Revolved Object tool to create 3D objects from these profiles. First, open the Revolved Object dialog box and change the # of Steps to 24. By increasing the # of Steps, you will produce a smoother model. Click on the Surface Object from Open Source option (figure 9-58). Selecting the Surface Object from Open Source option is necessary in order to apply the Trim/Split and Stitch tools on these objects. Figure 9-59 shows the revolved flashlight.

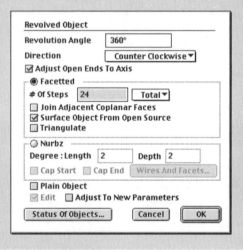

Fig. 9-58. Revolved Object dialog box.

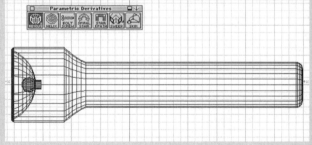

Fig. 9-59. Revolved flashlight.

Creating the Thumb Switch

To create the thumb switch, you will trim out a portion of the flashlight surface. On the resulting piece you will perform an extrusion and deformation.

1. Create a 3D Extruded rectangle intersecting the handle of the flashlight. It should pass through the handle where you want the button to be, but should not pass through the entire object.

2. With the Line of Intersection tool active, click on the flashlight body, then select the extruded solid. The portion of the body where these two overlap will be detached from the flashlight body (figure 9-60).

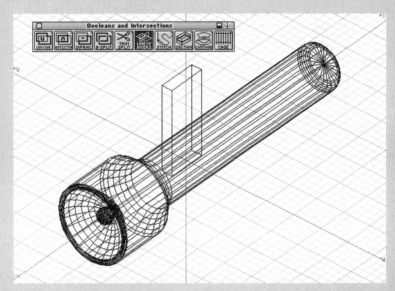

Fig. 9-60. Create a 3D object intersecting the flashlight body.

3. Use the 3D Extrusion tool from the Derivatives tool palette to create a 3D solid from the piece you separated from the flashlight body. When you finish you should have a solid piece similar to that shown in figure 9-61.

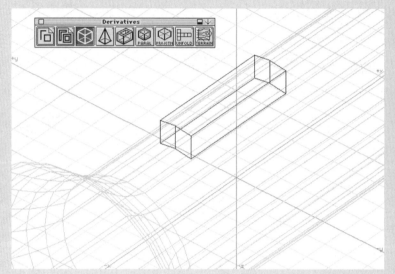

Fig. 9-61. Extruded thumb switch.

4. Select the top face of the button using Face Topological Level and apply a 1/8" mesh using the Mesh tool (figure 9-62). The mesh size is set in the Mesh tool options.

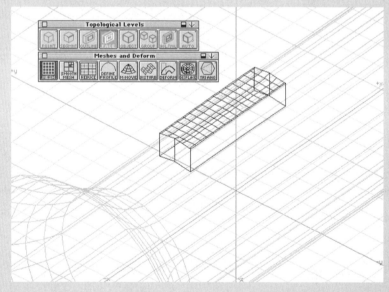

Fig. 9-62. Mesh the top face of the switch.

5. To shape the button to fit a thumb, you will need to use the Deform tool. In the Deformation Options, chose Radial Shear and set the Base Reference plane to YZ. At the Topological Level of Object, use the Deform tool to bend the button downward. The top faces are the only portions of the object meshed; therefore, they will be the only faces affected by this operation (figure 9-63).

Exercise 9-4: Modeling a Flashlight

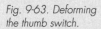

Fig. 9-63. Deforming the thumb switch.

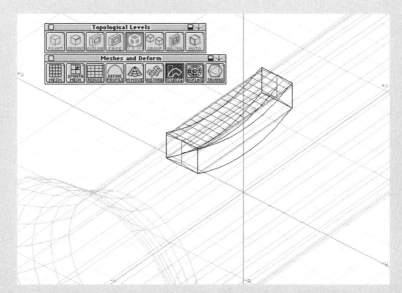

6. The last step in shaping the flashlight's thumb switch is to taper the edges. The easiest way to accomplish this is to select the top faces and scale them "inward." This will make the top face smaller than the bottom face and consequently taper the sides "downward" (figure 9-64).

Fig. 9-64. Scaling the top faces of the switch to add a taper to the sides.

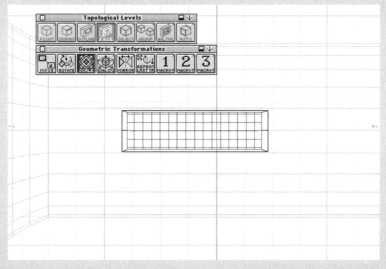

7. If you want add a little more realism to your flashlight, you can create a glass lens. To do this, switch the active reference plane to YZ. Using the Point Snap, create an extruded circle touching the edge of the lamp (figure 9-65).

Fig. 9-65. Addition of a lens to the lamp.

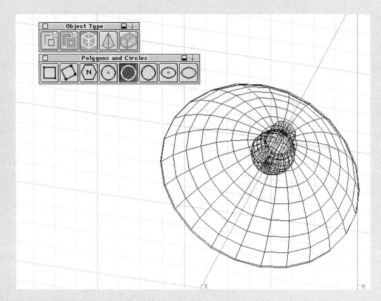

Adding Textures and Transparency

The remainder of this exercise will delve into the texturing and lighting effects necessary to make the model look as if it were a working flashlight. The first phase is to texture the flashlight body, lamp, lens, and light bulb. After texturing the flashlight, you will create a couple of "light" objects to produce the light's halo and glare.

1. Start by creating a Surface Style for the flashlight body by clicking on a blank space in the Surface Style palette. The Surface Style Parameters dialog box will open automatically. Name the texture Metal and set its color to a dramatic bright red. Select Metal, Accurate as the reflection type from the Reflection pull-down menu. Click on the Options button next to Reflection to set the Metal, Accurate options (figure 9-66). Leave all of the settings at their default settings except for Reflectivity, which should be lowered to 10 to 15 to prevent the object from being too shiny. Assign this surface style to the flashlight body using the Color tool.

2. Create a second surface style for the light bulb and lens. Name this surface style *Lens*. Make the Color white and set the Reflection to Glass, Accurate. You may want to use the Edit Simple Display Color option. This will allow you to set the wire frame color to a color other than white. Assign this surface style to the light bulb and lens using the Color tool.

> **NOTE:** *If you leave the display color as white and you are using a white background in your modeling window, the white objects will blend into the background, making them difficult to read.*

Exercise 9-4: Modeling a Flashlight

Fig. 9-66. Accurate Metal Options dialog box.

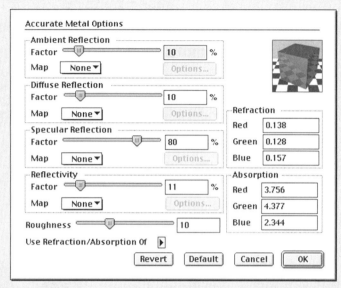

3. For the lamp, you will need to make a third surface style. Call this surface style Lamp and set its Color to gray. Choose Metal, Accurate from the Reflectivity pull-down menu and click on Options. In the options box, make the Reflectivity 100%, which will make the surface seem mirrored and highly reflective. Assign this texture to the lamp.

4. You now need to create a few objects to be used for the visual effects of the halo and glare that come from the light. To create a halo around the light bulb, you need to make a small sphere surrounding it. This can be done with the Spherical Object tool or by drawing a semicircle and using the Revolved Object tool (figure 9-67).

Fig. 9-67. Creating a halo object around the light bulb.

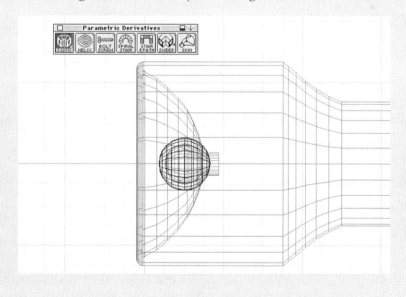

5. To make the spherical halo glow, you need to give it a glowing surface style. Create a new surface style and name it *Glow*. Set its Color to be a bright yellow and select Neon as its transparency type from the Transparency pull-down menu. Under the Transparency options, set the Intensity to 75%. This will make the object appear as a transparent cloud. Assign this surface style to the sphere around the bulb using the Color tool (figure 9-68).

6. To create a glowing cone of light that emanates from the flashlight, you will need to model a solid cone coming out of the flashlight and then give it the Glow surface style you created in the previous step. Change the reference plane to ZX. Using the Circle tool and the Converged Solid tool, create a long cone terminating inside the flashlight, as in figure 9-69. Assign the Glow texture to this object using the Color tool. You can use the Edit Cone Of Vision windows under the View pull-down menu to position lights.

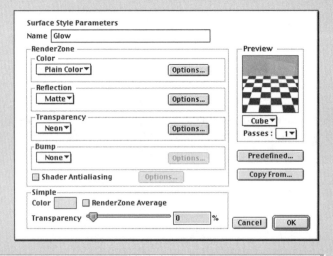

Fig. 9-68. Surface Style parameters for the glowing texture.

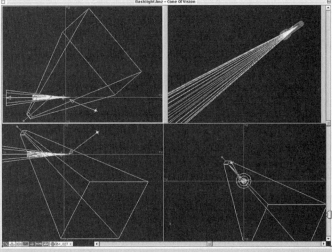

Fig. 9-69. Use the Edit Cone of Vision windows to position the lights.

Creating the Lights

If you render your model now it may begin to look like an operating flashlight, but it will not be convincing until you add the final touch, the lights! In this last phase you will add some lights to create a convincing scene.

1. Create a Directional light to be the ambient light in the scene. Double click on its icon in the Lights palette. This will open the Light Parameters. Because the primary light source should be the flashlight itself, shadows created from another light source would be confusing. For this reason, you should turn off the Shadows for this Directional light. In addition, you should turn down its intensity by setting Basic Intensity to 70%.

2. The next light to add to the scene is a Cone light. This will be the actual light of the flashlight. Click on an empty space in the Lights palette to create a new light. Double click on its icon to open the Light Parameters and select Cone from the Type pull-down menu. Turn on the Shadows for this light and in the parameters for the Shadows set the Outer Angle to 15° and the Inner Angle to 10°. This will make your cone light a narrow beam. Position this light at the center of the light bulb, shining out.

3. A third light is necessary to give the appearance of a bright point at the center of the light bulb. Create a new light and set this light's type to Point. Leave the Shadows off, as it is not necessary to have this light cast shadows. It will also reduce rendering time to have only one shadow-casting light. Change the Point Light Radius option to 1" to create a concentrated glow around the light bulb and add to the dramatic effect. Position this light to be at the center of the light bulb.

4. The last light to edit is the Sun (Light 1). The Sun is the default light that is automatically in your form•Z scene. Double click on its icon to open the Light Parameters dialog box, and make sure that Shadows are turned off. This light should be positioned to shine on the glowing cone and give it illumination. The glowing cone needs a light source shining at it from the site to give it a glowing quality. You may need to test several lighting positions until you get the desired effect.

5. You will need to change the light positions, depending on how you view the model. Figure 9-69 shows the positions of the view and the lights for the final rendering (figure 9-70). If you try a rendering looking from behind the flashlight, as in figure 9-55, you will need to turn off the point light because it will add too much light to the scene and "wash out" any objects in view.

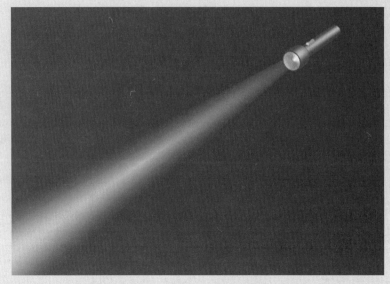

Fig. 9-70. Finished flashlight.

6. Place an object in the path of the Cone light to receive light and shadows. It will appear to be illuminated by the flashlight. Now render your model. Use RenderZone and be sure to turn on the Shadows, Transparency, and Reflectivity options. To best emphasize the glow of the flashlight, set the Project Color to black in the RenderZone options. Figure 9-70 shows the rendered flashlight.

Further Exercises

The following illustrations show the exercise models contained on the companion CD-ROM. For instructions on using the CD-ROM, see the Introduction.

Further Exercises

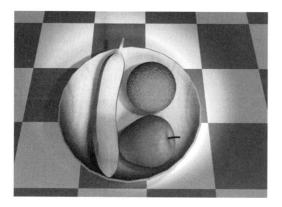

Fruit Bowl

Rowboat

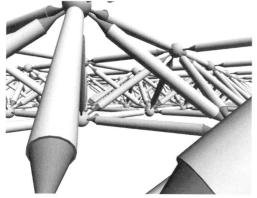

Toothpaste Tube

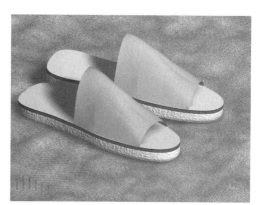

Sandals

Space Frame

Construction System

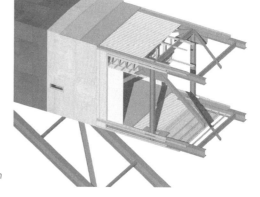

Chapter 9: Advanced Exercises

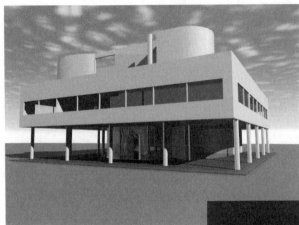

La Villa Savoye, Model Analysis

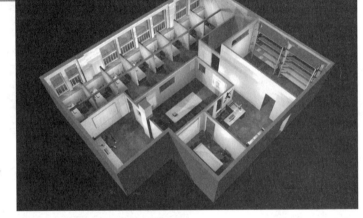

Complicated Model

Office Interior

part 3

Rendering, Drafting, Animation, and Tips

chapter 10

Rendering

Introduction

One of form•Z's greatest strengths is its rendering capabilities. form•Z has a variety of rendering options (available under the Display menu), which you can select from depending on the level of detail and realism you require in the view of the model. Among these rendering options, RenderZone stands out as the most sophisticated rendering tool, allowing you to easily produce photorealistic rendering of materials and lighting environments. Even though RenderZone is easy to use, it is rewarding to take the time to master its intricacies and subtle effects.

RadioZity is yet another powerful feature of form•Z that allows the accurate scientific calculation of lighting for environments to simulate realistic effects, as well as analysis of lights on materials in a model. A simple example model with a collection of rendering features has been created to graphically illustrate and guide you through the RenderZone options, manipulation of lighting and shadows, and the features of RadioZity. This section is organized so that you can grasp the various rendering options by simply comparing the images. Figure 10-1 shows the example scene rendered with RenderZone.

Fig. 10-1. Scene rendered with RenderZone display.

Rendering Display Types

There are a wide variety of display types available in form•Z under the Display menu. Each option has its own set of features and limitations. The following is a brief description of each display type and its features, and when you would most likely use each display type.

Wire Frame

This display option shows all lines that constitute a model, as if each component were completely transparent. This is useful when you want to see how models or components relate to each other or determine where correction is necessary after a tool or a command is executed. This is the fastest display option. It is useful during the modeling phase because it allows unobstructed views of all elements in a scene, as well as offering the fastest feedback after a command has been performed. Wire Frame is also the only mode that shows non-rendering modeling elements, such as points, lines, lights, and cameras. Figure 10-2 shows a scene displayed in Wire Frame.

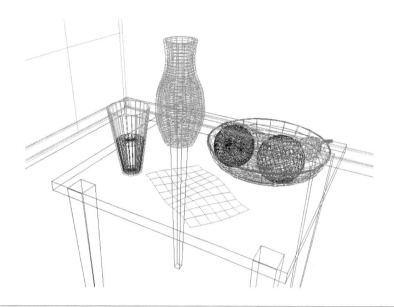

Fig. 10-2. Scene in Wire Frame display.

Quick Paint

This display option provides you the quickest preview of how a model will look with solid properties. However, it is often inaccurate when trying to resolve intersections between objects (figure 10-3).

Fig. 10-3. Scene rendered with Quick Paint display.

Hidden Line

The Hidden Line display option eliminates the back faces of objects and displays only lines in the model that would be visible to the viewer, giving the appearance of solidity. It is useful for generating projection and axonometric views of a model that are to be taken into the drafting module for layout and dimensioning (figure 10-4).

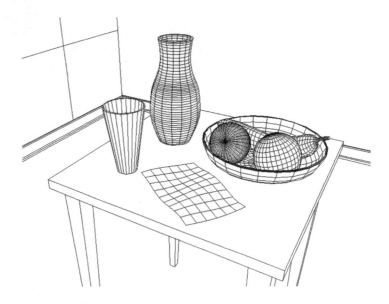

Fig. 10-4. Scene rendered with Hidden Line display.

Surface Render

This display option uses an algorithm that renders surfaces of a model from farthest in the particular view to closest, with the closer surfaces covering whatever is behind them. When used in conjunction with its shadow casting option, this option is useful to architects who want to create simple massing and daylight shadow studies of buildings. In figure 10-5, the scene was rendered with the Show Color option off and the Render With Shadow option turned on.

Shaded Render

This option produces high-quality, accurate renderings but lacks some of the sophisticated features of the RenderZone display option, such as materials, textures, reflections, depth, and background effects. It is useful for checking the integrity and placement of objects in a scene, as well as the scene's lighting effects, without wasting valuable time calculating realistic material properties of the objects (figure 10-6).

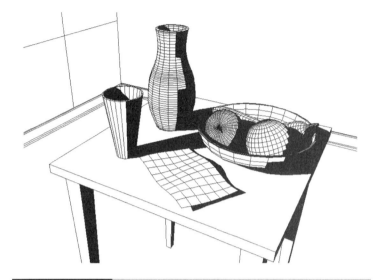

Fig. 10-5. Scene rendered with Surface Render display with shadows, but without color.

Fig. 10-6. Scene rendered in Shaded Render display.

RenderZone

RenderZone is the full-featured rendering option that provides realistic renderings of the highest quality in form•Z. Detailed descriptions of its features and options are provided in the section "RenderZone Options." Figure 10-7 shows a scene rendered with RenderZone.

Fig. 10-7. Scene rendered in RenderZone display.

QuickDraw 3D/OpenGL

Both QuickDraw 3D (Mac) and OpenGL (Mac, PC) are display options that allow the computer to rapidly display 3D rendered scenes through the use of algorithms that take advantage of hardware acceleration video cards that are either QuickDraw 3D or OpenGL enabled. They allow for fast interactive manipulation of the scene while the scene remains rendered. This is a very useful tool for previewing and composing a 3D scene without waiting for time-consuming renderings, presentations, and interactive walk-throughs using the Window tools. QuickDraw 3D and OpenGL can also be used to preview an animation, as discussed in Chapter 12. Basic options are shown in figures 10-8 and 10-9. Figure 10-10 shows a scene rendered with QuickDraw 3D.

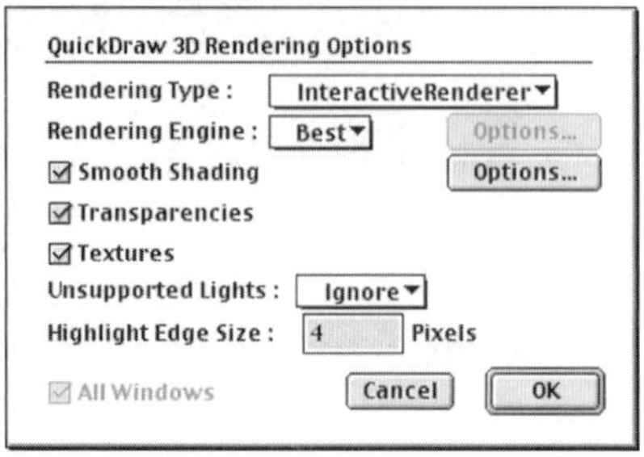

Fig. 10-8. QuickDraw 3D Rendering Options dialog box.

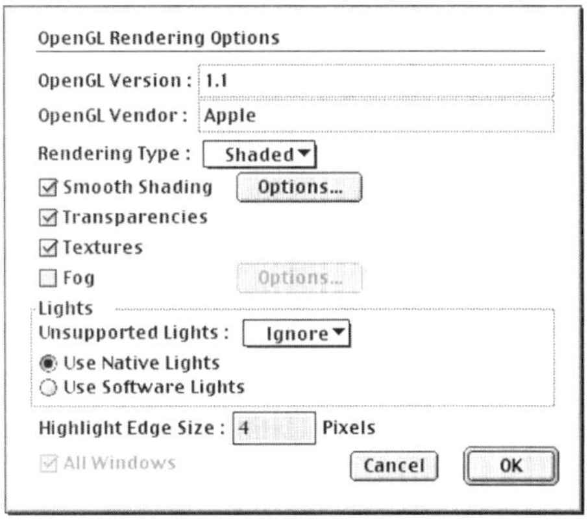

Fig. 10-9. OpenGL Rendering Options dialog box

Fig. 10-10. Scene rendered in QuickDraw 3D display.

RenderZone Options

The beginner, after the thrill of producing a first rendering with materials and lighting in RenderZone, may find the RenderZone Options dialog box (figure 10-11) a little intimidating because it appears to contain a great number of options to master. However, do not be intimidated; there are really only a few major options that need to be adjusted to produce stunning effects. Other advanced options are used to add subtle details to a scene. The sections that follow will take you step by step through most of the options in RenderZone.

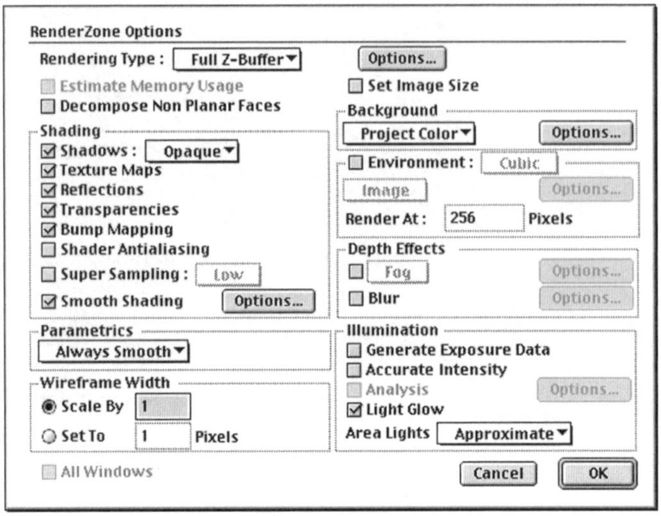

Fig. 10-11. RenderZone Options dialog box.

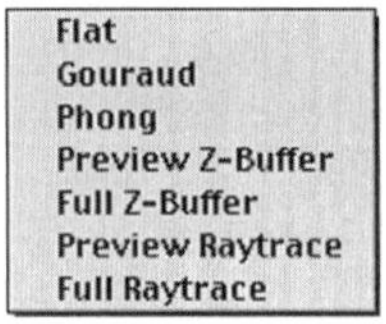

Fig. 10-12. Rendering Type pull-down menu.

Rendering Types

The Rendering Type pull-down menu is located at the top of the RenderZone Options dialog box (figure 10-12). It is important to know when and how to use the various rendering types offered in RenderZone. There are seven Rendering Types and the lower down an option appears in the Rendering Type menu, the more rendering features it contains and consequently the longer it will usually take to produce a rendering. The Full Raytrace option represents the highest-quality rendering type and the most time consuming to employ.

Flat, Gouraud, and Phong

The Flat option provides only the most basic rendering of the component shapes of objects, without any rendering of materials or smoothing. The Gouraud and Phong options differ only in that they provide some level of smooth shading to smooth out facets on an object's surface. These options can be used to generate renderings quickly, when material designation is not an essential part of the rendering. They are especially useful for viewing modeled objects and their arrangement without the addition of surface styles or lights that may prove distracting. Figures 10-13, 10-14, and 10-15 show scenes rendered with these three options.

RenderZone Options

Fig. 10-13. Scene rendered with Flat rendering type.

Fig. 10-14. Scene rendered with Gouraud rendering type.

Fig. 10-15. Scene rendered with Phong rendering type.

Z-Buffer and Raytrace

Full Z-Buffer and Full Raytrace provide high-quality renderings with the full set of RenderZone features. The two differ in the way they determine the visibility of a surface and in the way they anti-alias an image. These differences are in many cases minor and unnoticeable. Full Raytrace takes more time to render, but produces sharper textures, reflections, and shadows, which provide a more accurate rendering of texture-mapped and transparent objects (figures 10-16 and 10-17). Z-buffer, although not as accurate as Full Raytrace, can be custom configured to produce a range of antialiasing effects through the use of the Shader Antialiasing option in RenderZone. Refer to the section on Shader Antialiasing in this chapter for additional information.

> **NOTE:** *Antialiasing is the smoothing of jagged edges that appear in a scene. In figures 10-16 and 10-17, the advantages of Full Raytrace over Full Z-Buffer can be most easily recognized in the text quality on the letter and the antialiasing of shadow edges.*

Preview Z-Buffer and Preview Raytrace are used to render a quicker preview of a scene, at a lower quality than Full Z-Buffer and Full Raytrace. These two options provide you with most of the options available for Full Z-Buffer and Full Raytrace.

Fig. 10-16. Scene rendered with Full Z-Buffer rendering type.

Fig. 10-17. Scene rendered with Full Raytrace rendering type.

Shading Options

The Shading options box is used to turn on or off surface style properties and rendering effects on a global level in a scene when the image is rendered. The following sections describe shading options and the effects achieved with them.

Shadows

The Shadow option is used to globally switch on or off shadows in a rendered scene. Toggling this option will only produce noticeable changes in your scene if you have lights that are shadow casting. The shadow casting status of lights can be set in the Lights palette. In the Lights palette, each light can be individually adjusted for lighting options as well as for shadow type. Adding shadows to a scene enhances its 3D depth, especially in a perspective view (figures 10-18 and 10-19). See the "Lighting and Shadows" section in this chapter for more details.

Fig. 10-18. Scene rendered without Shadows option selected.

Fig. 10-19. Scene rendered with Shadows option selected. When the Opaque option is selected, shadows will always be rendered in black.

The Transparent option will allow the shadows to inherit the color of the object that produces the shadow when the object has transparent properties (figure 10-18 and 10-19). See the section "Lighting and Shadows" for a more detailed description of these features.

Texture Maps

This option allows you to turn off surface style properties that are based on imported images. Color image maps, transparency maps, and bump maps that use images as their sources will be rendered in an average color (figures 10-20 and 10-21). This is a useful option for saving time when previewing a scene because texture maps can substantially increase rendering time, especially when the image files are large. The realistic effects on the apple and banana in the following illustrations were achieved though the use of texture maps. Refer to the Fruit Bowl exercise in Chapter 9 for instructions on how to produce realistic texture maps on objects.

Fig. 10-20. Scene rendered without Texture Map option selected.

Fig. 10-21. Scene rendered with Texture Map option selected.

Reflections

The Reflection option toggles on and off the reflectivity of objects that have reflective surface style properties. Figures 10-22 and 10-23 show the effect of this option on the mirrored reflection in the metallic vase and glass.

Fig. 10-22. Scene rendered without Reflections option selected.

Fig. 10-23. Scene rendered with Reflections option selected.

Transparencies

Objects that have a transparent surface style assigned to them will be rendered only if the Transparencies option is selected. Figures 10-24 and 10-25 show a glass of water rendered with and without the Transparencies option selected.

Fig. 10-24. Scene rendered without Transparencies option selected.

Fig. 10-25. Scene rendered with Transparencies option selected.

Bump Mapping

Bump maps can be added to a surface style to conveniently produce 3D texture effects on the surface of a model without having to model them. Textures are specified in the Bump option of the Surface Style Parameters dialog box. In figures 10-26 and 10-27, you can see the rough bump map applied to the orange, making it appear more realistic. There is also a bump map that creates the wood plank pattern on the table.

Fig. 10-26. Scene rendered without Bump Mapping option selected.

Fig. 10-27. Scene rendered with Bump Mapping option selected.

Shader Antialiasing

Rendered objects sometimes produce jagged lines, especially around their edges and in their texture maps. Antialiasing smoothes these jagged edges by softening the color transition between edge pixels and background pixels. This creates the perception of smoothness and makes the rendering look less harsh and consequently more realistic. The Shader Antialiasing option in RenderZone gives the user a high level of control over the way individual objects in a scene are antialiased.

This option is used in conjunction with the Shader Antialiasing option in the Surface Style Parameters dialog box. The Shader Antialiasing option only works with the Full Z-Buffer rendering type and provides additional control over texture-based anti-aliased effects. By default, Full Z-Buffer will only anti-alias the pixels along the edge of objects.

Figure 10-28 shows the Shader Antialiasing Options dialog box, accessed through the Surface Style Parameters dialog box. There are three levels of Quality that can be assigned for each category. The higher-quality settings will result in better image quality but longer rendering time.

Fig. 10-28. Shader Antialiasing Options dialog box.

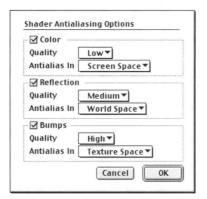

Note that the Shader Antialiasing option in RenderZone only applies to those objects that have the Shader Antialiasing option set in the Surface Style Parameters dialog box for their surface style. Shader Antialiasing option is useful in eliminating moiré patterns that often occur when fine-checkered patterns are rendered. Figures 10-29 and 10-30 show a close-up of the fruit bowl with the Shader Antialiasing options toggled on and off.

Fig. 10-29. Scene rendered without the Shader Antialiasing option selected.

Fig. 10-30. Scene rendered with the Shader Antialiasing option selected with Full Z-Buffer rendering type.

Super Sampling

Super Sampling increases the overall quality of the rendered scene by over-sampling each pixel in an image. Over-sampling is a process in which an image is rendered at a higher resolution than specified to achieve better detail. This typically produces better antialiasing in an image and is especially useful in eliminating moiré patterns and flickering in animations. When the Low option is selected, there is a 2 by 2 over-sampling of each pixel. Medium and High options have over-samplings of 3 by 3 and 4 by 4, respectively.

Although Super Sampling produces superior results, it will drastically increase rendering time. In the High Super Sampling option, for example, the rendering needs to produce 16 times more pixel and hence the rendering will take up to 16 times longer to render! Figures 10-31 and 10-32 show scenes rendered with no Super Sampling and with Super Sampling set to High.

Smooth Shading

All objects in form•Z are generated with facets. In general, the more facets, the smoother the model will appear during rendering. You can economize by creating an object with a minimal number of facets and then using the Smooth Shading option to smooth out its facets, thus giving the impression of a smooth surface (figures 10-33 and 10-34).

> **NOTE:** *Although smooth shading will enhance the smoothness of an object's faces, it does not alter an object's actual geometry. The rim of the fruit bowl in figures 10-33 and 10-34 clearly illustrates this point.*

Fig. 10-31. Scene rendered with no Super Sampling.

Fig. 10-32. Scene rendered with Super Sampling set to High.

Fig. 10-33. Scene rendered without Smooth Shading option selected.

Fig. 10-34. Scene rendered with Smooth Shading option selected.

You can manually set an object to be smooth shaded by using the Smooth Shade tool in the Attributes tool palette. The Smooth Shading Options in RenderZone allow you to globally adjust which objects, whether or not they have smooth shading properties assigned, are smooth shaded. This is important because you may sometimes want certain objects to be faceted when rendered.

Parametrics

This option allows you to globally adjust how objects defined by parametric surfaces are shaded. You have the choice of the Use Smooth Shading option from RenderZone, Always Smooth, or Use Smoothing Attribute (which allows the Smoothing Attribute to be individually assigned to each object). Objects with parametric properties include objects defined in the Primitives tool palette as well as Nurbs, Patches, and C-Meshed objects.

Wireframe Width

You can set an object to Render As Wire Frame by using the Render Attributes tool in the Attributes tool palette. There are two settings in the RenderZone options with which you can globally adjust the line width of objects set to Render As Wireframe. You can adjust the Wire Frame Width either to a relative scale or an absolute pixel width. Figure 10-35 shows an object rendered as wireframe with a 2-pixel line width.

Fig. 10-35. Scene rendered with an object wireframe width of 2 pixels assigned to the glass.

Fig. 10-36. Background options pull-down menu.

Background

Using the Background option (figure 10-36), you can introduce a variety of background effects and images into a model. Of the many Background options, those worth special mention are the Project Background, Image, Environment, and Alpha Channel options.

Project Background

In the Project Background option, you can set the rendering to show the grid and axes by selecting the Render Grid And Axes As 3D Lines option (figure 10-37).

Image

The Image option, under the Background pull-down menu, allows you to import an image file as a background to a model (figure 10-38). After importing the image, you can use the Match View tool to match the model to the perspective of the image. Refer to the Match View tool in Chapter 6 for more details on how to achieve this effect.

Fig. 10-37. Scene rendered with Project Background selected under the Background pull-down menu with the Render Grid And Axes As 3D Lines option selected.

Fig. 10-38. Scene rendered with Image option selected under Background.

Environment

This option is grayed out until the Environment section is selected with either the Cubic or Spheric option. The Environment option will render the background assigned in the Environment section. A background environment is projected onto the interior of a cube or sphere surrounding the model. The advantage of using this option becomes apparent when the view of the model changes. The background image will shift with the view according to what is visible on the environment cube or sphere.

This option is especially useful when creating an animation wherein the view of the background needs to move along with the movement of the camera. Setting up the environment is discussed in "Environment" section that follows.

Alpha Channel

The Alpha Channel option allows you to save a rendering with alpha channel information. The alpha channel retains the transparency information of a model when you save the file as an image file. Image file formats that support alpha channels are TIFF, PICT, Targa, and GIF. You can later use the alpha channel to place the model into an image using an image manipulation program such as Photoshop, keeping all transparencies in the model intact.

Environment

This section in RenderZone allows you to set up reflection and background environments. An environment is an image or a texture map projected onto the interior of a cube or a sphere so that reflective objects in the scene will not only reflect other objects but the simulated environment surrounding the scene. As mentioned in the previous section, once an environment has been set up, it can also be used as a background.

When the Cubic option is selected, the reflective objects in the scene will reflect an environment mapped onto an imaginary cube surrounding the model. The images mapped onto the six interior sides of this cube are loaded by pressing the Options button. This opens the Environment Maps dialog box (figure 10-39). Here, you can load the six individual images that depict the surrounding environment. Figure 10-40 shows the vase reflecting the cubic environment with the imported image maps.

RenderZone Options

Fig. 10-39. Environment Maps dialog box.

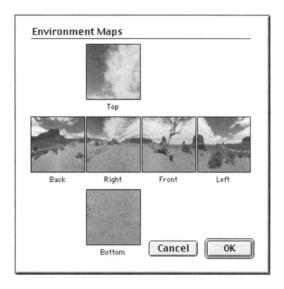

Fig. 10-40. Scene rendered with Cubic Environment showing imported image maps.

Spherical Environment is similar to Cubic Environment except that Spherical Environment maps an image onto an imaginary sphere. An image can be loaded for this option in the same way as a cubic environment. When the Options button is pressed, the Spherical Environment Image Options dialog box (figure 10-41) opens. Figure 10-42 shows a scene rendered with the Spherical Environment option.

Fig. 10-41. Spherical Environment Image Options dialog box.

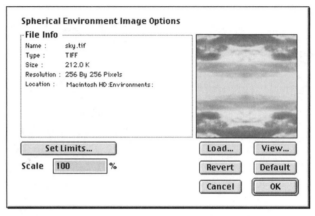

By default, the image map is wrapped around the sphere so that is extends once around the equator and from one pole to the other. This mapping method creates substantial distortions as the map extends toward the poles. The distortion is least at the equator. If the scene or the animation does not need to view the entire sphere up to the poles, the Set Limits option in the Spherical Environments Image Options dialog box allows you to adjust the mapping of an image around the sphere.

Fig. 10-42. Scene rendered with Spherical Environment selected for Reflections.

In the Environment Image Limits dialog box (figure 10-43), the Horizontal setting adjusts where the image map starts and ends and its rotation along the equator. Vertical adjusts how far the image map extends toward the poles.

When the Rendered option is selected, objects with the Reflection option set to Environment in Surface Style Parameters will have a simulated reflection of the scene. The reflection appears as though it were projected onto an imaginary cube that surrounds the model. This option reduces rendering time, but at the cost of accuracy of the reflections. Figure 10-44 shows a scene rendered with Rendered Environment.

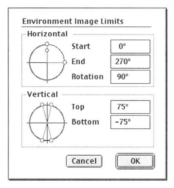

Fig. 10-43. Environments Image Limits dialog box.

Fig. 10-44. Scene rendered with Rendered Environment selected in Reflections.

Depth Effects

The following sections discuss depth effect options available in form•Z. Depth effects allow you to enhance the three-dimensionality and realism of an image by mimicking natural phenomena such as atmospheric perspective, focus, fog, and snow.]

Depth Cue

Depth Cue uses the Hither and Yon clipping planes to gradually blend the model into a specified color, thus achieving an impression of depth. In the Depth Cue Options dialog box, the mix and intensity of the blend can be adjusted (figure 10-45). When using Depth Effects, the Hither and Yon clipping planes should be adjusted in Edit Cone Of Vision prior to using these effects.

Fig. 10-45. Scene rendered with Depth Cue option selected.

Fog

The appearance of fog can be simulated in RenderZone with the Fog option. The Fog Options dialog box allows you to create and control the fog density and color. The farther the Distance From Hither, the weaker the fog effect becomes. The Background Mix controls the amount the color of the depth effect will be mixed with the background (figure 10-46).

Fig. 10-46. Scene rendered with Fog option selected.

Snow

The Snow option generates simple snowflakes in the scene that have an aspect of depth associated with them (figure 10-47). The Near/Far Scale option adjusts the size of the flakes from closest to the farthest from the viewer. The Flake Density and Flake Color settings adjust the density of the snowfall and the color of the flakes. Noise and # Of Impulses control the level of randomness of the flakes.

Fig. 10-47. Scene rendered with Snow option selected.

Blur

This option adds a depth of field to your rendering, blurring out objects too close to or too far from the designated range, much in the same way that a camera may blur out areas that are not in focus (figure 10-48).

Fig. 10-48. Scene rendered with Blur option selected.

Illumination

The following sections describe options used to analyze and make adjustments to the illumination in the scene being rendered.

Generate Exposure Data

This option is used in conjunction with the Correct Exposure When Present option in the Image Options dialog box under the Display menu. When the Generate Exposure Data option is selected, it generates exposure data that can be used to manually correct the exposure of the scene. You do this in the Image Exposure Options dialog box, which is accessed through the Options button in the Image Exposure Options dialog box (figure 10-49).

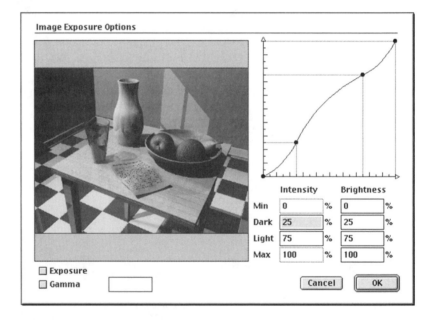

Fig. 10-49. Image Exposure Options dialog box.

Use Accurate Intensity

This option gives you the choice of using the Basic Intensity or Accurate Intensity for the lights in the scene. This is set for each light in the Light Parameters dialog box. Basic Intensity is expressed as a percentage, whereas Accurate Intensity is expressed in scientific units. If a radiosity solution for the scene exists, accurate intensities will always be used.

Analysis

This option is used to analyze the light in a model. This is done by rendering only the light intensity in the scene through a system of color coding. This analysis can be used to see if lighting criteria are met in a given space. The analysis can be performed either with a radiosity solution or by using the Use Accurate Intensity option (figure 10-50).

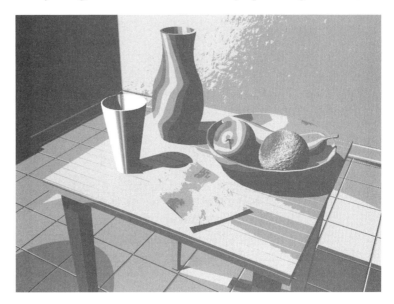

Fig. 10-50. Scene rendered with Analysis option selected.

The Illumination Analysis Options dialog box (figure 10-51) is opened when the Options button is clicked next to the Analysis option. Color/Value shows an editable list of the Lux value and the corresponding color used in the analysis rendering.

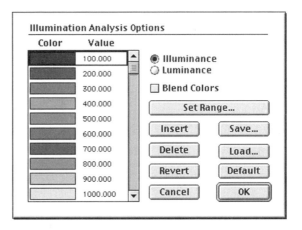

Fig. 10-51. Illumination Analysis Options dialog box.

Illuminance is the amount of light intensity received on a square unit area measured in Lux. Luminance is the amount of light intensity reflected by a surface, also measured in Lux. Illuminance is commonly used to determine the appropriate lighting conditions for a space, whereas Luminance is used to determine areas of glare.

The Blend Colors option will create smooth transitions between the colors in the rendering. The Set Range option allows you to customize the minimum, maximum, or increment of the Illuminance/Luminance values.

Light Glow

This option allows you to globally turn on or off the light glow properties of all lights in the scene. Light Glow can be set individually for each light in the Light Parameters dialog box. For more information about Glow, refer to the "Lighting and Shadows" section that follows.

Area Lights

This option provides three choices as to how area lights are handled in the scene. The Ignore option bypasses all area lights in the scene and renders the scene without them. The Approximate option substitutes area lights with point lights. The Use option will use the available area lights in the scene. This option will increase the rendering time substantially.

Lighting and Shadows

The Light Parameters dialog box (figure 10-52) is accessed by double clicking on the name of the light in the Lights palette. The Light Parameters dialog box contains various options that control the parameters of the specific light named at the top of the window. The following sections explain the various Light Parameters options.

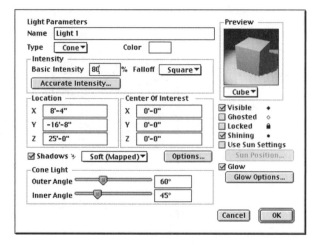

Fig. 10-52. Light Parameters dialog box.

Name

New lights can be created either in the Lights dialog box or by clicking in a blank space in the Lights palette. All lights are given a default name of "Light N," where N is the next number that has not been taken by another light. This name can be changed in the Light Parameters dialog box. When you make a copy of any light in the modeling window, form•Z will automatically rename the copied lights by adding the word *Copy* and a unique number after the original name to each additional light. Note that any changes to the original light are not reflected in the corresponding copies; that is, each light must be modified individually.

Light Types

The following sections describe the options available for lighting types. These include Distant, Point, Cone, Projectors, Area/Custom, and Point Light/Radius.

Distant

The default Type setting for a new light is Distant, which is primarily used to mimic the sun. The simplest type of light to use, the Distant light casts constant light with parallel light rays over the entire model, regardless of how close or far away the light is from the scene. In other words, its Intensity remains constant throughout a scene (figure 10-53). The most important factor governing how a Distant light will illuminate your model is its orientation relative to your model, which can be changed in a variety of ways.

Fig. 10-53. Scene rendered with Distant light.

One thing to remember is that although you may turn any light on or off, form•Z requires that one of your lights be designated as the Sun and remain a distant light. This is the only light used by Quick Paint and Surface Render to illuminate a scene.

Point

The point light is much like an exposed light bulb because it casts light in every direction from a single point in space (figure 10-54). This makes it a quick solution for illuminating interior spaces. Unlike the Distant light, the Intensity of a Point light can be set to decrease over a distance, allowing it to behave more like an actual point light source (e.g., a candle or a light bulb). This behavior is called Falloff. Refer to the "Falloff" section that follows for more details on this option.

Fig. 10-54. Scene rendered with Point light.

Cone

The Cone light emits light in a cone-shaped beam (figure 10-55). This cone can be adjusted with the slider controls at the bottom of the Light Parameters dialog box. In addition to controlling the overall angle of the light cone, the Inner and Outer Angle settings control the transition of the edges of the light's beam from light to dark. The closer the angles are to each other, the sharper the beam and the shadow will be. Similar to Point lights, Falloff can be adjusted to control the decrease in intensity over a distance.

Fig. 10-55. Scene rendered with Cone light.

Projector

The Projector light allows you to project an image of your choice over a model much like a slide projector (figure 10-56). Selecting Projector in the Type menu invokes the Projector Light options at the bottom of the Light Parameters dialog box. The options in this dialog box control the Angle of the projected light pyramid and the Spin or rotation of the image being projected. Clicking on the box containing the default image invokes the Projector Map Options dialog box, which allows you to Load an image to be used with the Projector light.

Fig. 10-56. Scene rendered with Projector light.

Area/Custom

Area and Custom lights, originally designed to be used solely for radiosity solutions, can now be used without radiosity. An Area light is a light associated with a physical object whose surfaces are emitting light, such as neon and florescent lighting. Refer to the RadioZity section for a discussion of how Area lights are used.

Color

Clicking in the Color box allows you to adjust the color of the light source. See the "Tips and Getting Best Results" section for tips on light color.

Intensity

The Basic Intensity is what all of the rendering methods (except radiosity solutions) use to determine how much light a light source is emitting. The default Basic Intensity setting for any new light is 100%. After entering your desired Basic Intensity, press the Tab key or click in another field to update the Preview image. Although the Preview image estimates the light's current Intensity, it is best to monitor your light adjustments by doing quick test renderings in the model window. See "Setting Up Quick Rendering Previews" in the "Tips and Getting Best Results" section of this chapter.

The Accurate Intensity button is designed for adjusting lights to be used in generating a radiosity lighting solution. However, Accurate Intensity settings can be used in a non-radiosity rendering if the Accurate Intensity option is selected in the RenderZone dialog box. Refer to the RadioZity example for a discussion of how the Accurate Intensity options are used.

Falloff

Falloff determines how the Intensity of the light will decrease as the distance between the light source and objects increases. Falloff is an extremely useful option in that it gives you a high level of control over the effect of lights. See the "Getting Best Results" section of this chapter for tips on using Falloff to quickly create realistic lighting conditions. This option is only available for Point, Cone, and Projector lights. The Constant option makes a light's Intensity independent of distance and will light objects equally, regardless of the distance from the light (figure 10-57).

Fig. 10-57. Cone light with Constant Falloff.

The Linear option decreases the Intensity of a light linearly with distance. For instance, if object A is twice as far from a light source as object B, object A will receive half the light object B receives (figure 10-58).

Fig. 10-58. Cone light with Linear Falloff.

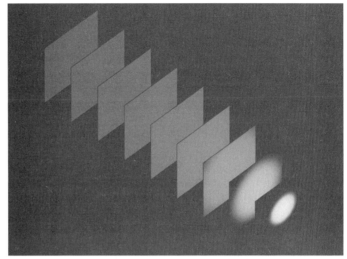

The Square option decreases the Intensity of a light exponentially with distance. For instance, if object A is twice as far from a light source as object B, object A will receive a quarter of the light object B receives. This option most closely simulates how a real light behaves (figure 10-59).

Fig. 10-59. Cone light with Square Falloff.

➤ **NOTE:** *When Linear or Square Falloff is selected, Intensity within the light increases above the given percentage and objects within the light cone or light radius may be overexposed. If this effect is not desired, you can decrease the size of*

the light cone or radius, lower the overall intensity of the light, or revert to a Constant Falloff.

Location/Center Of Interest

The Location determines the position of the light source in space. The Center Of Interest determines the position where the light is directed and is only available with Distant and Cone lights. In the modeling window, the Location of the light source is denoted as a solid circle with a cross hair and the center of interest denoted as an arrowhead. This position can either be entered numerically using x,y,z coordinates in the Light Parameters dialog box or positioned graphically in the modeling window. You can also use Edit Cone Of Vision to adjust the Location of the light source and the Center Of Interest.

Point Light/Radius

The Radius option is only visible when a Point light is selected in the Type menu. The value entered here (with a default setting of 10) determines the size of the point light sphere.

Shadows

In order to generate a rendering with shadows, it is first necessary for a light source to be designated as shadow casting. In addition, the Shadows option must be selected in the RenderZone Options dialog box. There are a number of options that can be used to adjust the quality of the shadows in the Lighting Parameters dialog box. These options are discussed in the following section.

Hard and Soft Shadows

In the Light Parameters dialog box, form•Z provides you with the option of having a light cast either Hard (raytraced) or Soft (mapped) shadows. Hard shadows are generally used to simulate shadows cast by a single distant source such as the sun and are ideal for exterior scenes (figure 10-60). Soft shadows, on the other hand, are better for simulating the multi-source lighting conditions of an interior space (figure 10-61). The level of blurring for Soft Shadows can be adjusted individually for each light in the Soft Shadow Parameters dialog box, accessed with the Options button in the Light Parameters dialog box.

Fig. 10-60. Scene rendered with Hard (raytraced) Shadows.

Fig. 10-61. Scene rendered with Soft (mapped) Shadows.

Keep in mind that Soft (mapped) Shadows require significantly more memory, depending on the number of shadow-casting lights and the size of the rendering. It is common for form•Z to run out of memory when trying to calculate the shadows in a scene. The following are suggestions for dealing with memory problems.

- Allocate more memory to form•Z (Mac).
- Decrease the number of Soft Shadow Casting lights.
- Use Hard (raytraced) Shadows.
- Turn off the Shadow attributes of large objects (e.g., ground planes).
- Decrease the quality and resolution of the Soft Shadows in the Light Parameters dialog box.

Opaque Versus Transparent Shadows

If you have light-casting Hard (raytraced) shadows, RenderZone gives you the option of casting either Opaque or Transparent shadows. Opaque Shadows, the default option, calculates the shadows of all objects as if every object were completely opaque, allowing no light to pass through (figure 10-62). Transparent Shadows takes into account the transparency of each object and calculates how much light actually passes through the object and how that light may change color as it does so (figure 10-63). If your model contains transparent materials such as glass or transparency maps, you may want to use Transparent Shadows for a more realistic effect, even though the scene may render more slowly.

◆ **NOTE:** *Transparent Shadows may only be cast by lights with the Hard (raytraced) Shadows option selected.*

Fig. 10-62. Scene rendered with Opaque Shadows.

Fig. 10-63. Scene rendered with Transparent Shadows.

Glow

The Glow option is available only for Cone, Point, and Projector lights. When this option is selected, the path of the light is rendered as a glowing volume when using Z-buffer or Full Raytraced rendering types. The Glow Options dialog box is divided into two sections: Simple and Accurate. The Accurate options are only accessible if the Falloff is set to Square in the Light Parameters dialog box (figure 10-64).

The Simple options box contains relatively simple methods for specifying glow parameters such as Intensity, Falloff, and Distance. Each of these settings is similar to the corresponding settings of the light itself (described previously). However, these Glow options affect only the glow appearance and have no bearing on the actual light.

The Accurate Options box contains settings similar to those in the Simple Options box but adds a Quality pull-down menu and Noise Options. A higher Quality setting yields better volumetric lighting at the expense of rendering time. The Noise option gives the effect of dust particles in the glow region. The appearance of the dust patterns can be controlled through the Size, Type, # of Impulses, Contrast, and Detail options. A scene rendered with Accurate Glow is shown in figure 1-65.

Fig. 10-64. Glow Options dialog box.

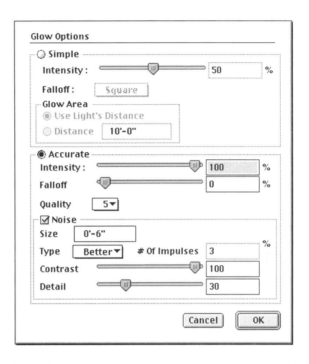

Fig. 10-65. Scene rendered with Accurate Glow option selected.

Radiosity

The following sections discuss radiosity-based rendering and the RadioZity feature. Radiosity is a remarkably realistic method of rendering based on the analysis of light reflections from surfaces. Radiosity excels in interior renderings and allows for soft, gradual shadows that cannot be achieved by ray tracing techniques alone. In form•Z, radiosity-based rendering is called RadioZity.

Lighting an Interior with RadioZity

Radiosity-based rendering provided by form•Z RadioZity is a powerful tool that allows for accurate simulation of lighting conditions. Radiosity takes into account the material qualities of objects and the nature of the light sources to reiteratively calculate light bouncing off surfaces. Reflected light is calculated independent of the location of the viewer, to generate a more accurate and convincing representation of ambient lighting conditions in a scene.

The following section introduces RadioZity by way of rendering an interior space containing the model used throughout this chapter. For an additional exercise in using RadioZity, refer to the interior lighting exercise in Chapter 9.

Radiosity can be rewarding when the final lighting condition is achieved, but may take a lot of testing, going back and forth between options and time-consuming renderings. This simple exercise is by no means comprehensive; rather, it provides an introduction to further work with radiosity-based rendering. Figures 10-66 and 10-67 show the same interior space rendered with RenderZone and RenderZone RadioZity.

Fig. 10-66. Scene rendered with Full Raytraced RenderZone.   *Fig. 10-67. Scene rendered with Full Raytraced RadioZity.*

Concepts of Radiosity

Radiosity is not a rendering tool but a calculation tool that subdivides the surfaces of a model into small polygons, called "patches." Radiosity calculates the amount of light absorbed and reflected by each of the patches to create more photorealistic lighting conditions when the scene is rendered. In order to render a scene with radiosity, it is necessary to perform a Shaded Render or RenderZone rendering after the radiosity solution has been generated.

The process radiosity uses to determine ambient reflected light consists of several steps. Radiosity first factors the lights from the primary sources (that is, the lights in the scene) and determines how each light hits each patch. It then uses the patch that reflects the most light and makes that a light source itself, called a secondary light source. The light from this source is then used again to determine how light hits each patch. The patch that receives the most light from this step becomes another secondary source. This becomes an iterative process that would go on forever if no limits were imposed on it.

Setting Up Renderings with RadioZity

The following is a quick run-through of the steps you need to render your model with RadioZity. You can perform these steps on the model provided on the companion CD-ROM in order to familiarize yourself with the various options.

In the Display pull-down menu (figure 10-68), there are four radiosity items: Radiosity Options, Initialize Radiosity, Generate Radiosity Solution, and Exit Radiosity. These items, in this order, represent the steps you need to take to generate a rendering with radiosity. Before executing any of these steps, however, the radiosity intensity of each light must be adjusted—that is, the Accurate Intensity setting of each light that has the most direct influence on how your radiosity rendering will look.

Fig. 10-68. Display menu with Radiosity commands highlighted.

Accurate Intensity

When calculating a radiosity solution, RadioZity ignores the Basic Intensity settings used by the typical rendering methods, using instead the Accurate Intensity settings obtained through the Lighting Parameters dialog box. All lights, except for Projector, have default Accurate Intensity settings.

✒ **NOTE:** *RenderZone can use the Accurate Intensity settings for previewing a scene without calculating a radiosity solution if the Accurate Intensity option is selected under RenderZone Options.*

Accurate Intensity is defined using either Radiometric or Photometric values (figure 10-69). Radiometric settings are expressed as watts and an efficiency percentage. Wattage is the measure of electrical power as denoted on most light bulbs. Efficiency is the percentage of that power that becomes visible light.

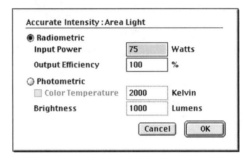

Fig. 10-69. Accurate Intensity dialog box for Point Light.

Photometric settings are expressed as lumens and temperature. Lumens define the total amount of light emitted by a source. A typical incandescent bulb of 60 watts has a value of about 730 lumens, and a 100-watt bulb has a value of about 1380 lumens. A candela is a unit of luminous intensity in a specific direction. Color temperature is measured in Kelvin and indicates how warm or cool a light will feel. Incandescent light bulbs have a value of about 2000 K and are considered "warm," whereas white fluorescent lamps have a value of about 3250 K and are considered "cool."

Two lights (figure 10-70) are involved in this example: a Distant light acting as the sun, and an Area Light that simulates the light from a window.

The Accurate Intensity of Distant Lights is set as Watts per Square Foot. In this example model, the Accurate Intensity of the Distant Light was set to 0.15 Watts/ft^2 (a suitable level determined through trial and error). This light will simulate the direct light from the sun streaming through the window onto the wall.

An Area Light is a light in the shape of an object or surface. It can be made visible or invisible and can cast light that is either the color of the object or the color selected in the Color field of the Light Parameters dialog box. An Area light is created by preselecting an object or surface and then creating a new light source. In the Light Parameters dialog box, the Type of light is then changed to Area. In this example, an Area light was created from a simple 2D plane that is the same size as the window. The Area light is set to be invisible by turning off the Render Light Object option. The Ignore Light Object Color option was selected and the Quality adjusted to Medium with the slider. The Accurate Intensity was set to 75 Watts. This light will simulate the daylight coming through the window.

Fig. 10-70. Axonometric view of model with one Distant and one Area light.

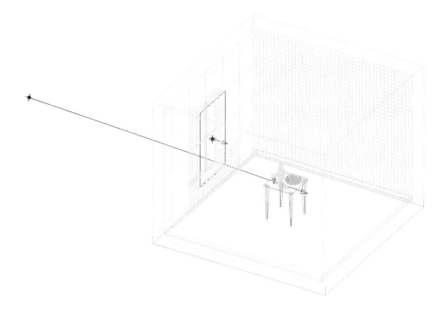

Radiosity Options

The Radiosity Options dialog box (figure 10-71) allows you to adjust the parameters of how radiosity subdivides surfaces and creates patches for the model. The settings in the Initialization box are used by the Initialize Radiosity menu command, whereas the settings in the Generation box are used by the Generate Radiosity Solution menu command.

Fig. 10-71. Radiosity Options dialog box.

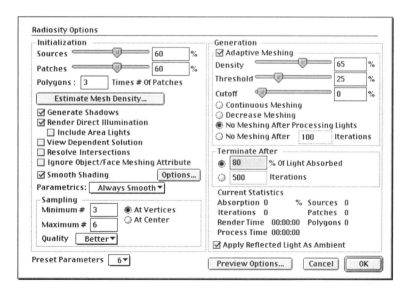

The easiest way to get started in using radiosity is to bypass the Radiosity options altogether and select one of the Preset Parameters located at the bottom of the dialog box. The Preset Parameters are a preconfigured set of options organized according to the quality of the radiosity solution. Simply select the quality level for your model. Bear in mind that the higher the quality, the more time will be spent calculating the mesh for the patches. After you get a scene set up correctly at a lower quality, you can increase the quality for final renderings or adjust any of the options described in the material that follows to individually "tweak" the solution to your liking.

In the Initialization settings of the Radiosity Options dialog box, the Sources slider controls the density of the mesh generated from Area light sources, whereas the Patches slider controls the mesh density generated from secondary sources. These two items remain fixed once the mesh has been initialized. The Polygons setting generates a multiple of additional subdivisions of the initial Area Sources and Patches as RadioZity processes the solution. Note that both the Patches and Polygons options can be set per object or face through the Rendering Attributes tool. The Estimate Mesh Density button will give you a rough estimate of the number of sources, patches, and polygons before a radiosity mesh is initialized.

The Generate Shadows option will incorporate shadows into the radiosity solution. Note that the lights must have their Shadows option turned on in their respective Light Parameters dialog box in order for shadows to be generated.

The Render Direct Illumination option is an extremely useful feature in that it allows crisp shadows to be created with a relatively light mesh. With the Render Direct Illumination option selected, form•Z splits up the illumination task by allowing the Raytrace and Z-Buffer algorithms of RenderZone to calculate direct illumination, and the radiosity process to calculate the secondary or ambient light.

By allowing each type of algorithm to do what it does best, you essentially get the best of both worlds, and a higher-quality solution in a shorter amount of time. If you look at the Radiosity Options dialog box for the example model in figure 10-71, you see that even with medium level settings a high-quality rendering with sharp shadows can be generated in a relatively short time. Figures 10-72 and 10-73 show scenes rendered with and without the Render Direct Illumination option.

The View Dependent Solution option makes the radiosity solution, which is typically view independent, into a view-dependent solution. In other words, radiosity typically computes a radiosity solution for the entire model, even areas that are not visible in the current scene. The benefit of the view independent method is that you can change the view you want to render at any time without having to compute another radiosity solution. By selecting the View Dependent Solution option, additional mesh subdivisions are not performed in areas not visible from the current view. This results in a shorter solution generation time. The drawback, of course, is that the optimal solution is calculated only for the current view.

Fig. 10-72. Scene rendered without Render Direct Illumination option selected.

Fig. 10-73. Scene rendered with Render Direct Illumination option selected.

The Resolve Intersections option is used to resolve shadow leaks that may occur when two objects are placed against each other. Shadow leaks are caused when there is not a sufficient level of information calculated to resolve accurate shadows (figures 10-74 and 10-75).

Fig. 10-74. Radiosity rendering with visible shadow leaks where objects sit on the table.

Fig. 10-75. Same scene with Resolve Intersections option selected.

The Ignore Object/Face Meshing Attributes option overrides any meshing parameters applied to an object or face through the Rendering Options tool in the Attributes tool palette. The Smooth Shading option determines which surfaces are smooth shaded when rendered. These settings override the settings in the Shaded Render, RenderZone, QuickDraw 3D, and OpenGL options when applied to a radiosity solution.

The Sampling Box is used by RadioZity to control the quality of the lighting environment. The recommended settings for Minimum/Maximum are between 1 and 5 for minimum and 10 and 20 for Maximum. These values are referenced by the Quality options menu when processing radiosity values. As usual, the trade-off for a higher-quality setting is a longer wait.

Generation Options

The Adaptive Mesh option subdivides the Patches into finer polygons according to the settings in the three sliders. The recommended value for the Density slider is between 70 and 80 percent for a high-quality rendering. Values above 80 percent may have detrimentally heavy mesh resolutions. You can also set radiosity Adaptive Meshing on a per object or per face basis through the Rendering Attributes tool.

Threshold compares the intensity of mesh resolutions between adjacent patches and, according to the threshold value, determines when subdivisions are generated in a patch. Lower threshold values will generate more subdivisions.

Cutoff suppresses subdivisions in darker areas of the scene because the details in these areas are less visible. Higher cutoff values produce lower mesh densities.

The Continuous Meshing option maintains the same parameters through the generation of the radiosity solution, whereas Decrease Meshing relaxes the parameters as the solution gets closer to termination. Continuous Meshing produces the highest quality but Decrease Meshing may be preferable because it speeds up the solution with little loss in quality. The No Meshing After Processing Lights option stops meshing after all primary sources have been processed. The No Meshing After n Iterations option stops the meshing after a set number of iterations, each iteration being the calculation of one source patch affecting the scene.

You can elect to terminate the radiosity solution either by the percentage of Light Absorbed or number of Iterations. The first option takes exponentially longer as larger values are selected. In general, a setting of about 85% produces a high-quality solution. The latter option can be used to control the time it takes to generate a solution because it gives more of an indication of how long the solution will take to generate (figures 10-76, 10-77, and 10-78).

The Current Statistics Area indicates the actual number of sources, patches, and polygons generated, as well as the time spent on processing and drawing the solution. The Apply Reflected Light as Ambient affects only how a solution appears while being generated. With this option selected, the scene will appear initially brighter than the complete radiosity solution. Conversely, with this option off, the scene will appear initially darker than the complete solution. Figure 10-79 shows the scene before the Initialize Radiosity command is executed.

Fig. 10-76. Termination set to 40% of Light Absorbed.

Fig. 10-77. Termination set to 60% of Light Absorbed.

Fig. 10-78. Termination set to 80% of Light Absorbed.

Fig. 10-79. Walls meshed prior to initializing Radiosity to force the radiosity mesh to be generated in a set direction to avoid jagged shadows.

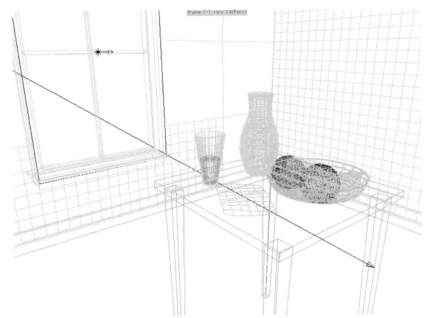

The entire wall on the right and certain regions of the wall on the left in figure 10-79 were meshed with the Mesh tool prior to starting radiosity. This is one way of controlling the orientation of irregular shadow patterns by forcing radiosity to generate its mesh along the pre-meshed orientation.

The Initialize Radiosity command under the Display menu generates mesh patches for all objects in a model based on the parameters designated in the Radiosity Options dialog box. At this point, form•Z enters radiosity mode and the model cannot be altered without exiting Radiosity. Lighting parameters may be adjusted, but the radiosity function will have to be re-initialized.

Generating a Radiosity Solution

The radiosity solution is generated (figure 10-80) by selecting the Generate Radiosity Solution menu item from the Display pull-down menu. Depending on the option selected in the Preview Options in the Radiosity Options dialog box, the scene is periodically updated with a preview showing the current solution. The rendering method used to preview the scene should be implemented before you select Generate Radiosity Solution. It is recommended that QuickDraw 3D or OpenGL be used, as they provide quick and relatively accurate previews of the image as the solution is generated. You can stop the generation of a solution midway through the Generate Radiosity Solution process and still retain the solution generated to that point by clicking on the Stop button.

Fig. 10-80. Radiosity solution generated.

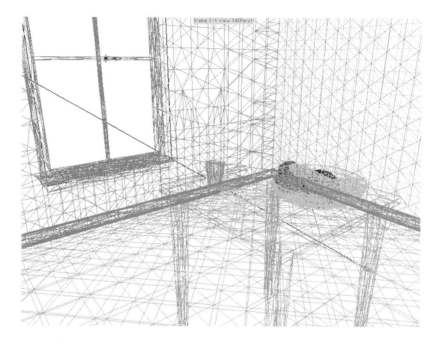

Once the solution is generated, you can render your model. The display options available are Wire Frame, Shaded Render, RenderZone, and QuickDraw 3D or OpenGL. You can save the solution by first making sure the Save Solution In Project File option is selected in the Radiosity Preferences dialog box (figure 10-81), accessed through the Preferences command under the Edit pull-down menu. Now you can save the form•Z file while in Radiosity mode and the radiosity solution will be saved along with the model.

Fig. 10-81. Radiosity Preferences dialog box accessed in Preferences under the Edit menu.

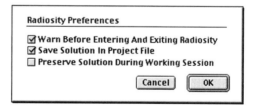

Exiting Radiosity

Selecting the Exit Radiosity menu item exits Radiosity mode. Once you have exited Radiosity mode, any rendering you perform will not use the radiosity solution to calculate the scene's lighting. If you exit Radiosity mode, you will have to repeat the process from the beginning to produce a radiosity solution and perform a rendering. It is recommended that you save your solution before exiting.

Tips and Getting Best Results

This section provides guidelines and hints that will help you achieve the best results when using RenderZone. The section is divided to provide specific hints in different aspects of the rendering process.

Handling Image Irregularities

The Decompose Non-planar Faces option at the top of the RenderZone Options dialog box automatically triangulates nonplanar faces that may otherwise produce irregularities during rendering. Selecting this option may represent an additional speed gain during rendering because form•Z does not have to check for nonplanar faces within the scene.

Using Quick Keys

As always, get into the habit of using quick keys. The Macintosh and Windows key combinations in the following table will have you working quickly in the RenderZone and Wire Frame display modes.

Function	MacOS	Windows
RenderZone	Command + K	Ctrl + K
Wire Frame	Command + W	Ctrl + W
Stop Rendering	Command + ,	Ctrl + ,

Image Options

The Image Options dialog box (figure 10-82) located in the Display pull-down menu is used to set the resolution shown in the modeling window and, more importantly, the resolution of the final rendering output when saved as an image file such as PICT or TIFF.

The Use Window Size option sets the image size as the resolution of your modeling window. Use Custom Size allows you to select a different image size for output. Once this option is selected, the image will be saved at this resolution regardless of what you may see on the screen. When the Maximize Window In Screen option is not selected, the window will match the actual resolution of the image. When the option is selected, form•Z displays the modeling window to fit the screen but saves the image at the resolution designated in the Use Custom Size dialog box.

Fig. 10-82. Image Options dialog box.

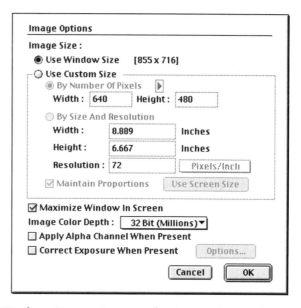

The Use Window Size option sets the image size as the resolution of your modeling window. Use Custom Size allows you to select a different image size for output. Once this option is selected, the image will be saved at this resolution regardless of what you may see on the screen. When the Maximize Window In Screen option is not selected, the window will match the actual resolution of the image. When the option is selected, form•Z displays the modeling window to fit the screen but saves the image at the resolution designated in the Use Custom Size dialog box.

Correct Image Exposure is used to make brightness corrections to an image after it has been rendered. See "Correct Image Exposure" in the "RenderZone" section of this chapter for more details.

form•Z Imager

form•Z Imager (figure 10-83) is a standalone program that can be used to batch-render views from any number of projects. The Imager application is also faster in rendering images because it saves computer resources by not displaying the image on the screen or offering modeling tools.

Fig. 10-83. form•Z Imager.

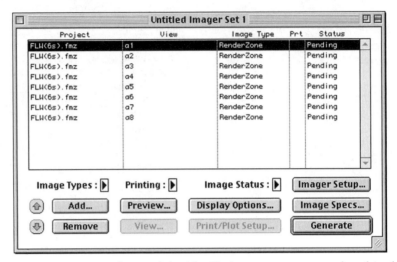

To use Imager, save the model with all views you want rendered in the Views palette. You can later load these views into Imager and assign the type of rendering, file format, and resolution for your views. This application is useful for performing time-consuming tasks, such as rendering a large number of views or large files, especially overnight.

Economizing Rendering Time

Setting Up Quick Rendering Previews

Rendering is always a time-consuming process because of the many adjustments that have to be made to get the right effect. You can save time by setting up a quick preview to view the changes you have made before doing a full rendering. The following are three ways to set up quick previews.

1. Use the Set Image Size option in the RenderZone Options dialog box to draw a box around a small portion of your model (figure 10-84). RenderZone will then proceed to render only this selected portion, saving you the time that would have been consumed by the generation of the parts of the rendering you do not need to see.

2. Use one of the preview Rendering Types, such as Preview Z-Buffer or Preview Raytrace. This provides you with a low-quality preview while retaining most of the RenderZone options.

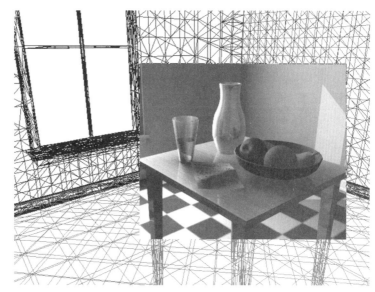

Fig. 10-84. Scene rendered with Set Image Size options selected.

3. Use a small window for previewing. Select New Model Window under the Windows menu. Resize the window to a Custom size of 320 x 240 or 240 x 180 in Image Options under the Display menu. You can use the larger window to make adjustments to the model and lighting while using the smaller window to render quick tests of the changes you have made to your model. You can later use the larger window to produce a full rendering when you are satisfied with the results.

Turn Off Texture Map Option

Texture Maps take up a substantial amount of rendering time. Therefore, switching off the Texture Maps option in the RenderZone Options dialog box during a preview rendering will reduce rendering time.

Lighting and Shadow Tips

The following sections present guidelines for establishing and manipulating light and shadow. When setting up lights for the basic illumination of a scene, it is almost always beneficial to use multiple light sources. For smaller scenes, a common technique known as triangular lighting is often employed. This consists of three lights, referred to as a "key" light, a "fill" light, and a "kicker." The key light is typically a shadow-casting light, shining at about a 45-degree angle to the scene, which serves as the primary source of illumination. The fill light shines at a complimentary angle and is used to illuminate extremely dark areas caused by shadows from the key light. This light is usually not shadow casting and rarely exceeds more than half the intensity of the key light. The kicker is typically placed high and to the back of the scene and serves to help separate the subject from its background and enhance specular highlights (figure 10-85).

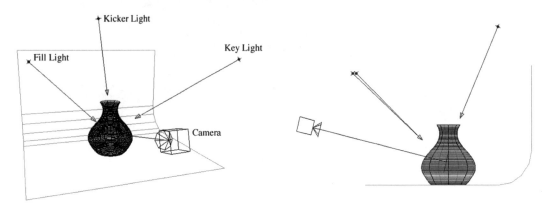

Fig. 10-85. Triangular lighting is often used to illuminate small scenes or objects.

For larger scenes, triangular lighting is often not sufficient and you will want to choose another strategy for illuminating the scene. Another method, which works equally well for large scenes or interior environments, is often referred to as zone lighting. This method involves breaking the scene down into smaller components or sections and illuminating each area independently. When using this method, pay special attention to the falloff rate for each light used so that the scene does not become overexposed. The Interior Lighting exercise from Chapter 9 uses this strategy.

Effecting Daylight Conditions

In order to simulate daylight conditions for a given time of day in a specific location in the world, use the Use Sun Settings option in the Light Parameters dialog box. Note that the default setting uses the positive y axis (+y) as the direction North. The Sun dialog box is shown in figure 10-86.

Positioning Lights with QuickDraw 3D and OpenGL

You can position lights and see immediate results by preselecting the light source, the point of interest, or the entire light with the Pick tool before engaging the QuickDraw 3D or OpenGL rendering modes. You can then use the Move command to change the light's position and see the effects of your changes immediately.

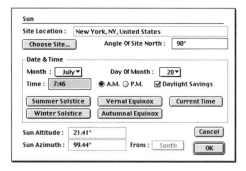

Fig. 10-86. Sun dialog box.

Lighting Single Objects

For dramatic renderings of solitary objects, first create a floor plane a few times larger than the object positioned on it. Make the background color black and turn the Ambient Light down to 0% (in the Lights Options dialog box) so that any part of the scene not directly lit will not be visible. Using shadow-casting cone lights, illuminate the object diagonally from narrow angles so that only the object and its immediate surroundings are lit and the rest of the scene fades to black.

Using Linear and Square Falloff Lights in Place of Shadow-casting Lights

In lighting large models, you may find that using a lot of shadow-casting lights will cause you to run out of memory or will make the rendering time for each scene unacceptably long. As previously mentioned, the first thing you may want to try is using Hard (raytraced) shadows instead of memory-intensive Soft (mapped) shadows. Another strategy is to use a few key shadow-casting lights, such as the sun, or a few large point lights and a number of smaller, non-shadow-casting lights to fill in dark areas or illuminate highlights in the scene. In order to make sure these smaller lights do not overexpose the scene, their falloff should be set to Linear or Square, and their Intensity turned down to a level that provides just the amount of light you need.

Light Color

Experiment with various light colors to add a more realistic ambiance to your renderings. For example, adding a hint of yellow helps simulate sunlight or incandescent bulbs, whereas a little orange simulates the warmer light of a sunset. Color adjustments can be made to the Ambient Light in the Light Options dialog box, and to the lights themselves in the Light Parameters dialog box.

Tweaking Material Reflective Properties

Once you have set up the lighting for a scene, it is often helpful to perform fine-tuning through the reflective properties of the materials rather than exclusively through the lights. Under Reflection Options in the Surface Style Parameters dialog box, you can adjust a material's various reflective qualities. The Ambient Light, Diffuse Light, and Reflection settings all control how much of the each particular light that strikes the material is reflected.

A particularly useful parameter under Matte and Plastic Reflection Options is the Glow setting. With a default setting of 0%, the Glow setting enables a material to appear self-illuminated, even if there is no light present. This setting is useful if you really want to make a certain material "pop out" in a rendering. It is also a great way of making white materials appear truly white without using a lot of light.

Changing Light Parameters with the Query Tool

Rather than going through the Lights menu to adjust a particular light, you can access a light's parameters through the modeling window by clicking on it with the Query tool.

Using Edit Cone Of Vision to Adjust Lights

You can adjust the location of lights graphically using the Edit Cone Of Vision option under the View pull-down menu. This is convenient because the lights can be viewed and adjusted in all three projections at once. You can also get immediate feedback on the effects of adjustments using the QuickDraw 3D or the OpenGL display option.

Avoiding Jagged Shadow Edges in Radiosity Mode

Jagged shadow edges in Radiosity mode are a common problem. This can be remedied a couple of ways.

- Use the Render Direct Illumination option to generate crisp shadows.
- Using the Mesh tool, mesh surfaces where jagged shadows occur so that the mesh pattern runs parallel to the shadow edges. This will force the radiosity mesh to follow the mesh pattern you have applied to the surface, thus creating a resolved shadow edge.

chapter 11

Drafting

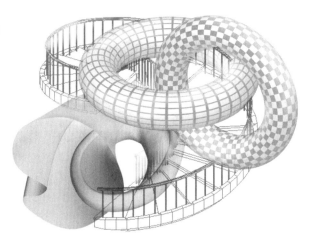

Introduction

This chapter reviews form•Z's 2D drafting features, focusing on its dimensioning capabilities. 2D drafting is neither the greatest strength of form•Z nor the main focus of this book. Therefore, if your only concern is producing 2D documentation, you should probably be using one of the dozens of other software products optimized for drafting.

form•Z, however, will let you bring your 3D designs directly into its drafting environment, which provides a substantial drafting capability. If this is not sufficient for your needs, you can save your files in standard formats for export to other drafting software.

➥ **NOTE:** *For information on importing and exporting files, see Appendix B.*

The drafting module is integrated into the form•Z environment so that you can easily move between 2D drafting and 3D modeling modes. You can generate shapes in the drafting module and paste them into the modeling module, where they can be further manipulated and turned into 3D forms. Conversely, you can cut and paste elements from modeling to drafting, where they can be prepared as working drawings or technical documents with dimensions, line weights, styles, hatching, title blocks and frames, and so on.

The Drafting Environment

The 2D module interface is quite similar to that of the 3D modeling module. You will recognize the Menus, Palettes, Command Tools, and Window Tools, but you will notice that certain familiar items such as the Heights menu are grayed out and unavailable. This is because they are

Chapter 11: Drafting

not applicable in a 2D environment. On the other hand, there are new features—such as dimension lines and the palettes that control hatching and line styles—that are meaningful only in 2D drawings.

You can enter the drafting module of form•Z at any time by selecting New (Draft) from the File pull-down menu at the top of the screen. Figure 11-1 shows the layout of the default Drafting screen with the Coordinates, Line Weights, Line Styles, View, and Layers palettes arranged at the right side of the window.

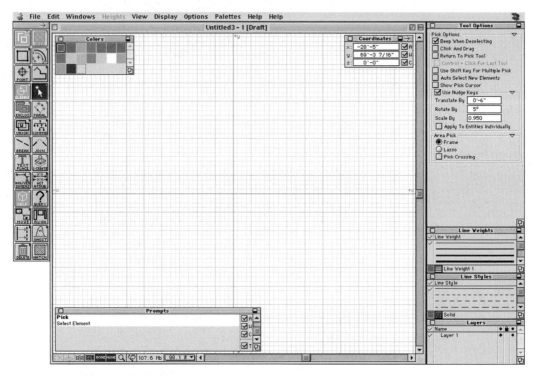

Fig. 11-1. Default screen of the Drafting module.

> **NOTE:** *The New (Draft) and New (Model) options are described in Chapter 3.*

An easy means of surveying the main features of the drafting module of form•Z is to select the Drafting Tools item in the Help menu at the top of the screen. As you drag the mouse over each command icon, a description of that command appears in the box at the bottom (figure 11-2). Note that the Menus item in the Help menu, when selected from within the drafting module, provides an explanation of each Drafting menu item. These differ slightly from the menus displayed when in Modeling mode.

The Drafting Environment

Fig. 11-2. Command tools of the drafting module as displayed by the Help function.

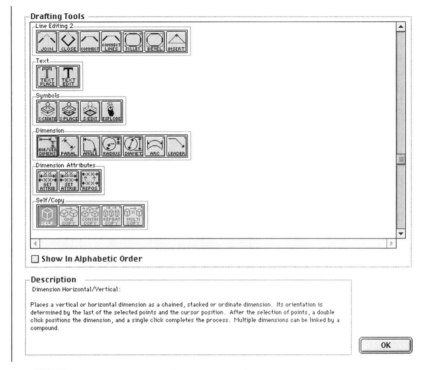

☞ **NOTE:** *See Chapter 2 for more information on getting help within form•Z.*

To demonstrate the main features of the drafting module, the sections that follow use the example of the screwdriver, which was developed as an intermediate exercise in Chapter 8 (figure 11-3).

Fig. 11-3. Four views of screwdriver rendered with the Hidden Line display option in the Modeling module.

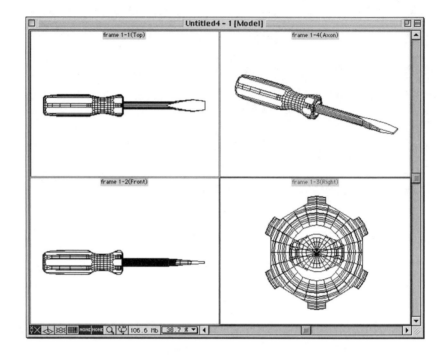

Pasting from Modeling into Drafting

A simple cut-and-paste operation is all it takes to import an image of a model from the modeling module to the drafting module. You should review discussions in Chapter 3 of the Cut, Copy, Paste, Paste From Drafting, and Paste From Modeling commands in the Edit menu.

Figure 11-3 shows four views of a screwdriver rendered in Hidden Line mode. In general, a Hidden Line view is superior to a Wire Frame view because it will eliminate the extra, unwanted lines that would complicate the view of the object meant to receive dimension lines.

You can cut and paste the top view of the screwdriver from the Modeling to the Drafting module. Perform the following steps for any model you want to transfer to the drafting module.

1. In the Modeling window, display the desired view of the object, with the Hidden Line display option under the Display menu (figure 11-4).

Fig. 11-4. Top view of the object in the Modeling environment, in Hidden Line mode.

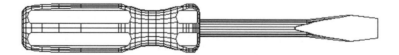

Pasting from Modeling into Drafting

2. Select the object with the Pick tool.
3. Use the Copy command from the Edit menu to store that view of the object in the system's clipboard.
4. Use the New (Draft) command in the File menu to open a new drafting window.
5. Use the Paste From Modeling* command in the Edit menu to place the object in the drafting module.
6. Use the Fit command in the Window Tools to adjust the view of the object.

Figure 11-5 shows the top view of the screwdriver after it is pasted into a new drafting window. Note that the rulers have been turned on with the Show Rulers command in the Windows menu to clearly show the size of the object. Window Setup has been adjusted to display major grid lines at 1-inch intervals, indicating that the screwdriver is just over 5 inches long.

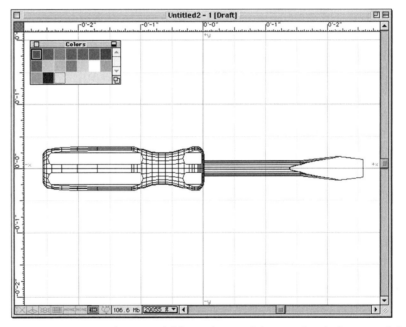

Fig. 11-5. Top view of the object is pasted into a drafting window.

In preparing to transfer a model from the modeling to the drafting module for dimensioning, a Wire Frame view may be preferred to a Hidden Line view in cases where it is desirable to "see through" the object. In this case, the Remove Duplicate & Overlapping Lines option should be selected from the Paste From Modeling dialog box, which can be accessed by pressing the Option key (Mac) or the Ctrl key (Windows) while selecting the Paste From Modeling* item in the Edit menu (figure 11-6).

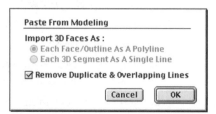

Fig. 11-6. Selecting Remove Duplicates & Overlapping Lines in the Paste From Modeling dialog box.

➥ **NOTE:** *The Paste From Modeling option is available only in the drafting module, and only if an object has been cut or pasted into the 3D Clipboard. See Chapter 3 for further discussion of cutting and pasting between 2D and 3D modules.*

Preparing to Add Dimension Lines

The screwdriver model was developed at full scale; that is, it was modeled at a length of 5-1/4 inches after making certain adjustments in the Window Setup and Numeric Accuracy settings. The dimensions of the object are remembered by the system when you cut and paste a view of the 3D model into the 2D drafting module. This allows you to obtain accurate measurements of the object using the dimensioning tools.

➥ **NOTE:** *The Window Setup and Numeric Accuracy settings are discussed in Chapter 3.*

Because the screwdriver is a relatively small object, all text, dimension lines, arrowheads, and so on will have to be set at a size smaller than the default settings, which are more appropriate for larger-scale architectural projects. These settings are adjusted in a series of dialog boxes accessed by double clicking on any of the Dimension tool icons in the Dimensions tool palette (figure 11-7). Note that the Tool Options palette provides the same controls as the dialog box, and is perhaps more convenient in that it instantly displays the options associated with each tool.

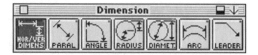

Fig. 11-7. Main dimensioning tool selected in the Dimensions tool palette.

Double clicking on the first icon in the Dimensions tool palette brings up the Dimensions: Horizontal/Vertical Options dialog box (figure 11-8). This dialog box allows you to constrain the dimension tools to horizontal or vertical dimension lines, or lines whose orientation is determined dynamically according to cursor movement. Choose the option that best fits your needs.

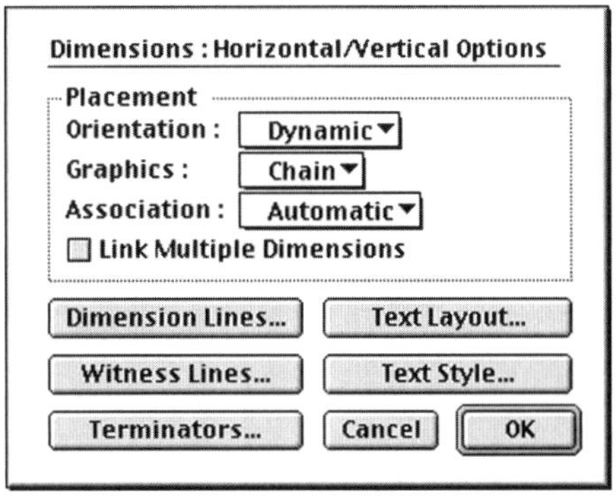

Fig. 11-8. The Dimensions: Horizontal/ Vertical Options dialog box.

Setting Numeric Accuracy

The measurements you obtain from your model with the Dimension tool will only be as accurate as the current Numeric Accuracy setting. To adjust the accuracy, select Working Units under the Options menu. This will bring up the Project Working Units dialog box, which has a Numeric Accuracy field.

If an object is small or has fine details, a finer setting, such as 1/64", may be appropriate. For a large object, such as an architectural model of a house, the default accuracy of 1/16" is sufficient, and most of the default settings for dimension lines will not have to be adjusted. The numeric accuracy for the screwdriver model is set at 1/64" because of the small size of the object and the need for precision in the description of the screwdriver blade (figure 11-9).

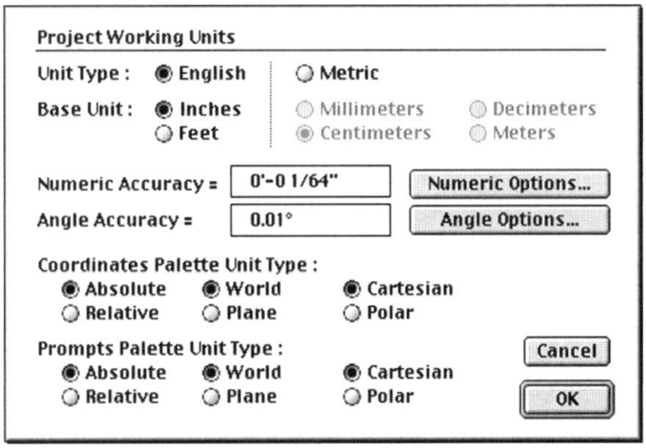

Fig. 11-9. Numeric Accuracy set to 1/64" in the Project Working Units dialog box.

In the Dimensions: Horizontal/Vertical Options dialog box, you can specify dimensions in three formations: Chain, Stacked, and Ordinate. These three arrangements, shown in figure 11-10, can be selected in the Graphics pull-down menu. Chain is the default and the recommended setting for most situations, and is used in the screwdriver example.

Fig. 11-10. Three arrangements of dimension lines: Chain, Stacked, and Ordinate.

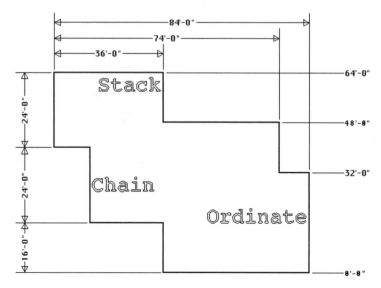

At the bottom of each dialog box associated with a dimensioning tool are five important buttons: Dimension Lines, Witness Lines, Terminators, Text Layout, and Text Styles (figure 11-8). These lead to dialog boxes that control the size and style of the three main components of dimension lines. In the case of the relatively small screwdriver model, all three components have to be adjusted to create a suitable dimension line style. The final settings selected for the screwdriver are shown in the dialog boxes reproduced here. The value of 0'-0 1/8" was pasted into each field in the dialog boxes (figure 11-11).

Fig. 11-11. Dimension Lines dialog box.

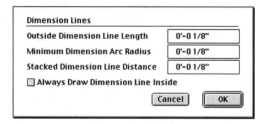

Setting Witness Lines

The Witness Lines button brings up the Witness Lines dialog box (figure 11-12). This lets you customize the guidelines drawn from the object to the dimension arrows. By default, witness lines appear at the start and end point of the dimension line; however, you can specify one condition or the other, or neither.

Normally the witness line stops short of the object, and you can adjust the Offset From Element parameter to specify just how short. Similarly, the witness line is usually extended beyond the dimension line, and you can adjust the Witness Line Extension to specify just how far beyond. If the Uniform Line Length option is selected, the witness lines will be based on the specified value rather than the distance from the dimension line to the object. In the example of the screwdriver, the witness line offset and extension are set at 1/8". The remaining options control the appearance of witness lines for arc dimensions.

Fig. 11-12. Witness Lines dialog box.

Setting Terminators

The Terminators button in the Dimensions dialog box brings up the Terminators dialog box (figure 11-13). You can select one of five styles for the start and end points of dimension lines: Arrow Head, Slash Mark, Dot, Circle, and Cross. The Fill option will place solid colors within these terminator shapes. In the Length and Height fields you can customize the size of terminators at the start and end points. In the example of the screwdriver, the terminator length and height are set at 1/8".

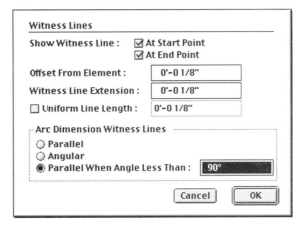

Fig. 11-13. Terminators dialog box.

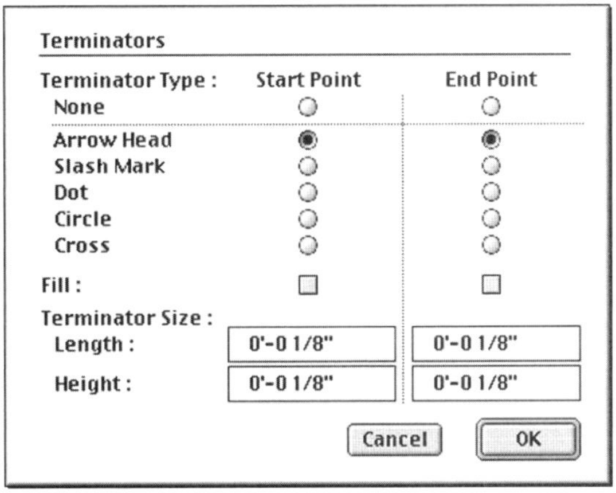

Setting Dimension Text

The Text Layout button brings up the Dimension Text dialog box (figure 11-14). This lets you adjust the appearance of the text component of dimension lines. The Drawing Read Direction setting of Bottom forces the text to be horizontal; Bottom & Right limits dimension text to vertical and horizontal positions; and Normal, the default and recommended setting, allows the text to follow the angle of the dimension line even if it lies on a diagonal.

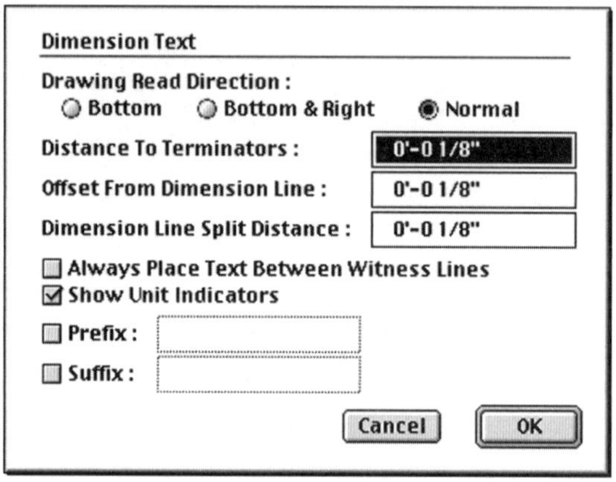

Fig. 11-14. Dimension Text dialog box.

The Distance To Terminator option determines the placement of the text when the text label is unable, due to lack of space, to fit between the terminators. The Offset From Dimension Line setting should be a bit less than the size of the letters so that the text label is always placed near the dimension line. In the example of the screwdriver, the Distance To Terminator option and the Offset From Dimension Line option are set at 1/8".

The Prefix and Suffix fields allow you to specify a text string to be added before or after the dimension text. When adding a suffix or prefix, you may want to turn off the Show Unit Indicators option. For example, if you were using metric units, you could define a suffix of "Centimeters" to replace the standard "cm" unit indicator.

The Text Style button brings up the Dimension Text dialog box (similar to the Edit Text tool), which can be accessed directly via the Text tool palette and can be used to modify the text component of dimensions after they are created. This lets you customize the font, height, style, and justification of the text, and choose between high-quality TrueType or Postscript fonts or low-quality, but fast, BitMap fonts. The Leading field controls the vertical spacing of rows of text. Note that the Width and Outline Smoothness options are applicable to high-quality fonts only. In the example of the screwdriver, the font size is set at 1/8" and the font is BitMap.

Do not expect to get the perfect size and style of dimension lines on your first try. It is always best to try a cycle of experimentation: adjust the settings, create some dimension lines, use the Undo command, and start again. After a few tries, you will get a feel for the settings that will work best with the particular object you are dimensioning. The Dimension Text Editor window is shown in figure 11-15.

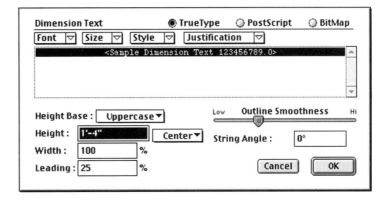

Fig. 11-15. Dimension Text Editor window.

Placing Dimension Lines

The dimension is placed with a series of mouse clicks. The following steps apply to the screwdriver example and are illustrated in figure 11-16.

1. Select the Point Snap tool from the Window Tools.
2. Select the Horizontal/Vertical Dimension tool in the Dimensions tool palette.
3. Click on the first point on the object to be dimensioned (point A, figure 11-16).
4. Click on the second point to be dimensioned (point B, figure 11-16).
5. Click on the last point to be dimensioned (point C, figure 11-16).
6. Double click at a point that indicates the direction in which the witness lines are to be extended (point D, figure 11-16).
7. Click at point E (figure 11-16) to define the exact position of the dimension line.

Fig. 11-16. Placing a chain of dimension lines with a series of mouse clicks at points A through E.

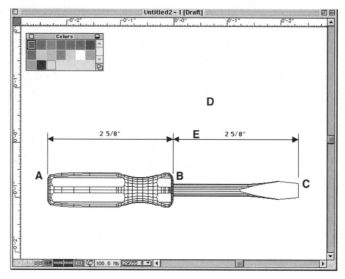

An overall dimension line can be placed using the same technique, minus step 4. Figure 11-17 shows the result of repeating the procedure using only points A and C. The final click positions the dimension line at a distance from the first dimension chain.

Fig. 11-17. Adding an overall dimension line.

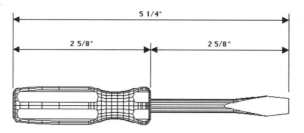

Vertical dimension lines are placed with the same command and technique. Figure 11-18 shows three vertical dimension lines whose text labels do not appear between the terminators because of lack of space.

Fig. 11-18. Adding vertical dimension lines.

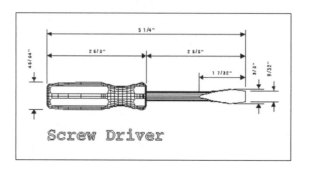

Placing a Text Label

Fig. 11-19. Selecting the Place Text tool in the Text tool palette.

You can add as many notes, labels, and titles to your drawing as you want. Use the Place Text command in the Text/Symbols tool palette (figure 11-19). For example, the label Screw Driver was added at the bottom of the drawing, as shown in figure 11-18.

Once text has been placed in a drawing, tow methods of modifying it are available to you. You can use the Undo command and reuse the Place Text command with different settings, or you can use the Edit Text tool in the Text tool palette. This will bring up the Drafting Text Editor (figure 11-20), which will allow you to change the attributes of the text as well as the words themselves.

Fig. 11-20. Drafting Text Editor.

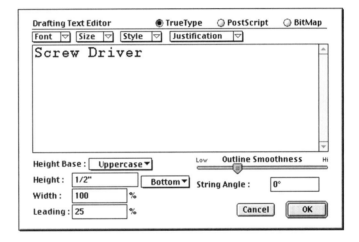

Draft Layout Mode

Previous to version 3.0, the drafting module restricted you to one scale (such as 1/4"=1'-0") per drawing. A new feature called Layout Space allows you to compose elements that can be at different scales. To use this feature, perform the following steps.

1. Select the Draft Layout Mode item in the Display menu.
2. Add "panes" for each drawing you want to add to the composition, using the Pane modifier in the Generation tool palette, located in the first row of the second column of the drafting tools.
3. Use any drawing tool to outline the shape of the panes.
4. Use the Pick Pane tool in the fourth row of the second column of the drafting tools.
5. Adjust the view, layer, and display settings for the selected pane. You can also set it to a fixed scale, such as 1/2"=1'-0", in the Display menu.

Chapter 11: Drafting

6. Once you are done adjusting settings inside the pane, click anywhere outside the pane, with the Pick Pane tool still selected, to deselect the pane.

7. Use the regular Pick tool to move and resize the panes to create the desired page composition.

Preparing to Print

In order to get a final product on paper, you will need access to a printer or plotter. In either case, the Plot/Print Setup option in the File menu will let you set the scale of the final output. Because the screwdriver is less than 6 inches long, it can be printed at full scale on a normal 8.5" x 11" laser printer. Figure 11-21 shows how full scale (1'-0"=1'-0") is defined in the Plot Scale field of the Plot/Print Setup dialog box. A horizontal, or landscape, page orientation can be selected from Page Setup in the File menu.

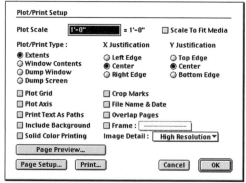

Fig. 11-21. Full-scale printing is set in the Plot/Print Setup dialog box.

You should always preview your work before sending it to the printer. The Page Preview command is located in the File menu and will display the exact appearance of final pages (figure 11-22).

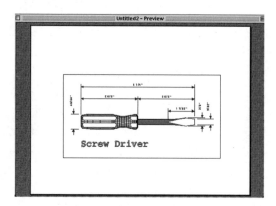

Fig. 11-22. Page Preview window.

> **NOTE:** *The Page Setup, Plot/Print Setup, Page Preview, and Print commands are discussed in Chapter 3.*

chapter 12

Animation

Introduction

This chapter explores the animation capabilities of form•Z 3.x. Animation is a powerful way to study and present a model by creating a simple path the camera can follow, through and around the model. In the case of architectural models, you can study what it would feel like to walk through the spaces you have designed. In the case of industrial design, you can circle around the object and get a better grasp of how the design works in 3D.

This chapter examines animation in form•Z through exercises and in-depth explanations of animation features. After a brief explanation of animation basics, and an overview of the animation commands in form•Z, the first exercise takes you step by step through the basic procedures needed to create a simple animation using the cityscape you created in Chapter 7. The second exercise explains how to use the more advanced features within animation through the eyes of an insect in a room. The last example explains some of the subtleties of creating an animated architectural presentation of a large-scale building. Figures 12-1, 12-2, and 12-3 show the three animation exercises covered in this chapter.

> ❖ **NOTE:** *The models for the first two exercises and the movie of the last exercises are contained on the companion CD-ROM, and can be loaded prior to going through the exercises.*

Chapter 12: Animation

Figs. 12-1, 12-2, and 12-3. The three scenes from animation examples covered in this chapter.

form•Z Animation Basics

There are a number of animation terms and concepts that need to be covered before proceeding. An *animation* is a sequence of rendered images (or frames in a film strip) that are played back at a certain frame rate to give the illusion of continuous motion. The basic frame rate at which the eye is fooled into thinking the animation is continuous is about 30 frames per second (fps).

In an animation there is a *camera* (or "eye point" as it is called in form•Z) and a *center of interest* (hereafter referred to simply as *COI*). The *eye point* is where the viewer is located, and the COI is the point the viewer is looking at. The eye point and COI both follow a path as the viewer (you, for example) walks through a building. These paths are called the *eye path* and the *COI path*.

In form•Z, an animation is generated through the creation of *keyframes*. Keyframes are strategic views that help define the eye and COI paths. form•Z produces the animation by first creating the eye and COI paths based on an interpolation between predefined keyframes you select. The program then generates all frames defined by these paths.

The overall length of the animation is defined by the number of frames between the start frame and the end frame, and the frame rate at which these frames are played back. Say, for example, the animation contains 300 frames, which will be playing back at 30 fps. You can calculate that this animation will be 10 seconds long. This will be too short to show a walk-though of a 3-storey building, but may be appropriate for a short tour around a small object. It is important to think about the duration of the animation before you generate it. This also helps gauge how much time you will need to render the frames. If, for example, you need 1,800 frames for a 1-minute animation, and each frame takes about 3 minutes to render, you will need about four days to render the entire sequence!

The time counter in form•Z animation uses SMPTE standard time code format: hh: mm:ss:ff. In this system, hh represents hours, mm represents minutes, ss represents seconds, and ff represents frames. If you have a frame rate of 30 fps, the ff counter will go up to 29 before each second advances. Now that you know the basics, it is time to make a movie!

Animation Quick Guide

The following is a quick guide through the basic steps needed to generate an animation in form•Z. You can try these steps for any model you may have already built and saved. Refer to figure 12-4 to find the screen locations of these five steps.

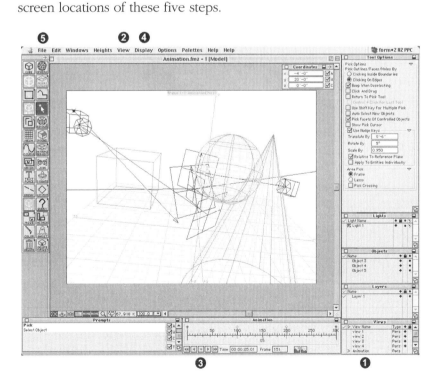

Fig. 12-4. Quick guide for creating an animation in form•Z.

Step 1: Use the Views Palette

Save all views you want as keyframes for the animation. Make the views you want to use visible by turning Visibility on in the Views palette. The diamond symbol is the visibility switch. Then select the views in the sequence you want the animation to be made, using the first column of the Views palette. (See number 1 in figure 12-4.)

Step 2: Create the Animation from Keyframes

Once the keyframe views have been selected, select the Animation From Keyframes option in the View menu. This opens the View Parameters dialog box. This is where you can manipulate the animation. Choose a name and use the default settings for now and select OK. This creates a new view in the Views palette containg the animation sequence you have created. (See number 2 in figure 12-4.)

Step 3: Use the Animation Palette

You can preview the animation by using the controllers in the Animation palette. You can choose to view the preview in Wire Frame, QuickDraw3D (MacOS), or OpenGL (Windows, MacOS). (See number 3 in figure 12-4.)

Step 4: Generate the Animation

After you have previewed the animation as a wireframe, you can choose to generate a rendered animation by selecting Generate Animation from the Display menu. Choose the Display Type, the number of frames, and a file name; then select Generate. This creates a .fan file you can play back using the Play Animation option, also located under the Display menu. (See number 4 in figure 12-4.)

Step 5: Export the Animation

You can export the animation in a more widely used movie format, such as QuickTime (MacOS) or AVI (Windows). Select Export Animation in the File menu. In the options box, select the rendered .fan file you want to export. (See number 5 in figure 12-4.)

Animation Tour and Exercises

Exercise 12-1: A Casual Walk Through a City

A casual walk though a city can be animated with ease using the animation tools in form•Z. It is as easy as capturing the views you like and then specifying these as an animation sequence. Of course, with the power of animation at your fingertips, why just create a walk to the end of the street. Why not take off and view the city from above!

LEARNING OBJECTIVE 12-1: This exercise covers the basic steps needed to set up an animation. Discussed are setting up keyframes, length of the animation, previewing, and generating a rendered animation.

Analysis and Method

The cityscape is a perfect example of an environment that calls for multiple views from different perspectives—from the view of a pedestrian to the view of a person in an office building to a person looking down from a helicopter. The animation sequence will be generated by setting up six strategic views as keyframes. These keyframes will be used to generate the eye path and COI path. The View Parameters options dialog box will be used to manipulate the length of the animation.

Before starting, first load the Cityscape example contained on the companion CD-ROM. The file is named *cityscape.fmz*. This will be used as the scenery for the animation. You can also view the model, including the keyframes and animation, in the file named *cityscape_anim.fmz*.

Exercise Steps

The following simple steps guide you through the process of setting up the necessary keyframes in order, putting them into an animation sequence, and finally rendering your animation.

Setting up Keyframes

The first step in creating an animation is to set up the keyframes from which the animation will be created. Keyframes are strategic views set up just like any other views saved in the form•Z Views palette.

1. Open View Parameters in the View menu, select Perspective as the View Type, and enter 70 degrees as the Angle in the Perspective Parameters box. This ensures that all subsequent saved views are perspectives with sufficient viewing angle to see the surrounding buildings. Click on OK.

2. Use the Edit Cone Of Vision option from the Views menu to create the views that will be used as the keyframes. Start with a view from the sidewalk at the end of the street. Click on the window bar at the top of the window to open the Cone Of Vision menu and select Save View to save the first keyframe, as shown in figure 12-5.

Fig. 12-5. First view saved as a keyframe.

3. Create two other views that follow the street to the end by clicking on the Line of Sight (this is the line that connects the eye with the COI) and dragging it to the appropriate locations.
4. At the fourth view, the camera rises as the COI turns, starting to look backward (figure 12-6).

Fig. 12-6. Fourth view turns and starts to look back at the city as the camera rises.

5. The fifth view is higher and looks back toward the city. Start moving away from the model. The sixth and final view should look down at the buildings, as shown in figure 12-7.

Exercise 12-1: A Casual Walk Through a City

Fig. 12-7. Final view, looking down at the city.

6. In the Views palette, turn on the visibility for all six keyframe views. This is the column with the diamond symbol at the top. This makes the camera for all views visible in the model window (figure 12-8).

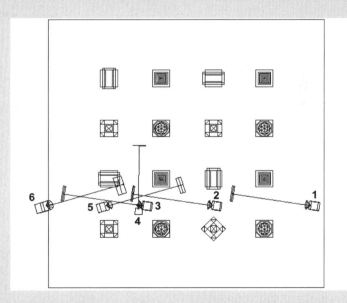

Fig. 12-8. Six keyframes shown in plan.

Making an Animation Preview

Now that you have saved the six keyframes, perform the steps that follow to create an animation path and preview. Be careful to follow the steps in exact order, because the animation path will not be created if the commands are out of sequence.

1. Select the views in the sequence in which you want the animation path to travel by checking the Selection column in the Views palette, as shown in figure 12-9. This is the column with the check mark on the top. The order in which the selection is made is very important, because the animation path will be created in this sequence.

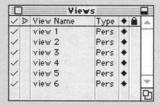

Fig. 12-9. Views palette with visibility turned on and views selected in order.

2. From the View menu, select the Animation From Keyframes option. This evokes the View Parameter dialog box. Make sure that Custom is selected in the Camera View box, and that 320 x 240 is used for the Width and Height. In the Animated View box, make sure you have the Frames set to 300, Frames Per Second set to 30, and Time Control set to Keyframes. Eye Path and COI Path should both be set to Spline. Enter a name for your animation in the Name field. The settings should be those shown in figure 12-10.

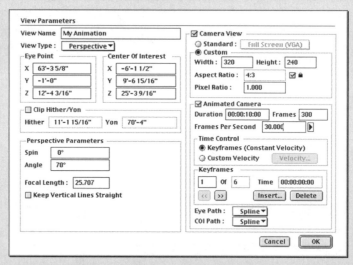

Fig. 12-10. View Parameters dialog box.

3. When you click on OK, the previously empty Animation palette is now highlighted. This palette shows the number of frames and the location of the keyframes in the sequence (figure 12-11). In addition, check to see that the animation is now a view in the Views palette. You can return to the animation at any time by selecting it in the View palette.

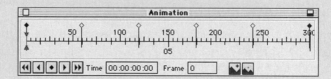

Fig. 12-11. Animation palette

Exercise 12-1: A Casual Walk Through a City 565

4. Click on the Play button in the Animation palette and see it go! You can stop the animation by pressing the Escape key or the red stop button on the Animation palette. If you have QuickDraw 3D or OpenGL display options, you can see a rendered animation preview by selecting either the QuickDraw 3D (MacOS) or OpenGL (Windows, MacOS) display option in the Display menu and then clicking on the Play button.

Generating a Rendered Animation

1. The final phase is to create the rendered animation. This takes substantially longer because each frame needs to be individually rendered. Be sure to save your model before you do this. From the Display menu, select Generate Animation. In the Animation Generation dialog box, for the Display Type, select RenderZone. Select All Frames for the Frames To Render option, as shown in figure 12-12. Click on New to create a new form•Z animation file, or ".fan" file. This will prompt you to enter the name and location of your file. The other options are explained in the following Fly Animation example.

Fig. 12-12. Animation Generation dialog box.

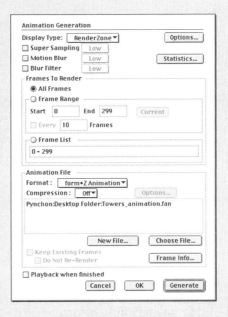

2. Click on Generate when you are ready. This starts the rendering, with a window showing the progress. Do not trust the Estimate Time Remaining. It has the tendency to change its mind when it hits a section of the animation with a lot of detail.

3. When the animation has been generated, select Play Animation in the Display menu to see the result: A Casual Walk Through a City.

Exporting Your File to QuickTime or AVI

You can export your animation to a more widely recognized movie file format such as QuickTime (MacOS) or AVI (Windows) either while you are generating your animation or after you have generated your .fan form•Z animation file. In the former case, you can select QuickTime or AVI in the Animation Generation dialog box in the Format option. This will create a QuickTime or AVI file directly while generating the animation without generating a .fan form•Z animation file.

In order to export an existing .fan file to QuickTime or AVI, select Export Animation from the File menu. You are first asked to select the .fan file you want to export. Then you can select the frames to export in the Animation Frame Export Options dialog box (figure 12-13).

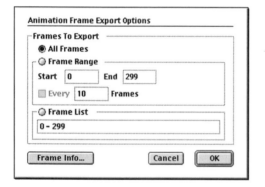

Fig. 12-13. Animation Frame Export Options dialog box.

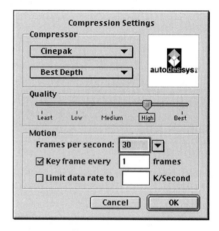

Fig. 12-14. Compression Settings dialog box.

After you select the name and the file format in which you want to export your file, select the Save button. You will be asked to choose a compression format in the Compression Settings dialog box (figure 12-14). Animation files tend to grow large quite rapidly if uncompressed. Compression allows you to efficiently reduce their size with only a slight loss in quality during playback.

Exporting Stills from Your Animation

You can also export stills of your animation as image files such as PICT, TIFF, or Targa. This can be also done through the Export Animation option under the File menu. Choose the frame or frames you want to export as stills in the Animation Frame Export dialog box. In the Save dialog box, select one of the image file formats, such as EPS, JPEG, PICT, PNG, Targa, or TIFF. This will export the selected frames as stills in the file format you have designated.

Exercise 12-2: A Day in the Life of a Fly

Trying to simulate a flight path of an insect presents interesting challenges for animation. Because the path of a fly may sometimes be inverted, with sudden changes in direction and speed, it becomes a complex exercise in not only setting up the initial path but in manipulating the animation in a way that expresses these movements. form•Z animation allows you to customize settings such as velocity of the Eye Point path, COI path, Spin, and View Angle using the Custom Velocity controls.

LEARNING OBJECTIVE 12-2: *Learning to use the advanced features of form•Z animation by which animation paths are made to look more natural.*

Custom Velocity Controls, when used skillfully, can recreate subtle motion effects that add a dynamic depth to your animation. This exercise will lead you through the manipulation of Velocity Controls Curves to achieve complex feats of acrobatics. Discussions continue into Motion Blur and Super Sampling toward the end of the exercise.

One word of warning for the beginner: This exercise may end up being quite time consuming, because setting up a path with velocity controls may require a lot of tweaking and running back and forth between previews. You will be happy to know that it took the authors a good many hours of adjustments to achieve the animation included on the companion CD-ROM. Patience, perseverance, and perfectionism will pay off in the end and the results will be very rewarding. Who said it would be easy?

Analysis and Method

The fly is in a room that may look familiar. It is the same model of a room with a window and table used for Chapter 10, on rendering.

The path of the fly is as follows. It starts out upside down on the ceiling. It then falls and gathers momentum and flies between the legs of the table. It heads straight for the window and tries to get out. After hitting itself twice on the window pane, it decides that the bowl of fruit on the table is enticing and heads over to the table and lands on a fruit. All of this happens within the space of 300 frames!

Custom Velocity Controls can be manipulated once the animation paths have been created in the View Parameters dialog box for the animation. In the Time Control box you can select Custom Velocity to access the Velocity Control dialog box. By manipulating graphs that represent the frames on one axis, and the percentage along the path on the other axis, you can adjust the rate at which the camera moves along the path.

Creating too many keyframes may make the Velocity Control curve difficult to manipulate. Therefore, the number of keyframes in the this exercise were kept to a minimum of eight frames.

Be sure to first view the fly animation movie contained in the companion CD-ROM before proceeding. There are two fly animation models on the CD. The first, titled *fly_anim.1.fmz*, contains just the keyframes. The second, titled *fly_anim.2.fmz*, is the model with the animation path and all Custom Velocity Controls set up. Use the former model when following the exercise (figure 12-15).

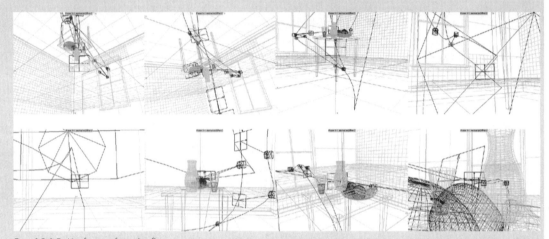

Fig. 12-15. Keyframes from the fly animation.

Velocity Controls Basics

Before setting up the fly animation, take a little while to experiment and get acquainted with the velocity control curves. In the following steps you will set up a small animation with three keyframe views to see what the curves do and to study the relationship between the two axes: the Percentage-along-path axis and the Frames axis.

Comparison Between Constant and Variable Velocity

1. Set up a quick animation using three keyframes (any model and any three views will do, or use the cityscape animation). Then select Custom Velocity in View Parameters for the animation. Open the Velocity Control dialog box by clicking on the Velocity button.

2. The Velocity Control dialog box shows graphs of the animation path. The vertical axis represents the "percentage along path" for the camera. The Horizontal axis represents the "elapsed number of frames." The curve represents the changes in velocity of various functions of the camera along the path.

3. Select the Per Track option. This shows the Velocity curve for the Eye Point. In this short exercise you will only be manipulating the curve for the Eye Point. In the pull-down menu, curves can be manipulated for Eye Point, Center Of Interest, Hither, Yon, Zoom/Pan, Spin, and View Angle (figure 12-16).

Fig. 12-16. Per Track editing options pull-down menu.

4. Click on the Constant option at the bottom of the dialog box. This resets the curve to be a flat line (figure 12-17). Exit the dialog box and play the preview of the animation. This Velocity Control Curve makes the camera travel along the path at a constant velocity from beginning to end, without variation. The camera will move abruptly at the beginning of the animation and stop equally abruptly at the end.

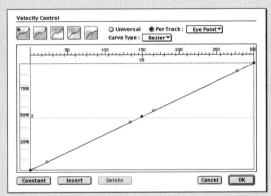

Fig. 12-17. Velocity Control Curve with Constant option selected.

5. Set up the curve again, this time using the control handles on the points to make the curve resemble figure 12-18. Play the animation preview.

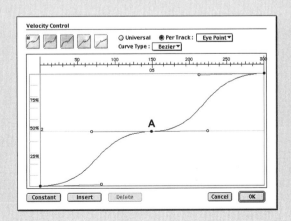

Fig. 12-18. Velocity Control Curve manipulated using the handles.

6. This second curve will start more smoothly and will slow down as it gets closer to point A. After it passes through point A, it will speed up and finally slow down as it reaches the end. You will notice that the camera travels faster between the points in the second animation. This is because the steeper the curve the faster the camera will be moving, because it is using fewer frames to travel along a certain percentage of the path.

Editing the Velocity Control Curve

1. The way the Velocity Control Curve meets a control point is important in determining the way the camera behaves. At the top of the Velocity Control dialog box are five icons showing predefined types of point-line conditions (figure 12-19). Try out various point types. These options are grayed out until you click on a point you want to edit. Click on the icon to apply the changes to the selected point in your curve.

Fig. 12-19. Curve Type options.

2. The first two boxes show curves meeting at both sides of the point. The difference between the two is that the first is a Continuous Smooth curve and the second is a Broken Smooth curve. This means that with the Continuous Smooth curve the entry velocity and the exit velocity into the point is the same, whereas with the Broken Smooth curve the velocities can be different. Figures 12-20 and 12-21 illustrate the two curves.

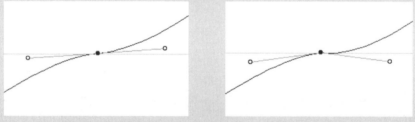

Figs. 12-20 and 12-21. Continuous Smooth curve and Broken Smooth curve.

3. Sometimes you want the animation to come to an abrupt stop at a certain point in your animation, and then start up again slowly. This would happen if there is an obstacle in your path, for example. You can achieve this effect by using the Flat-Smooth curve, which is the fourth option in the top of your Velocity Control dialog box. First select the point; then select the point type to change it. You can achieve the opposite effect by using the third icon, the Smooth Flat curve option.

4. You can abruptly change the velocity of the camera using the last icon, the Both Flat curve option.

5. You can insert a control point in the curve by using the Insert button at the bottom of the dialog box. First select a point. Then click on the Insert button. This will insert the point after the selected point. To delete a point, simply select the point and click on the Delete button.

Now you are ready to tackle the fly animation.

Fly Animation (Advanced)

Setting up the Fly Animation

1. Load the file titled *fly_anim.1.fmz* from the companion CD-ROM. This contains the scene with all keyframes. Make the views visible in the Views palette and select the eight views in order. Select the Animation From Keyframes menu item from the Display menu to create an animation path from the selected keyframe views.

2. Double click on the animation you have just created in the Views palette. This opens the View Parameters dialog box for the animation path (figure 12-22). Note that in the Time Control box, the Keyframes (Constant Velocity) option is selected. This is the default velocity, where the Eye and COI follow the animation path at a constant velocity.

Fig. 12-22. View Parameters dialog box.

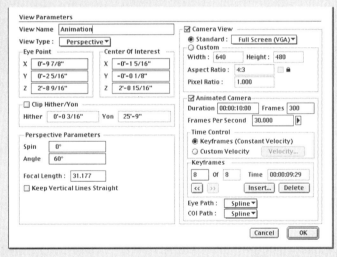

3. Select Custom Velocity. This will activate the Velocity options button. Click on this button to access the Velocity Control dialog box (figure 12-23).

4. In the Velocity Control dialog box, by default you should have Universal selected. This option allows you to globally manipulate all functions of the camera, such as the velocity of the eye point and COI. This option is suitable for adjusting the speed of the camera during a straightforward walk-through in which you only need to slow down and speed up. However, for the flight of a fly, a lot more control is needed over the path.

5. Select the Per Track option. This highlights a pull-down menu that shows the names of the curves you can individually manipulate.

Fig. 12-23. Velocity Control dialog box.

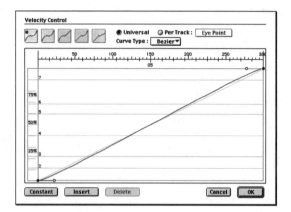

For the fly exercise, three of the curves will be adjusted: Eye Point, COI, and Spin. The number of Control Points on the curves is determined by the number of keyframes in the scene. In this exercise, there are seven points that can be manipulated.

Editing the Velocity Control Curve

In the following images, what should be happening in each keyframe is explained beside an image showing the Eye Point, COI, and Spin Velocity Control curves together at that frame. Use the images to set up the Velocity Control Curve for your fly animation and to study the effects achieved by each manipulation. Note that after the Velocity Control Curves have been manipulated, the frame resembles the original views used as keyframes when making the animation.

Keyframe 1

Here the fly is on the ceiling, ready to make its descent. The Eye Point curve shows that it starts slowly, then speeds up as it falls down. The COI curve stays consistent with the Eye Point curve (figures 12-24 and 12-25).

Keyframe 2

At this point the fly is starting to make its spin as it falls and gains speed. Both the Eye Point and the COI curve show that it is speeding up. The Spin curve shows that between approximately frame 65 to frame 80 it spins rapidly around to face right side up (figures 12-26 and 12-27).

Keyframe 3

Now that it is right side up, the Eye Point and COI curves show the fly is still gathering speed as it flies through the table legs (figures 12-28 and 12-29).

Exercise 12-2: A Day in the Life of a Fly

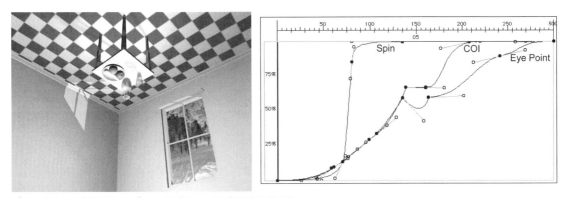

Figs. 12-24 and 12-25. Keyframe 1 / Frame 0 / 00:00:00:00.

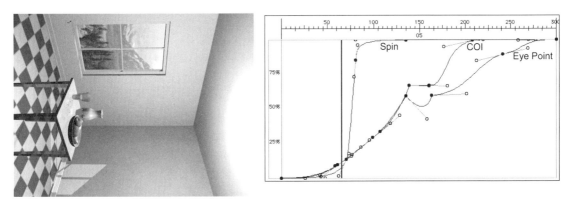

Figs. 12-26 and 12-27. Keyframe 2 / Frame 65 / 00:00:02:05.

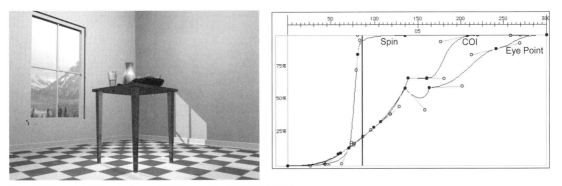

Figs. 12-28 and 12-29. Keyframe 3 / Frame 86 / 00:00:02:26.

Keyframe 4

The fly is still gathering speed as it innocently flies toward the bright light coming through the window (figures 12-30 and 12-31).

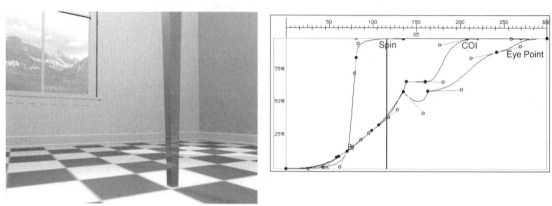

Figs. 12-30 and 12-31. Keyframe 4 / Frame 115 / 00:00:03:25.

Keyframe 5

At frame 135, the fly hits the window and bounces off. Then it tries to go out again, before turning back. The Eye Point curve does something unusual to achieve this movement. The curve comes to an abrupt point and then turns down. This means that the camera movement stops and follows the path in the opposite direction, the way it came! This is how the bounce-back movement of the fly is simulated.

It is interesting to see what the COI curve is doing during all of this. After the fly hits the window for the first time, you can see that the COI remains more or less fixed. This means that the fly is looking at a fixed point along the COI path, presumably a point outside the window (figures 12-32 and 12-33).

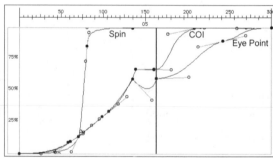

Figs. 12-32 and 12-33. Keyframe 5 / Frame 162 / 00:00:05:12.

Keyframe 6

The second, less severe bounce has an equally sharp velocity approaching the point of the bounce, but instead of bouncing back it slowly starts again along the path, toward the table again. The COI curve at this point speeds ahead of the Eye Point curve to look at the table and the fruit bowl. This has the effect of the fly turning to look at the fruit bowl right after bouncing off the window the second time (figures 12-34 and 12-35).

 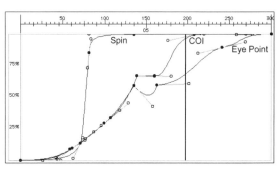

Figs. 12-34 and 12-35. Keyframe 6 / Frame 196 / 00:00:06:16.

Keyframe 7

The fly slowly gathers speed as it passes by the table and fruit bowl, but upon realizing it has passed it, slows down before heading back to the table. The point in the Eye Point curve corresponds to the point in the path the fly starts to turn back. The COI curve has reached its destination, which is the fruit bowl well ahead of time (figures 12-36 and 12-37).

 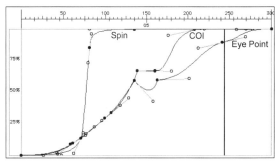

Figs. 12-36 and 12-37. Keyframe 7 / Frame 244 / 00:00:08:04.

Keyframe 8

The fly slows down to finally come to a rest on the fruit. All of this happens in just 10 seconds, or 300 frames (figures 12-38 and 12-39).

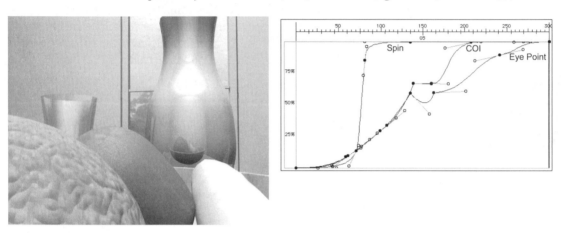

Figs. 12-38 and 12-39. Keyframe 8 / Frame 299 / 00:00:09:29.

Once you have finished setting up the path and are satisfied with the results, there a couple of finishing touches you can add to the animation, as explained in the following section.

> **NOTE:** *Before you generate the animation, it is a good idea to make the Camera View in the Views Parameters dialog box a smaller size. Select Custom, and a resolution of 240 for Width and 180 for Height. This is because generating the animation at the full setting of 640 x 480 will take a substantial amount of time. In addition, playback may be very jumpy at 640 x 480 because most computers still cannot handle full-screen playback unless they have additional video acceleration hardware installed.*

Adding Motion Blur to Your Animation

Often in an animation, when there are fast camera movements, the animation tends to be jumpy. This is due to the fact that there is too much of a difference between each frame when the camera is moving rapidly. The Motion Blur filter smooths out the animation by adding a blur in the frames that have rapid camera movements relative to the motion of the camera. This is a digital effect that simulates what your movie camera will naturally do when you are panning too fast.

The Motion Blur effect is achieved by rendering additional frames between two scenes. The more rapid the camera movement, the more frames are required to create a convincing motion blur.

You can set an animation to have motion blur by selecting the Motion Blur option in the Animation Generation dialog box. This is accessed when you select the Generation Animation menu item under the Display menu. There are three quality settings: Low, Medium, and High. This setting determines the number of extra frames that will be generated to achieve the motion blur effect. Note that motion blur increases the time to generate the animation.

The fly animation movie file contained on the companion CD-ROM was generated with motion blur. Figures 12-40 and 12-41 show selected frames rendered without and with motion blur.

Fig. 12-40. Animation frame generated without motion blur.   *Fig. 12-41. Animation frame generated with motion blur.*

Super Sampling

Super Sampling renders each pixel in the scene at a higher resolution and then shrinks it to the original size to achieve better quality and better anti-aliasing in the image. This is especially effective in reducing flicker or moiré patterns that may occur during the animation generation.

The Super Sampling option is set in the Animation Generation dialog box. When the setting is set to Low, each pixel will be rendered at a 2 x 2 over-sampling resolution; at Medium, 3 x 3; and at High, 4 x 4. At the High setting, expect your rendering to take up to 16 times longer that it normally takes, because each pixel needs to be rendered at 16 time its original resolution. The fly animation on the companion CD-ROM was generated with Super Sampling set to Low.

Adding a Background Image to Your Animation

You can add a background image to your animation that changes according to your eye and COI paths by using the Background option in the RenderZone display option. In the RenderZone Options dialog box, use the Environment option for the Background. Then in the Environment box, select Spherical and Image in the pull-down menus. When you click on the Options button, you can load an image and have it map around an imaginary sphere that surrounds your model. This image will change according to the camera's angle of view.

If you use a regular background with your animation, you will find that the background image will be the same regardless of the location of the camera. For the fly animation, however, a simpler technique was used. Because only a restricted view outside one window needs to be seen, a rectangle was placed close to the window, with the image texture mapped onto it. This is not unlike how Hollywood set designers place a large canvas with scenery outside a window for an interior scene (figure 12-42).

Fig. 12-42

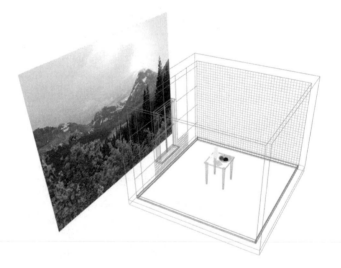

Fig. 12-42. Perspective View of the scene showing rectangle with texture mapped outside scenery.

Previewing Your Animation in OpenGL/QuickDraw 3D

Preview a rendered version of the entire animation using the OpenGL (Windows/MacOS) or QuickDraw 3D (MacOS) before you generate the animation. Select OpenGL or QuickDraw 3D from the Display menu in your animation view. This option will give you a better gauge of what the final product will look like without having to render a single frame.

Exercise 12-3: Advanced Animation Example

This exercise explores Frank Lloyd Wright's National Life Insurance Building, a project never built. This 30-second animation is included on the companion CD-ROM. This section will review the techniques used to create the model, placing cameras for the fly-through and generating the animation.

LEARNING OBJECTIVE 12-3: *Working with advanced animation.*

Creating the Model

Figures 12-43 and 12-44 show both a wireframe and a rendered view of the completed model. Although the model may look extremely complicated, it is actually much simpler than one might think. A few important techniques allowed for the creation of realistic looking effects, yet kept the model size and modeling time to a minimum.

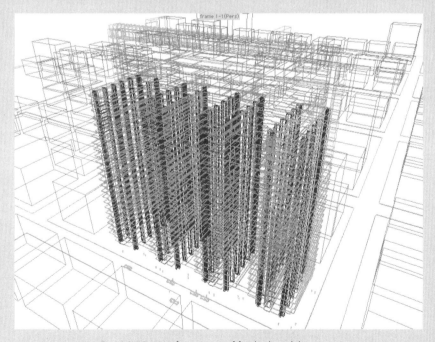

Fig. 12-43. Wireframe view of finished model.

Fig. 12-44.
RenderZone view of finished model.

Underlays

An underlay is an excellent tool for quickly and accurately generating a 3D model from existing drawings or sketches. In this model, one of Wright's hand-drawn plans was scanned and imported into form•Z using the Underlay command under the Windows menu.

Once the underlay image had been scaled correctly, the columns and floor slabs were drawn over the image as surfaces and then extruded to the correct height and thickness according to Wright's section drawing. The typical floor slab was then copied vertically using the Repeat Copy command. The wider floors at the bottom and the narrower floors at the top of the tower are typical slabs, with their end points adjusted accordingly. Figure 12-45 illustrates the structural model of one tower.

Custom Bump Maps

Figure 12-46 illustrates one of the ornamental concrete blocks Wright specified in his design. Modeling this block would not only be extremely time consuming, but would also involve many polygons, especially when copied hundreds of times throughout the model. The best way to approach a problem like this is to import a grey-scale image to be used as a custom bump map. A custom bump map is used rather than an image map because it will behave more like a textured block when it interacts with light, and allows for the adjustment of color separately. This bump map was combined with a procedural Mist color, as shown in figures 12-47 and 12-48.

Exercise 12-3: Advanced Animation Example 581

Fig. 12-45. Axonometric view of one tower's structure.

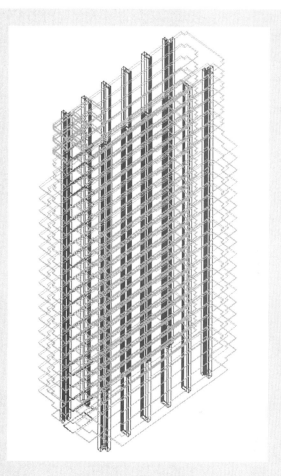

Fig. 12-46. One textured block rendered with a custom bump map.

Fig. 12-47. Mist Color settings for textured blocks.

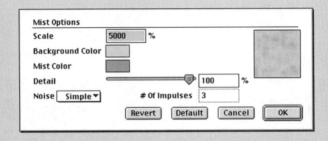

Fig. 12-48. Bump Map options for textured blocks.

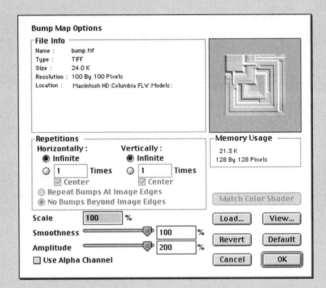

Grid Transmission Map

Figure 12-49 is a close-up of the building's facade, showing the glass and mullion system. As you can see, the mullions are not actual objects but a gridded map on the surface of the glass. Because this model is designed for use in an animation and not for close-up images, you can get away with using a mapped surface rather than a polygon and time-intensive mullion system. Unlike the custom bump map, you do not have to scan in a gridded map. form•Z has a procedural gridded map, which affords a lot of flexibility in the grid size, spacing, and scale. Figures 12-50 and 12-51 show the parameters of both the grid transmission map and the glass.

Exercise 12-3: Advanced Animation Example

Fig. 12-49. Frame 360 from included animation.

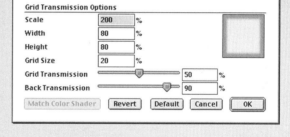

Fig. 12-50. Grid Transmission options for mullioned glass.

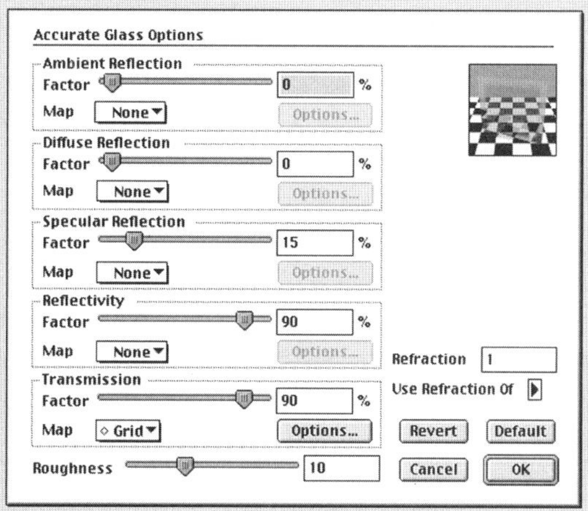

Fig. 12-51. Accurate Glass options for mullioned glass.

Lighting

When lighting a model, it is important to understand how the choices you make will affect the amount of memory and time it will take to render a scene. This is especially true if you plan on using the model to generate an animation. As discussed in Chapter 9, the use of shadow-casting lights can significantly increase the demand on your computer. Generally, the use of Soft Shadows increases the amount of memory required to generate a scene, whereas Hard Shadows require more processing time. Using more than one shadow-casting light multiplies these effects accordingly. Therefore, it is important to use your lights as efficiently as possible by limiting the use of shadow-casting lights. In this model, only six of the 400 lights cast shadows. These six lights are shown in igures 12-52 and 12-53.

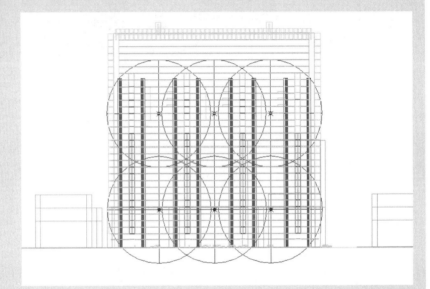

Fig. 12-52. Front view of model with six shadow-casting point lights.

These six shadow-casting lights will cast shadows both across the building's facade and into its immediate surroundings. These lights will approximate the shadows generated by the building as if it were illuminated at night, resulting in a believable final image. Figure 12-54 shows the model illuminated with these six lights exclusively.

The next step is to illuminate the building internally with a number of small point lights that would mimic the hundreds of individual desk and table lamps that would have been present in a building of the 1920s. Because these lights are going to be very small and inside the building envelope, you can save a lot of rendering time by making them non-shadow-casting lights. Because you are using a large number of these point lights, you want their effect to be contained within a small area. Otherwise, the scene would become overexposed. Confining the influence of these lights is achieved by changing the Falloff to Square option and setting the Radius and Intensity, as shown in figure 12-55.

Exercise 12-3: Advanced Animation Example

Fig. 12-53. Top view of model with six shadow-casting point lights.

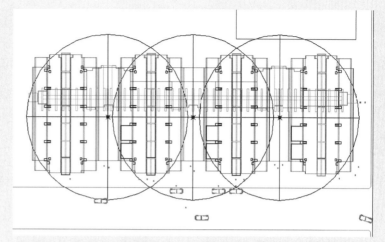

Fig. 12-54. Rendered view of model with six shadow-casting point lights.

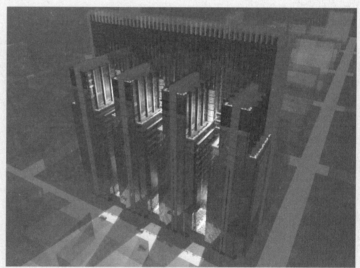

Fig. 12-55. Light Parameters of an interior non-shadow-casting point light.

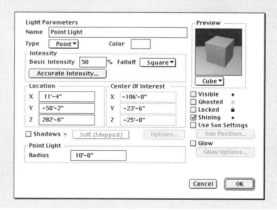

Because these lights are not shadow casting, you are not as concerned about the number of these types of lights in the model. In fact, a large number of these small lights will help give the model the massive sense of scale it needs to look believable. Therefore, the lights are placed in a pattern that completely covers one floor and copied to every other floor up the entire building, as seen in figures 12-56, 12-57, and 12-58.

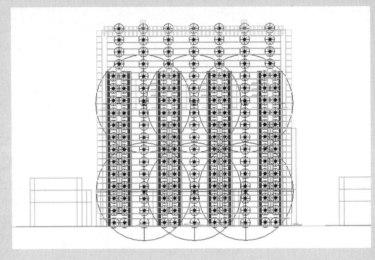

Fig. 12-56. Front view of model with six exterior shadow-casting point lights and numerous interior non-shadow-casting point lights.

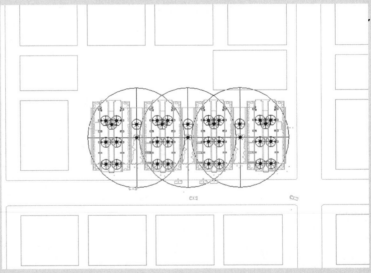

Fig. 12-57. Top view of model with six exterior shadow-casting point lights and numerous interior non-shadow-casting point lights.

The final lighting step is to add a series of cone lights that will highlight the textured blocks along the front and side faces of the building. In order to reveal the texture of the blocks as much as possible, the lights will be placed at a shallow angle relative to the face. Allowing the light to "rake" across the surface like this will increase the contrast of light on the surface of the blocks, making the ornamental pattern more visible.

Fig. 12-58. Rendered view of model with six shadow-casting and 400 non-shadow-casting interior lights.

As seen in figures 12-59 and 12-60, there are 12 cone lights in all: four below the surface of the street, illuminating the front face of each of the towers; four smaller lights focused on the concrete blocks on the upper half of each tower; two lights highlighting the top of the elevator cores; and a light on each end of the building, illuminating the concrete block stair towers.

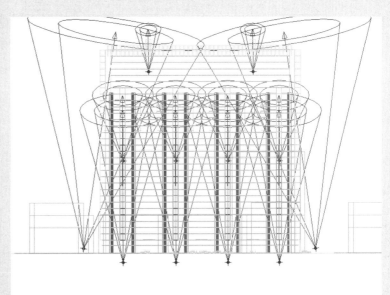

Fig. 12-59. Front view of model with non-shadow-casting exterior cone lights.

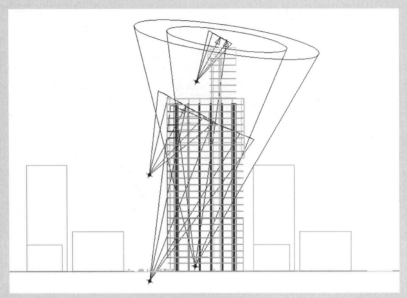

Fig. 12-60. Right view of model with non-shadow-casting exterior cone lights.

Because all of these cone lights point upward, there is no surface for them to project shadows against and, therefore, no need for them to cast shadows. However, like the point lights, there is a Falloff assigned to these lights so that they behave more like an actual light source. The parameters for one of the front cone lights are shown in figure 12-61.

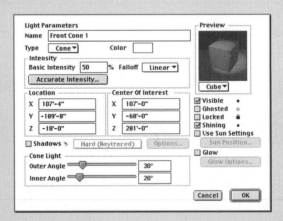

Fig. 12-61. Light Parameters dialog box for one of the front cone lights.

The completed model, shown in figure 12-62, contains over 400 light sources. This sounds like a lot of lights, and one would expect that it would require a lot of computing power to complete a rendering. However, the following choices were made during the lighting process that kept the required computing resources to a minimum.

- Out of the 400 lights in the model, only six cast shadows.
- All six of the shadow-casting lights cast Hard (raytraced) Shadows, keeping memory requirements to a minimum

- Much of the building detail was simulated through the use of a custom bump map and a procedural transparency map.

Fig. 12-62. Rendered view of model with six shadow-casting point lights, 400 non-shadow-casting interior point lights, and 12 exterior non-shadow-casting cone lights.

Placing Cameras for an Animation

When setting up an architectural animation, it is important to have in mind the types of views you want before setting up the animation keyframes. In the animation of Wright's National Life Insurance Building, there were four key views that were desired right from the start: a pedestrian view from the street level, two "birds-eye" views from either side of the building, and a close-up of the building's facade. Because it was known that these were views to be covered in the fly-through, they were set up first, as shown in figures 12-63 and 12-64. Figures 12-65 through 12-68 show other views of the building.

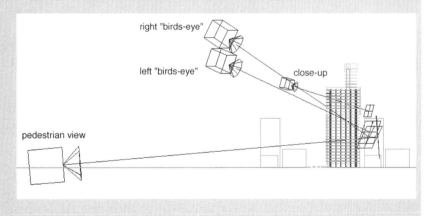

Fig. 12-63. Right view of model with four "anchor" views.

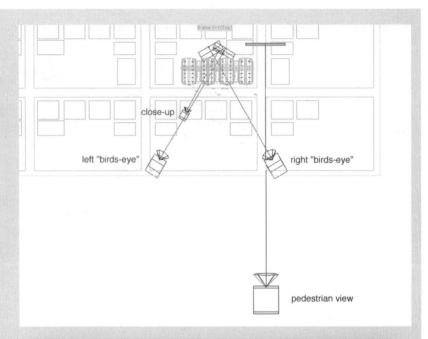

Fig. 12-64. Top view of model with four "anchor" views.

Fig. 12-65. Pedestrian view (frame 0).

Exercise 12-2: Advanced Animation Example   591

Fig. 12-66. Right "birds-eye" view (frame 135).

Fig. 12-67. Left "birds-eye" view (frame 225).

Fig. 12-68. Close-up view (frame 360).

The views shown in figures 12-65 through 12-68 serve as "anchors" in that they ensure that all of the required views will occur in the animation. Any additional views created for the fly-through sequence are used to create smooth transitions between the four anchor views, ensuring that the animated camera remains focused on the building through these transitions. Figures 12-69 and 12-70 show the placement of four additional cameras (a2, a4, a7, and a8) as transitions between the four anchor views (a1, a3, a5, and a6).

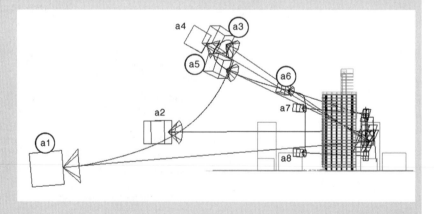

Fig. 12-69. Right view of model with cameras for the eight keyframes visible.

Exercise 12-2: Advanced Animation Example

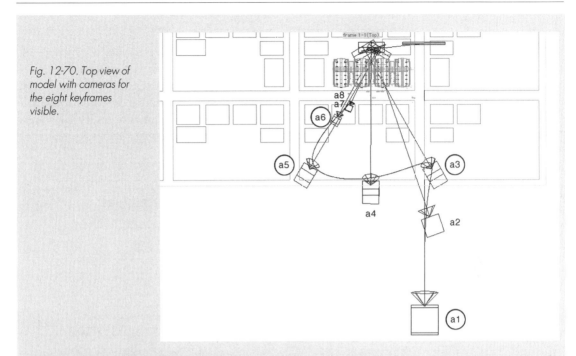

Fig. 12-70. Top view of model with cameras for the eight keyframes visible.

As you can see, the Centers of Interest for all views are directly behind and in the center of the building. This will ensure that the building remains in the center of the window throughout the fly-through sequence.

Using Walkthrough View to Create an Animation Path

As of Version 3.1 of form•Z, there is a very useful tool that helps you to set up quick animation paths. In the Windows tools, the Walkthrough view has an option that allows you to save the walkthrough view as an animated camera path (figure 12-71).

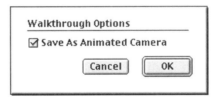

Fig. 12-71. Walkthrough Options in the Walkthrough View allow the view to be saved as an animated camera path.

This tool may be used to save an animation path that can be used to see a quick animation through your model. It can also be used to set up a mockup of animation paths that can be used later as a framework for setting up more elaborate animations.

Generating an Animation

Once the views have been set up correctly, generating the animation is relatively easy. As explained earlier in this chapter, all of the views are first made visible and then selected in the sequence desired for the animation. The Animation From Keyframes command under the View menu brings up the dialog box shown in figure 12-72.

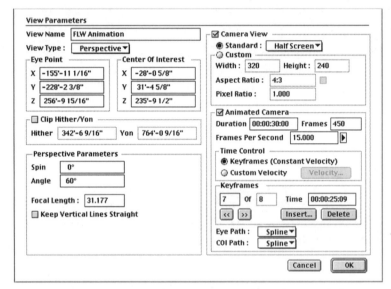

Fig. 12-72. View Parameters dialog box for the animation.

In the View Parameters window you can adjust the pixel dimensions, the duration, the frames per second (fps), the velocity, and the path shape of the animation. In this case, you would select a Half Screen size of 320 x 240 pixels. This is a size that will show a decent amount of model detail and remain playable on most computers. The duration of the animation was set to 30 seconds, with a frame rate of 15 fps. The 30-second duration allows for a relatively slow camera velocity throughout the fly-through and a slower frame rate, of 15 fps. The Time Control was set to Constant Velocity and the animation paths were constructed with splines. These can be seen in figures 12-69 and 12-70.

chapter 13

Fifty-five form•Z Tips

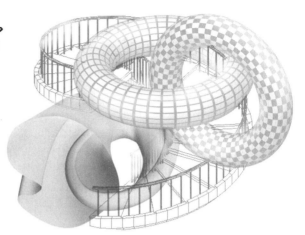

Introduction

These 55 tips are helpful hints that will make working in form•Z even easier. If you are a form•Z expert, knowing these will only make you better. If you are a beginner, mastering these techniques and tricks will make people think you have been using the program for years. The tips contain everything from rendering to modeling tips, organizational tricks, and hot-key manipulations.

➥ **NOTE:** *To access tips by subject, see the Subject Index to Chapter 12 at the back of the book.*

Fifty-five Tips

Fig. 13-1. Grid Snap tool.

The sections that follow present 55 tips for using form•Z that range from quickly setting Grid Snap to previewing plotter and printer setups before printing.

Tip 1: Snap To It! Quickly Setting Grid Snap

To quickly set your grid snap, double-click on the Grid Snap icon to access its options dialog box (figure 13-1). There you will find the Match Grid Module toggle. This will automatically match your grid snap spacing to your Window Grid spacing (figure 13-2).

Fig. 13-2. Match Grid Module option.

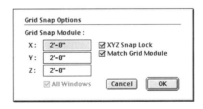

Fig. 13-3. Layers palette.

Tip 2: Setting Layer Attributes Quickly

When you turn on or turn off the visibility of multiple layers, form•Z always processes one layer at a time and refreshes the screen between operations. This refresh can be time consuming with complex models. To turn off multiple layers without the screen refreshing, hold down the Shift key while toggling the Visibility option in the second column of the Layers palette. If you continue holding down the Shift key, you can turn off several layers without the screen redrawing (figure 13-3).

Tip 3: How Far? Using the Measure Tool

Fig. 13-4. Measure tool.

Fig. 13-5. Measure options.

If you want to check a measurement quickly without the hassle of using the dimensioning tools, you can use Measure. Double click on the Measure tool to access its options box and toggle one of the many distance calculation options, such as Point to Point (figures 13-4 and 13-5).

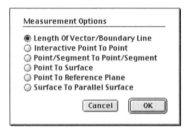

Tip 4: Rendering with Wire Frame

Sometimes a RenderZone or Shaded Render image is not legible enough, which would be improved if the wireframe of the model showed through. Instead of compositing your rendered image with a wireframe image in Photoshop, form•Z includes a special function that will do it for you automatically.

Fig. 13-6. Query Attributes tool.

If you query an object using the Query Attributes tool (figure 13-6), a dialog box will appear (figure 13-7) in which you can toggle the option to Render As Wireframe. This option has to be set for each object, but you can copy attributes between objects using the Get Attributes and Set Attributes tools.

Fifty-five Tips

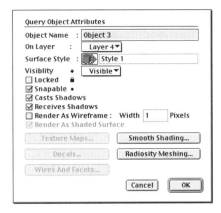

Fig. 13-7. Query Object attributes.

Tip 5: Closing Lines in the Prompts Window

As you draw an object, you can type coordinates into the Prompts window to guarantee the accuracy of your work. To close a line, type *c* in the Prompts window (figure 13-8). To end a line, type *e* in the Prompts window.

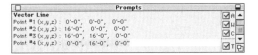

Fig. 13-8. Closing a line in the Prompts window.

Tip 6: Zoom to the Zoom Using Hot Keys

It takes a lot of time to go to the bottom of the window every time you want to zoom in, zoom out, or fit your model in the window. Use hot keys instead. Command + F will fit your model in the window. Command + [zooms in. Command +] zooms out. Use Shift Command + [or] to zoom in and out by window.

Tip 7: Printing Tip—Getting Good-quality B&W Printouts

If you are trying to get legible renderings out of a black-and-white printer, try rendering your images in only black and white. Change the Surface Render options by toggling off the Show Color option and turning on Render with Shadows (figure 13-9).

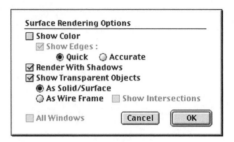

Fig. 13-9. Surface Rendering Options dialog box.

Tip 8: Returning to the Pick Tool

To prevent accidentally performing functions, select the Pick tool between operations. To have the system perform this function automatically, set the Pick Tool options to Return To Pick Tool (figure 13-10).

Fig. 13-10. Pick Tool Options dialog box.

Tip 9: Resetting Undo/Redo to Conserve Memory

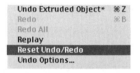

Fig. 13-11. Reset Undo/Redo tool.

If you are working on a large model, you may run out of disk space on your hard drive because every action you do, undo, or redo is recorded. If you are starting to run out of disk space on your hard drive and do not have enough room to save, try using Reset Undo/Redo under the Edit pull-down menu at the top of the screen (figure 13-11). This will erase the hidden file that records all of your actions.

➭ **NOTE:** *After you Reset Undo/Redo, you will not be able to undo changes made before resetting.*

Tip 10: Clear Rendering Memory

To make rendering more efficient, form•Z remembers certain calculations used to perform a rendering so that it will not have to do as much recalculation if you change views in a scene and rerender. This "memory" of a rendering is called Rendering Memory. Sometimes you will start to run out of RAM and will not have enough memory to work. Often, much of your RAM is being occupied by Rendering Memory. To free up this

Fig. 13-12. Clear Rendering Memory.

Fig. 13-13. form•Z Imager.

RAM, reset Rendering Memory under Clear Rendering Memory in the Display pull-down menu at the top of the screen (figure 13-12).

Tips 11: Using Imager

Sometimes a model is simply too complicated to attempt to render in form•Z. If you find you are running out of memory (RAM) while rendering big projects or images, try using the form•Z Imager (figure 13-13). This program is designed solely for rendering and is much more efficient and less memory intensive.

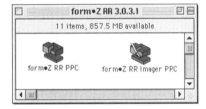

Tip 12: Clear All Ghosted

When you perform many functions in form•Z, such as Boolean operations and Derivative object functions, your original objects are not actually deleted, merely ghosted or hidden from view. These ghosted elements take up as much memory as visible elements. Therefore, if your model is getting large and unwieldy, try removing them. The Clear All Ghosted function in the Edit pull-down menu will delete all ghosted elements in your model (figure 13-14). Objects ghosted on a layer will not be affected.

Fig. 13-14. Clear All Ghosted.

Tip 13: Placing Scale Figure Views

If you are modeling a scene and want to place the view as if it were at eye level, try using View Parameters. The View Parameters window is available in the Views pull-down menu. This window allows you to numerically set the location of the view. To place it at eye height, you would set the z parameter under View Point to the proper eye height of a standing or seated person, such as 5'-0" (figure 13-15).

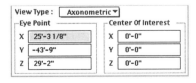

Fig. 13-15. Eye Point positioning.

Tip 14: Edit Display Color

Objects with a surface style whose color or texture matches or approximates the background can be difficult to find. To make your objects more visible while working, but retain their rendering qualities, edit the Simple Color. In the Surface Style palette, toggle off the RenderZone Average check box at the bottom of the window (figure 13-16).

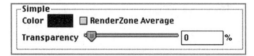

Fig. 13-16. RenderZone Average check box.

Once you toggle off the RenderZone Average check box, you can click on the patch of color in the Simple Color section and edit the color of your object. This color is applied to the wireframe only and has no effect on the color of the object in a rendering.

Tip 15: Nudging

If you want to quickly adjust an object's position, orientation, or scale, try using the nudge keys. With an object selected, use the arrow keys to move the object around, Control + arrow keys to rotate it, and Option + arrow keys to scale it. All of these functions can be done without selecting the Move, Rotate, or Scale tools. The move distance, rotation degree, and scaling percentage for the nudge keys can be set in the Pick tool options (figure 13-17).

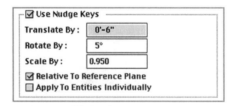

Fig. 13-17. Nudge keys setup.

Tip 16: Inserting a Point

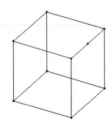

Fig. 13-19. Using the Insert Point tool.

To insert a point in a line, use the Insert Point tool in the Line Editing palette (figure 13-18). Click anywhere on a 2D line and a point will be inserted. Use the Midpoint or Segment Part snap to insert the point accurately (figure 13-19).

Tip 17: Extending the Cursor

form•Z includes a tool similar to AutoCAD's cross-hair cursor called Extend Cursor (figure 13-20). You can turn this tool on in the Windows pull-down menu at the top of the screen. This option turns on a set of cross hairs that extend to the edges of the window, parallel to the axes. This is especially useful when used in conjunction with the Rulers tool. You can turn on the Rulers tool by selecting Show Rulers from the Windows pull-down menu.

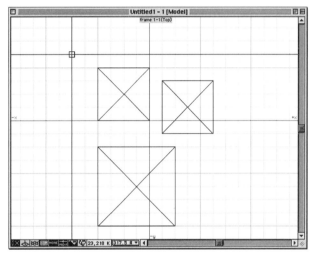

Fig. 13-20. The Extend Cursor option makes alignments easier.

▶ **NOTE:** *For more information on customizing the cursor, see the discussion of the Edit menu in Chapter 3.*

Tip 18: Grouping in the Object Palette

Groups can often be made more easily using the Objects palette instead of the Group tool (figure 13-21). Click on an open space in the Objects palette and an empty group will be automatically created. You can then select objects and add them to the group by clicking on their names in the Object window and dragging them into the group. To remove them from the group, simply drag the object names out of the group column.

▶ **NOTE:** *See Chapter 4 for further discussion of the Objects palette.*

Fig. 13-21. Objects Palette with nine objects collected in three groups.

Tip 19: Editing Point Types

To gain more control over C-curves, form•Z users can use the Edit Point Type option. With the Option/Alt keys held down, click on a point and you will be able to edit how the curvature is calculated at that point. The curvature can be set as Corner, Flat > Curve, Curve > Flat, or Curve, giving you greater control over the behavior of control curves (figure 13-22).

Fig. 13-22. The Point Type pop-up menu shows how a curve is constructed.

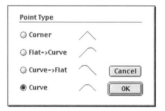

MacOS/PC NOTE: *Hold down the Option key (Mac) or the Alt key (PC) when editing points with the C-curve edit tool.*

Tip 20: Go Somewhere Sunny

If you want to set the sun position as if your model were in a particular location on Earth at a particular time, use the Sun Position item in the View menu at the top of the screen. You can specify the latitude and longitude of a location on the globe, or select from a list of cities. You can also specify the time of day and the year.

If you already know the proper sun altitude and azimuth values for your site and time, the sun can also be positioned by specifying them in the Geographic position options accessed via the Choose Site button in the Sun options dialog box (figure 13-23).

➤ **NOTE:** *For more information, see the discussion of the Sun Position options in the View menu section of Chapter 3.*

Fig. 13-23. Sun position options.

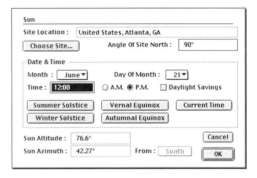

Fig. 13-24. Hiding the Palettes.

Tip 21: Run and Hide

To increase the viewable area on your screen, it is often easiest to hide the visible palettes. Instead of closing them all individually, it is possible to hide them all using the Hide Palettes function in the Palettes pull-down menu (figure 13-24). To reveal the palettes again, toggle off the Hide Palettes function. See Chapter 4 for a complete discussion of form•Z palettes.

Tip 22: Thinking Time

If you want to know how much memory your rendering will take, use the Estimate Memory Usage option in the RenderZone options found under Display Options in the Display pull-down menu (figure 13-25). This will calculate how much memory (RAM) it will take to render your image at the current settings. This way, you can find out before you start rendering if you have enough RAM to finish.

Fig. 13-25. Estimate Memory Usage option.

☑ **Estimate Memory Usage**

Tip 23: How Many Polygons?

If you want to find out how many polygons, faces, objects, groups, and points you have in your model, use the Project Info tool (figure 13-26). Select Project Info from the Help pull-down menu and it will count and display how many of each entity you have in your model.

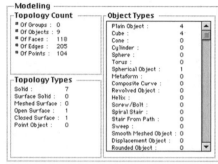

Fig. 13-26. Project Info tool.

Tip 24: Changing the Modeling Window Background

Some people find it is easier on the eyes to work with a dark background and lighter lines instead of a white background in the modeling window. To change the background color of the modeling window, open the Project Colors options from the Options pull-down menu (figure 13-27). Here, you can set the Background color, as well as other project colors such as the Select and Ghost colors and the colors of the Grid and Axes.

➥ **NOTE:** *If you change the background color, you may need to edit the Simple Display color of your Surface Styles so that your objects will show up in Wire Frame mode (see tip 14).*

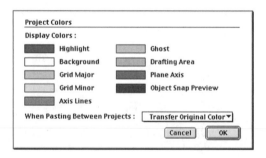

Fig. 13-27. Project Colors options.

Tip 25: Key Shortcuts

The Key Shortcuts tool in form•Z is an essential time-saving device, letting you perform functions quickly with just a keystroke instead of using the mouse.

Fifty-five Tips

MacOS/PC NOTE: *For example, Control + 6 (PC) or Command + 6 (Mac) are the Key Shortcuts for Top view. Pressing these two keys together will select Top from the Views menu. Using Control (PC) or Command (Mac) with any of the number keys will let you quickly switch between views.*

Shortcuts for the functions in the Menu bar are written next to their names in the pull-down menus. Key Shortcuts in the Edit pull-down menu contains a list of all key shortcuts. You can also create your own shortcuts with this option. The following table contains some of the most useful Key Shortcuts.

Function	MacOS	Windows
Deselect	Cmd + Tab	Ctrl + Tab
Undo	Cmd + Z	Ctrl + Z
Zoom In	Cmd + [	Ctrl + [
Zoom Out	Cmd +]	Ctrl +]
Fit View	Cmd + F	Ctrl + F
RenderZone	Cmd + K	Ctrl + K
Wire Frame	Cmd + W	Ctrl + W

Tip 26: Mouse Tracking

It is often necessary to type in the location of a coordinate instead of defining it by using the mouse. You can do this by typing dimensions and coordinates directly into the Prompts window. When using the Prompts palette to type dimensions, it is often easier if you turn off the palette's Mouse Tracking option.

Mouse tracking is a function of the Prompts palette that displays the current mouse coordinates (figure 13-28). This is usually desirable, but when tracking is on and you are typing in coordinates, they can be erased accidentally if the mouse is moved before you select Return or Enter. To prevent this, toggle off Mouse Tracking by clicking on the box labeled T on the right side of the Prompts palette.

Fig. 13-28. Mouse tracking.

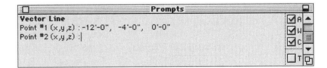

Tip 27: Rendering a Grid

Sometimes, for illustrative purposes, you want a grid to be displayed in a RenderZone rendering. To render the grid, you need to change the RenderZone options.

In the options, select Project Background from the Background pull-down menu and then click on the Options button next to the pull-down

MacOS/PC NOTE: *To open the RenderZone options, hold down Ctrl + Shift (PC) or Option (Mac) while selecting RenderZone from the Display pull-down menu.*

Fig. 13-29. Background pull-down menu.

menu (figure 13-29). There is an option to Render Grid And Axes As 3D Lines. This option displays the grid in a rendering similar to an object with its Render Attributes set to Render As Wire Frame (figure 13-30).

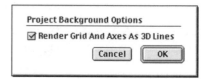

Fig. 13-30. Project Background Options dialog box.

Fig. 13-31. Join tool.

Fig. 13-32. Difference tool.

Tip 28: Performing Multiple Boolean Operations

If you have several objects in your project that need to be subtracted from a single object using the Boolean Difference tool, it is not necessary to subtract them one at a time. To subtract a group of objects, select all elements to be subtracted and join them using the Join tool from the Join and Group palette (figure 13-31).

Now when you perform a Boolean Difference with the Difference tool (figure 13-32), and click on the object to be subtracted, all of the objects you joined will be automatically selected and subtracted at once. This trick also works for the other Boolean functions: Union, Intersection, and Split.

Tip 29: Previewing Your Work

To save time, it is often a good idea to make a preview rendering before doing a full-scale rendering of your model. To preview a rendering, make the window size smaller. This will drastically cut down on rendering time. Select Image Options from the bottom of the Display pull-down menu and select Use Custom Size. Click on the arrow next to the option By Number of Pixels and select 320 x 240 from the pull-down menu. This will change your window screen size to 320 x 240 pixels and decrease your rendering time.

To perform a full-screen render, go back to the Image Options dialog box and select the Use Window Size option. Click on the bottom right-

Fifty-five Tips

hand corner of the modeling window and drag it to make the window large again (figure 13-33).

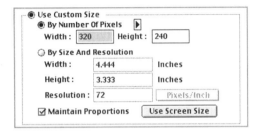

Fig. 13-33. Preview size options.

Tip 30: Saving Preferences

Once you have started to get familiar with form•Z, you will probably want to change some of the default settings to preferred settings. Almost all of the options (e.g., Working Units, Arc/Circle Resolution, Status Of New Object options, and Project Colors) you change in form•Z can be saved as personal preference files and used as default settings. To make a personal preference file (figure 13-34), set all of the options the way you prefer and then select Preferences from the Edit pull-down menu.

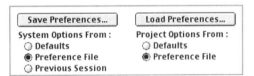

Fig. 13-34. Personal preference file.

Click on the Save Preferences button, which will save your settings to a file. Then set Systems Options From and Project Options From to Preference File. Now whenever you start up form•Z it will use your personal preferences instead of the default settings (figure 13-35).

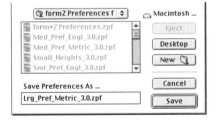

Fig. 13-35 Applied personal preference file.

Tip 31: Nonplanar Faces

If you deform objects or edit their points and then attempt to perform a Boolean operation, you will often get an error that reads "Non planar face(s) encountered." A nonplanar face contains four or more points,

Fig. 13-36 Triangulate tool.

where all of the points are not in the same plane. The remedy for this is to triangulate your object. Triangulating an object converts all faces into triangles that are, by definition, planar. To triangulate an object, select the Triangulate tool from the Meshes and Deform tool palette and click on the offending object (figure 13-36).

If the object has no nonplanar faces, form•Z will inform you that all faces are planar and will not triangulate the object. Once you have triangulated your object, you can proceed with the Boolean operation.

Tip 32: Rendering a Portion of a Window

MacOS/PC NOTE: Open the RenderZone options by either holding down Ctrl + Shift (PC) or Option (Mac) while selecting RenderZone from the Display pull-down menu.

Often you will not want to wait for an entire model to render to see what a portion of a model will look like. To render a portion of the screen, use the Set Image Size option.

Select the Set Image Size option at the top of the option box and click on OK. Now every time you access RenderZone instead of the entire window being rendered, the cursor changes to a paintbrush, which you can use to draw a box around the area you want to be rendered. To return to rendering full screen, turn off the Set Image Size option (figure 13-37).

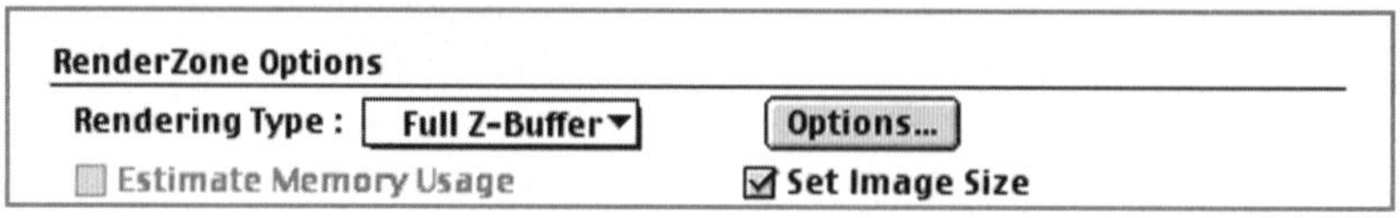

Fig. 13-37. RenderZone Set Image Size options.

Tip 33: Viewing Relative to a Plane

When using an arbitrary reference plane, it is often the case that you want to view an object relative to the arbitrary plane instead of the World reference plane. This means that instead of looking directly down at the World *xy* plane, the Top view is one of looking directly down at the arbitrary *xy* plane.

To see a Top view relative to an arbitrary reference plane, make the arbitrary plane active and select Plane Projection from the Views pull-down menu (figure 13-38). When you select Plane Projection, a submenu will appear, from which you can select any of the projection views (i.e., Top, Right, or Left).

Fig. 13-38 Plane Projection.

Tip 34: Insurance?

When you save a project, the Undo/Redo record is reset, "freezing" your model so that you can no longer undo the operations you have performed. This is usually not a problem. However, when you want to experiment, you want a little more insurance.

You can get unlimited Undos (as long as you have the disk space to store it) by changing the Undo options (figure 13-39). Select Undo from the Edit pull-down menu and turn off the Reset After Saving Project option. If you want even further insurance, you can turn on the Save Undos In Project option so that even after you Save and Quit, when you reopen your model, you will be able to Undo.

Fig. 13-39. Undo options.

Tip 35: Instant Replay

A great tool available in form•Z for presenting your modeling process is the Replay tool (figure 13-40). This function clears the screen and plays back all operations you have performed since Undo/Redo was last reset. Replay responds with a simple step-by-step animation of your model being built, which is ideal for client presentations and demonstrations. The Replay tool can also be used in a rendered mode, in which you render your model and then select Replay. If you save your Undo/Redo records with your model (see tip 34), you can play back the entire process at any time.

Fig. 13-40. Replay tool.

Tip 36: Separating an Object from a Group

Sometimes you want to separate one object from a group. Instead of ungrouping all objects and then making a new group minus that one object, you can remove it from the group using the Remove Object

Fig. 13-41. Ungroup tool.

From Group option. Double click on the Ungroup tool (figure 13-41) to open the Ungroup Operations dialog box (figure 13-42) and select the Remove Object From Group option. Select the Ungroup tool, and set the Topological Level to Object. Now you can select the object to be removed and it will be separated from the group.

Fig. 13-42. Ungroup options.

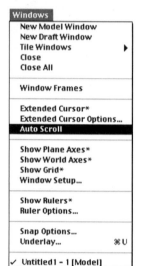

Fig. 13-43. Auto Scroll tool.

Tip 37: Auto Scroll

If you have ever tried to model an object that is larger than the window frame, you will appreciate this option. Instead of having to click on the scrolling slider to scroll around a window, the Auto Scroll tool will do it automatically (figure 13-43). Toggle on Auto Scroll by selecting it from the Windows pull-down menu. Now when you are performing a function that goes off the screen, the window will scroll itself.

Tip 38: The Pick Parade

Pick Parade is a function of the Pick tool that makes it easy to select an object lost in a crowd. When several objects are stacked on top of one another and you want to select just one, hold the Shift key down and click on the object. If the wrong object is selected, keep the Shift key held down and click again with the mouse. This will pick the next object. Continuing to hold down the Shift key and clicking will "parade" through the objects. When the object you want to select is highlighted, let go of the Shift key and click again with the mouse; the object will be picked.

Tip 39: Click and Drag

MacOS/PC NOTE:
Hold down the Option key (Mac) or Ctrl + Shift keys (PC) to click and scale an object

Turning on the Click and Drag option for the Pick tool adds a lot of functionality beyond simply selecting objects. Open the Pick Options dialog box by double clicking on the Pick tool and checking off the Click And Drag box (figure 13-44). You will then be able to move objects with the Pick tool alone by clicking on the object and, with the mouse button still held down, dragging the object around the screen. If you hold the Control key down while clicking on the object, you will be able to rotate it dynamically.

Fifty-five Tips

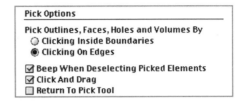

Fig. 13-44. Pick Options dialog box.

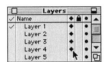

Fig. 13-45 Layers palette.

Fig. 13-46 Extend Plane tool.

Tip 40: Dragging in the Layers Palette

To quickly toggle on or off several layers in the Layers palette, it is possible to click and drag. Just click with the mouse in the Visibility column in the Layers palette next to a layer name (figure 13-45). This will toggle its visibility on or off. While continuing to hold down the mouse button, drag the cursor up or down within the column and all of the layers will be turned on or off as the cursor passes by them. Furthermore, this applies to any column in any palette.

Tip 41: Extending a Grid

To quickly extend a grid so that it reaches the extents of your model, select the Extend Plane Grid tool from the Reference Planes palette in Window Tools (figure 13-46). Hold the Shift key down and click anywhere in the modeling window and the grid will be automatically enlarged and moved to encompass your entire model.

Tip 42: Modeling with a Green Thumb

There is quick and easy trick to making trees. Instead of modeling out every leaf and branch, use the Transparency Maps tool. To make some convincing trees, take two simple rectangular planes standing on their short ends and intersecting each other like an X on a plane (figure 13-47). On these planes, apply a surface style with an Image Map of a tree with a black for its Color and a Transparency Map of the same image for its Transparency option background (figure 13-48).

If you select the option, Use Alpha Channel, and then render the model, you will have a tree that looks good from almost any angle (figure 13-49.) You may have to adjust the size of the texture map with the Texture Map tool if it is rendering too small. You can quickly create a forest by copying it randomly. For added realism, try using more than two planes.

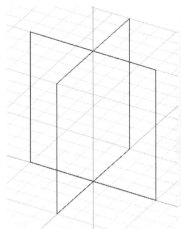  

Fig. 13-47. Two intersecting planes to be texture mapped.

Fig. 13-48. Image of tree with black background to be used for Image Map and Transparency Map.

Fig. 13-49. Rendered tree.

Tip 43: Pasting on Layers

Often when working with a large model, you find yourself cutting and pasting objects from one layer onto another. By default, objects are pasted onto the layer from which they came. Instead of pasting an object and then changing its layer with the Layers tool, it is often easier to use the Paste On Active Layer option. Open the Layer Options dialog box and check the Paste On Active Layer option in the bottom right corner (figure 13-50). This will paste objects directly onto the active layer.

Fig. 13-50. Paste On Active Layer option.

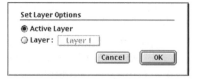

Tip 44: Relative Rules

The rulers in form•Z, when turned on, appear in the margins of the modeling window. This is a useful tool when you need to model quick, yet accurate, models. By default, the rulers show the distance measured from the origin (0'-0",0'-0",0'-0"). If you change the Coordinate mode to Relative in the Rulers Options, all measurements will be shown relative to the last mouse click (figure 13-51). This will make it easy to sketch models, and you will always know how far you are from the last point, intersection, or corner.

Fig. 13-51. Turning on the Relative Coordinate mode for rulers.

Tip 45: Snapping To Objects in Plan

Using object snaps such as Point Snap in projection views can often cause unexpected results. If you are looking down in Top view on a 3D object, such as a cube, and trying to snap to its corner points, you never know if you are snapping to the top or bottom point until you switch to a 3D view and see the results. To avoid unexpected results, toggle on the Lock Drawing To First Point or the Project Onto Reference Plane option located in the Object Snap options (figure 13-52).

Fig. 13-52. Turning on the Lock Drawing To First Point and Project Onto Reference Plane options.

Lock Drawing To First Point will create an imaginary reference plane parallel to the view at the first point you click. Every subsequent point drawn will be created in that plane. The Project Onto Reference Plane option will guarantee that you are modeling on the current reference plane no matter where in 3D space the point is to which you are snapping.

Tip 46: QuickDraw 3D

The QuickDraw 3D renderer is an amazing tool for getting instant feedback while modeling (figure 13-53). If you turn on the QuickDraw 3D mode and start modeling, the objects can be moved around, created, copied, and manipulated all while being rendered in real time. Using this allows you to see your model fully rendered while you work on it, giving you instant feedback. Be careful, though, as this renderer has its limits. Too many polygons in your model will slow down rendering time.

Fig. 13-53 Selecting the QuickDraw 3D renderer.

Tip 47: Fisheye

When you are viewing your model in perspective, sometimes you want to see a little more of the model, or tighten up the view to frame the project better. If you need to create a better cinematic effect with the perspective, try changing the lens on the camera. Open the Perspective options and change the Angle, located in the bottom left of the Perspective Options dialog box (figure 13-54).

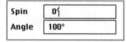

Fig. 13-54. Changing the Perspective Angle.

This angle defines the cone of vision. Making the Angle larger than the default of 60 degrees will make your field of vision wider, producing a "fisheye" effect. Making the Angle smaller will tighten the camera view, framing your scene more. Changing the Angle drastically can create dramatic or distorted camera effects.

Tip 48: Align and Scale Views

The Tile Windows are often a blessing when working with complicated 3D objects. They allow you to view your model in three projection views at once, as well as in a 3D view. However, after working with Tile Windows for a while, these views can become uncoordinated and no longer show corresponding portions of the project in all views. Try the Tile Windows function of Align & Scale Views, which is accessed from the Windows pull-down menu (figure 13-55). This function will change the view of each window to fit all of the objects in the project, effectively re-coordinating the windows in your modeling environment.

Fig. 13-55. Using the Align & Scale Views function for Tile Windows.

Tip 49: Have Model, Will Travel

If you have modeled your project with symbol libraries and need to move to another computer or send your files to someone via e-mail, it is best to package all of the libraries and the model in one folder. The Save Symbol Libraries option in the Save A Copy As options will package your model and all of your symbol libraries for you, ensuring that nothing important gets accidentally left behind (figure 13-56). The All Attached Libraries option will put any library that has been used in your project, even if it is no longer in use, into the project folder. The Libraries With Referenced Definitions Only option will put only those libraries with symbols in use into the project folder.

Fig. 13-56. Packaging symbol libraries.

Tip 50: Plot/Print Setup

When printing your form•Z models for portfolios, brochures, or handouts, often the scale of the drawing is not important, only that it fits the page. The Scale To Fit Media option in Plot/Print Setup prints your model as large as possible to fill the page specified in Page Setup (figure 13-57). If you have an 8-12" by 11" printer and do not use this option, you may accidentally print your model at a scale too large for one sheet of paper. form•Z will use as many sheets as necessary to print your model, even if it is a couple hundred. Therefore, if you are not using this option, be sure to preview your print first.

Fig. 13-57. Using the Scale To Fit Media option.

Tip 51: Updating Libraries

An easy method of editing and updating form•Z symbol libraries is to open them directly using the form•Z Library format when using Open in the File pull-down menu (figure 13-58). form•Z libraries that are opened can be edited as normal models and the changes saved to the library. With the library open it is possible to browse the symbols by clicking on them in the Symbols palette.

Fig. 13-58. Symbols Library file format.

Tip 52: Flipbook

You can make fun and simple flipbooks of your animations by exporting your animations as a set of independent images by selecting this option under Export Animation in the File pull-down menu. With individual images of each frame, you can print them and bind them into an animation that can be played by flipping through the pages.

Tip 53: Saving as Previous Version

form•Z now offers the option of saving a model file for use in a previous release. You can select the Project Options from the Save a Copy As menu choice in the File menu, and choose to save a copy of your release 3.1 model that will be compatible with versions 2.9.5, 2.9, or 2.8 (figure 13-59). This can be useful if work on a model is being shared by more than one computer, or if the model needs to be sent out and you are not certain of the available release number.

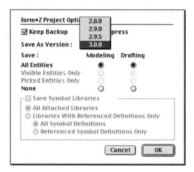

Fig. 13-59. Save a Copy As Project options box.

Tip 54: Recording Your Walkthrough

The Walkthrough tool in the View palette provides a convenient navigation method for exploring your models with real-time interaction either in wireframe, QuickDraw 3D, or OpenGL render. New for version 3.1, a Walkthrough Options dialog box has been added that contains the option to Save As Animated Camera, which when selected will save the walkthrough as an animated camera in the Views palette. These saved views can then be used later to generate a full animation.

Tip 55: Interactive Rendered Modeling

The use of the QuickDraw 3D and OpenGL rendering option is always valuable in supplying quick information regarding a model's composition and lighting characteristics from multiple views, by panning and rotating about the scene. Now in version 3.1, interactive modeling can be done while still in QuickDraw or OpenGL.

The Edit Controls and Edit Surface tools provide instant confirmation of the modeling procedures, interactively. If you make a modeling error, Undo and Redo work as well while in these rendering modes. You can observe the manipulation of the model itself geometrically, as well as see the effects of the editing on the shaded characteristics of the objects.

appendix A
form•Z on MacOS and Windows

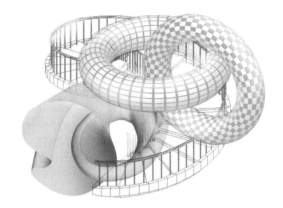

Introduction

Aside from a few minor differences, the PC (Windows) version and the MacOS version of form•Z are identical. The PC user will find no difficulties using this book, and moving from one platform to the other is almost seamless. The following sections describe the most apparent differences between the Mac and PC versions of the software.

Keyboard Commands

The most noticeable difference is in keyboard commands. Because of the differences between the Mac and PC keyboards, several of the key combinations are different. The equivalent of the Command key on the MacOS keyboard is the Ctrl key in the Windows version.

The Option key in MacOS does not exist on a PC keyboard. The equivalent in the Windows version of the software is Ctrl + Shift. For example, the keyboard command for Grid Snap On/Off in the MacOS version of form•Z is Option + G, whereas in the Windows version the same command is performed by pressing Shift + Ctrl + G. The Shift and Option key combination in MacOS is substituted with the Ctrl and Alt keys in Windows.

On the MacOS version, holding down the Option key while selecting an icon with a red dot in the corner, or a menu item with an asterisk, opens the Options dialog box. Because the Option key does not exist on the PC, Ctrl + Shift is used in the Windows version.

> **NOTE:** *You can view the full list of default keyboard commands by selecting the Key Shortcuts item under the Edit pull-down menu.*

Save As Options

There are some Save As options available to Windows users that are not supported in the MacOS version. The Windows version of form•Z supports one additional 3D data format, the 3D Studio file format. There are two additional image file formats for the PC version, BMP and Metafile. The PICT image file format is a Macintosh image format that is only available on the Mac and is not supported by the PC version of form•Z. For animation, the Quicktime Movie format is replaced by the AVI format, a more commonly used PC animation format.

OpenGL

OpenGL is a Windows-only option that makes use of OpenGL acceleration if you have the proper OpenGL hardware installed on your machine. OpenGL provides hardware rendering acceleration that allows for interactive manipulation of a model when rendered with the OpenGL display option. OpenGL gives you the option to either render as Shaded or Wireframe, as well as the option of rendering transparent objects as solids or transparencies. Smooth Shading is also supported under OpenGL. If you do not have an OpenGL card, the OpenGL Display option will still work, but you will not get the performance speed you would get with an OpenGL card.

Pick Tool/ Nudge Keys

As with the keyboard commands, the modifier keys (Shift, Ctrl, Alt) used with the Pick tool differ between the MacOS and Windows versions of form•Z. Geometric transformations (such as Move, Rotate, and Scale) can be performed on an object using the Pick tool when the Click and Drag option is selected in the Pick Options dialog box. This allows the user to simply pick an object by pressing the mouse button and, with the mouse button held down, to perform various geometric transformations by moving the mouse with a combination of modifier keys.

Using the Click and Drag option without any modifier keys simply moves the model parallel to the active reference plane. Using the Shift key constrains the movement to lines parallel to axes.

Using the Ctrl key on MacOS, or Shift + Ctrl keys in Windows, performs a rotation relative the active reference plane. Using the Command key in MacOS, or the Ctrl key in Windows, will rotate the object relative to the screen plane.

Scaling of objects is performed using the Option key in MacOS, or the Shift + Ctrl keys in Windows.

The Use Nudge Keys option in the Pick tool allows the user to perform geometric transformation to a selected object by using the arrow keys in conjunction with modifier keys. The modifier keys differ between MacOS and Windows versions.

NOTE: *You can view the full list of keyboard shortcut combinations under the Key Shortcuts item in the Edit pull-down menu.*

Loading Image Files into form•Z

form•Z allows for image files to be loaded within a number of tools. Images files can be loaded as texture maps, displacement maps, or background images, to name a few. When loading image files there are file formats that are either MacOS or Windows specific, and formats that can be used on both platforms. PICT files are MacOS specific, whereas BMP and Metafile formats are more Windows specific. PNG, JPEG, TIFF, and Targa formats are supported by both platforms.

appendix B

File Formats

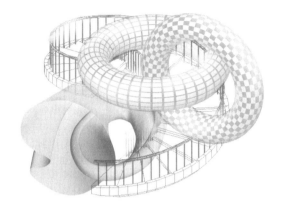

There are numerous file formats available for opening and saving files in form•Z. The file formats can be divided into image file formats, 3D/2D data file formats, animation file formats, and PC-only file formats. The file formats that fall under these categories are presented in alphabetical order within each category in the sections that follow. The following are the image file formats.

- EPS
- PICT
- Targa
- Quicktime Image
- JPEG
- PNG
- TIFF

When you save files using image file formats, only the image displayed on screen is saved. None of the 3D model information is saved. These file formats are for saving rendered images; the specific format depends on the intended use of the image. The following are the animation file formats.

- form•Z Animation
- Quicktime Movie

Animation file formats will save animations rendered in form•Z. They are similar to image file formats in that they do not save any 3D modeling data, only image data.

☛ **NOTE:** *Animation file formats are covered in detail in Chapter 12.*

The following are the 3D/2D data file formats.

- CIE
- DEM Data
- DXF
- CIBSE
- DWG
- FACT

- form•Z
- form•Z Library
- IES
- Illustrator
- OBJ
- SAT
- 3DGF
- VRML
- form•Z Imager
- form•Z Susp. Render
- IGES
- Lightscape
- RIB
- STL
- 3DMF

3D/2D data file formats will store the model data, but not the image displayed on screen. For example, if you render an image and save as form•Z, only the model data will be saved, not the rendered image. The method in which the model is represented and saved depends on the file format. Certain formats do not allow some types of objects to be saved; therefore, your choice of file format depends on the intended final use.

Another factor in deciding which format to use is whether or not a file format is platform specific. For example, the PICT file format is generally a Macintosh file format, whereas the TIFF format can be used on PC or Mac.

Three formats that are specific to PC: BMP, Metafile, and 3DS (3D Studio). These are discussed at the end of this appendix. The following are the PC-only data file formats.

- AVI
- Metafile
- BMP
- 3D Studio

The following sections contain brief explanations of each file format under the three categories of file formats. The descriptions include the applications for which each file format is best suited.

Image File Formats

The following sections describe the image file formats: EPS, JPEG, PICT, PNG, Quicktime Image, Targa, and TIFF. These are followed by the "3D/2D File Formats" section.

EPS

The acronym EPS stands for Encapsulated PostScript. Postscript is the language in which printers define the appearance of a page. If you are sending files out to be printed or plotted, this is the best format to use to guarantee your final print results. To guarantee your text will all appear properly, it is best to specify the Export Draft Text As Paths option, though this may create a larger file.

JPEG

The Joint Photographic Experts Group (JPEG) file format is usually used for saving images for use on the Internet. This file format is capable of high levels of compression, creating small files ideal for Internet use. Because this format always involves some form of compression, there is some quality loss in the image and it is not recommended for print use.

PICT

The PICT format is an image file format for the Macintosh. This format uses compression; therefore, if you are saving a lot of images, it may be a good choice. However, because of the compression it is generally a lower-quality image format than TIFF. If you are producing renderings for print, it is best to use the TIFF file format.

PNG

Portable Network Graphics (PNG) is a generic, cross-platform image format. This format is designed to be used for Web graphics; therefore, it is a very compact image format. This format is not widely in use as of yet, but has many improvements over other Web-based image formats such as GIF and JPEG. This image file format is not as high quality as TIFF and should not be used if you are intending to use your images for print.

Quicktime Image

Quicktime is a proprietary Apple format, and is available only if Quicktime 3.0 is installed on your machine. If installed, images can be saved in this format and viewed with a Quicktime software. If you will be transferring images between platforms and machines, this format is not recommended because it would require the Quicktime plug-in to be installed on a machine before the image could be viewed.

Targa

The Targa file format is a universal image format. Most graphic and video programs support this image file format. The file format is designed for use in video and is best used when producing sequences of images to be used for video.

TIFF

The Tagged-Image File Format (TIFF) is a universal image format. This file format is the industry standard for graphics and digital imaging. TIFF is popular because it retains the quality of the original image and can be read by almost all image programs on all computer platforms.

3D/2D File Formats

The following sections describe the 3D/2D file formats: CIE, CIBSE, DEM Data, DWG, DXF, FACT, form•Z, form•Z Imager, form•Z Library, form•Z Susp. Render, IES, IGES, Illustrator, Lightscape, OBJ, RIB, SAT, STL, 3DGF, 3DMF, and VRML.

CIE

The International Commission of Illumination (CIE) file format is for lighting data, which can be loaded and used in RadioZity renderings. To load files of this type, click on the Load button in the RadioZity Intensity Settings of the light source for which you would like to use the data.

CIBSE

The Chartered Institution of Building Services Engineers (British) file format (CIBSE) is for lighting data, which can be loaded and used in RadioZity renderings. To load files of this type, click on the Load button in the Radiosity Intensity Settings of the light source for which you would like to use the data.

DEM Data

The Digital Elevation Model (DEM) file format, developed by the U.S Geological Survey, is used for storing geographical elevation and position data. form•Z will import DEM files and interpret them as a surface objects. This format is only available for importing files and cannot be used for saving data.

DWG

DWG is the AutoCAD file format. When transferring your work back and forth between AutoCAD and form•Z, this is the best format to use. The method in which DWG stores objects is very similar to DXF, and has many of the same limitations. DWG is slightly more efficient than DXF; therefore, DWG file sizes are not as large as DXF file sizes. However, the DWG file format is less universal than DXF.

DXF

DXF is an acronym for Drawing Exchange Format. Although it is very inefficient, this format is probably the most universal file format. Almost all modeling, drafting, and illustration programs support this format. This format is also useful because it supports symbols and layers. However, when using this file format, most surface style information will be lost and only some types of C-curves will retain their controls. The main problem with DXF files is their size. These files can often become very large, making them difficult or impossible to open in form•Z.

FACT

This is the Electric Image file format. When going back and forth between Electric Image and form•Z, select this file format. FACT supports colors and image maps, as well as some of the other Surface Style settings, including Reflection and Bumps.

form•Z

form•Z file format is the native file format of form•Z. This is the default format. All of your files should be saved as form•Z unless you are transferring your work to a different modeling or drafting program.

form•Z Imager

The form•Z Imager file format will allow you to open form•Z Imager files using the form•Z Open dialog box. If you open form•Z Imager files, the Imager will automatically be launched and the file will be viewed within the Imager. The Imager is a standalone rendering program for form•Z.

form•Z Library

form•Z permits you to open a form•Z symbol library as if it were a model file. After opening a form•Z Library file you can browse the objects using the Symbols palette. Any changes made to a symbol library object when it is opened in this manner will be saved to the library and will affect any models that reference the library. This is a powerful method of editing and maintaining symbol libraries.

form•Z Suspended Render

The form•Z Suspended Render format is for saving renderings that are in the process of rendering and you would like to postpone the completion of the rendering process. This is useful for situations in which you begin a rendering without realizing how long it will take and then wish to postpone the rendering to work on other things on your computer.

A suspended rendering saves all of the rendering memory. Therefore, when you resume rendering, the program does not have to redo any of the computation. To suspend a rendering, hold the Shift key down as you cancel the rendering. Use Command + Period in Mac or Ctrl + Period in Windows to cancel a rendering. To resume a rendering, open the form•Z Susp. Render file.

IES

The Illumination Engineering Society (IES) of North America file format is for lighting data. This data can be loaded and used in RadioZity renderings.

To load files of this type, click on the Load button in the Radiosity Intensity Settings of the light source for which you would like to use the data.

IGES

The Initial Graphics Exchange Specification (IGES) format is, next to DXF, one of the most common and supported file formats. If the exact file format for the software you are transferring between is not available, this is usually the best file format to use. IGES supports most object types in form•Z, including C-curves and C-meshes, as well as layers and symbols.

Illustrator

This file format is for going between form•Z and Adobe Illustrator. This format supports lines and C-curves, which are converted to paths in Illustrator.

Lightscape

Lightscape is a radiosity rendering and animation application. This file format exports model information and lighting information to be used in Lightscape renderings.

OBJ

The Wavefront Object (OBJ) format is a 3D data format. This format is similar to DXF and 3DGF in that it represents an object as polygons. This format has been around awhile and is widely accepted, although it is not as universally used as DXF.

RIB

The RIB file format is the format used by Pixar's Renderman. This is a high-end rendering program used primarily in the movie industry. A RIB file stores all of the data necessary to render a file, including lighting and viewing parameters. RIB has little application outside exporting files to be used in Renderman.

SAT

The SAT file format uses a special library of objects marketed by Spatial Technology called ACIS. The ACIS library is used with several modeling programs. form•Z allows you to import SAT format files and convert them to objects that can be used in form•Z. However, to convert these files you need to install the form•Z ACIS module on your machine. ACIS data can also be embedded in DXF and DWG files. At this time, form•Z can only import SAT files.

STL

STL (an acronym for stereolithography, a popular rapid prototyping process) is the most commonly used format for rapid prototyping. The format polygonizes your model and saves it as a set of triangles. This format is only intended for rapid prototyping and does not support splines, layers, or colors. With this format, there are often several limitations imposed on your model by the fabrication process. Therefore, you should be familiar with the rapid prototyping process you will be using before beginning a model.

3DGF

3DGF is another 3D data file format. This file format, similar to DXF, represents a model as a list of polygons. Therefore, the controls for most parameter-based objects such as C-curves and C-meshes are usually lost.

3DMF

3D Movie Format (3DMF) is primarily a presentation format. This format saves files that can be viewed with a 3DMF viewer, which allows you to pan and orbit a lighted and rendered model for presentations and reviews. This file format can be used for transferring 3D data between software programs, but it is not well suited for the task.

VRML

Virtual Reality Modeling Language (VRML) is a 3D model format for the World Wide Web. This format represents models as a list of polygons or primitive objects. The format supports textures and image maps. Although it is not possible to do in form•Z, the VRML format also supports audio and video references, as well as hyperlinks and programming with Java or JavaScript.

Animation File Formats

The following sections describe the animation file formats: form•Z Animation and Quicktime Movie.

form•Z Animation

form•Z Animation is an animation file format specific to form•Z. That is, it can be opened using form•Z only. The file consists of a sequence of rendered images: one for every frame of your form•Z animation. form•Z animation files are created when you select Generate Animation from the Display pull-down menu. Once a form•Z animation file is created, it can be turned into a Quicktime Movie, AVI, or individual images using the Export Animation function under the File pull-down menu.

Quicktime Movie

The Quicktime Movie file format is an Apple format for movies and QTVR (Quicktime Virtual Reality) files. Movies and QTVR files created in form•Z and saved in this format can be opened and viewed using form•Z. QTVR files can be generated using the Save Quicktime VR function in the File pull-down menu. Quicktime Movie files can be generated using the Export Animation tool and are generally recommended as an animation format because they are more universally acceptable than the form•Z Animation file format.

➬ **NOTE:** *For more information on animation file formats, see Chapter 12.*

PC-only File Formats

The following four formats are specific to the PC (Windows) version of form•Z: AVI, BMP, Metafile, and 3D Studio. These file formats are described in the material that follows.

AVI

The Audio Video Interleave (AVI) format is the most commonly used video format on the PC. The standard for AVI format is defined by Microsoft and is consequently supported by the Windows operating system. The AVI file format is available on form•Z for exporting animations produced with form•Z's animation tools. There are a variety of possible compression formats available for AVI movies, with each format offering a different quality and compression ratio.

BMP

The Windows Bitmap (BMP) file format is a PC-only image format. This file format can be used in the form•Z drafting environment, as well as for renderings, underlays, and texture maps.

Metafile

The Windows Metafile format is a standard Windows image file format. This file format is supported by the PC version of form•Z only. The Metafile format supports pixel-based images, as well as vector-based lines. Therefore, it is very effective for saving images of drafted drawings or wireframe images.

3DS

3D Studio file format (3DS) is a 3D data format used by 3D Studio. This file format is specific to the PC version of form•Z and is used to transfer files to and from 3D Studio. 3D Studio is a modeling, rendering, and animation program. Models made in form•Z can be transferred to 3D Studio using the 3DS format, and can be animated.

appendix C
Keyboard Shortcuts

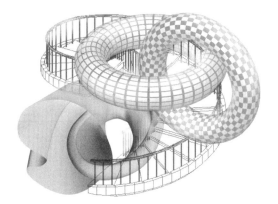

The following two tables constitute a complete list of keyboard shortcut commands for form•Z. The first table lists keyboard shortcuts for the Macintosh user. The second table lists shortcuts for the PC (Windows) user. In both tables, the command (in alphabetical order) is listed on the left, and the keyboard shortcut is listed on the right.

Mac Keyboard Shortcuts

Command	Keyboard Shortcut
Angle/Slope Snap	Option + Shift + D
Arbitrary Reference Plane Active	Option + A
Back	Command + 0
Bottom	Command + 7
Cancel	Command + .
Cancel	Escape
Capture Digitizer Mesh	F4
Close Drawing Thought Point	Shift + C
Close Drawing	C
Copy	Command + C
Cut	Command + X
Decrease Scale, X Direction	Command + Arrow, left
Decrease Scale, Y Direction	Command + Arrow, down
Decrease Scale, Z Direction	Command + Shift + Arrow, down
Delete Palette Item	Command + Delete
Deselect	Command + Tab

Command	Keyboard Shortcut
Duplicate	Command + D
Edit Cone of Vision	Command + E
End Drawing Thought Point	Shift + E
End Drawing	E
Exit Radiosity	F8
Face	Option + F
Fit All	Command + F
Front	Command + -
General Help Dialog	Command + /
Generate Radiosity Solution	F7
Grab Image	Command + G
Grid Snap	Option + G
Group	Option + U
Hidden Line Display	Command + H
Hole/Volume	Option + H
Import	Command + I
Increase Scale, X Direction	Command + Arrow, right
Increase Scale, Y Direction	Command + Arrow, up
Increase Scale, Z Direction	Command + Shift + Arrow, up
Initialize Radiosity	F6
Left	Command + 9
Move Negative X	Arrow, left
Move Negative Y	Arrow, down
Move Negative Z	Shift + Arrow, down
Move Positive X	Arrow, right
Move Positive Y	Arrow, up
Move Positive Z	Shift + Arrow, up
New Project, Modeling	Command + N
No Directional Snap	Option + Shift + X
No Object Snap	Option + Shift + N
Object	Option + O
Open GL Display	Command + =
Open	Command + O
Ortho and Diagonal Snap	Option + Shift + A

Command	Keyboard Shortcut
Ortho Snap	Option + Shift + O
Outline	Option + C
Paste	Command + V
Perpendicular Switch	Option + V
Point	Option + P
Preferences	Command + ,
Preset View 1	Command + 1
Preset View 2	Command + 2
Preset View 3	Command + 3
Preset View 4	Command + 4
Preset View 5	Command + 5
Previous View	Command + Shift + ,
Print Page	Command + P
Project Info	Command + Shift + /
Quick Paint Display	Command + T
QuickDraw 3D Display	Command + J
Quit/Exit	Command + Q
Radial Snap	Option + Shift + R
Radiosity Options	F5
Redo	Command + B
RenderZone Display	Command + K
Replot Drafting Window	Command + W
Reset View	Command + \
Right	Command + 8
Rotate Negative X	Control + Arrow, down
Rotate Negative Y	Control + Shift + Arrow, down
Rotate Negative Z	Control + Arrow, left
Rotate Positive X	Control + Arrow, up
Rotate Positive Y	Control + Shift + Arrow, up
Rotate Positive Z	Control + Arrow, right
Save	Command + S
Screen Digitizer Input	F2
Segment	Option + S
Select All Unghosted	Command + A

Command	Keyboard Shortcut
Select Previous	Command + `
Set View	Command + ;
Shaded Rendering Display	Command + L
Snap to Center	Option + Shift + C
Snap to Endpoint	Option + Shift + E
Snap to Face	Option + Shift + F
Snap to Intersection	Option + Shift + I
Snap to Line	Option + Shift + L
Snap to Midpoint	Option + Shift + M
Snap to Point	Option + Shift + P
Snap to Segment	Option + Shift + S
Surface Rendering Display	Command + R
Top	Command + 6
Underlay	Command + U
Undo	Command + Z
Uniform Scale Decrease	Option + Arrow, left
Uniform Scale Increase	Option + Arrow, right
Wire Frame Display	Command + W
World Digitizer Input	F3
XY Reference Plane Active	Option + X
YZ Reference Plane Active	Option + Y
Zoom In by Frame	Command + Shift + [
Zoom In Incrementally	Command + [
Zoom Out by Frame	Command + Shift +]
Zoom Out Incrementally	Command +]
ZX Reference Plane Active	Option + Z

PC (Windows) Keyboard Shortcuts

Command	Keyboard Shortcut
Angle/Slope Snap	Control + Alt + A
Arbitrary Reference Plane Active	Control + Shift + A
Back	Control + 0
Bottom	Control + 7

Command	Keyboard Shortcut
Cancel	Control + .
Cancel	Escape
Capture Digitizer Mesh	F4
Close Drawing Thought Point	Shift + C
Close Drawing	C
Copy	Control + C
Cut	Control + X
Decrease Scale, X Direction	Control + Arrow, left
Decrease Scale, Y Direction	Control + Arrow, down
Decrease Scale, Z Direction	Control + Shift + Arrow, down
Delete Palette Item	Control + Delete
Deselect	Command + Tab
Duplicate	Control + D
Edit Cone of Vision	Control + E
End Drawing Thought Point	Shift + E
End Drawing	E
Exit Radiosity	F8
Face	Control + Shift + F
Fit All	Control + F
Front	Control + -
General Dialog	Control + /
Generate Radiosity Solution	F7
Grab Image	Control + G
Grid Snap	Control + Shift + G
Group	Control + Shift + U
Hidden Line Display	Control + H
Hole/Volume	Control + Shift + H
Import	Control + I
Increase Scale, X Direction	Control + Arrow, right
Increase Scale, Y Direction	Control + Arrow, up
Increase Scale, Z Direction	Control + Shift + Arrow, up
Initialize Radiosity	F6
Left	Control + 9
Move Negative X	Arrow, left

Command	Keyboard Shortcut
Move Negative Y	Arrow, down
Move Negative Z	Shift + Arrow, down
Move Positive X	Arrow, right
Move Positive Y	Arrow, up
Move Positive Z	Shift + Arrow, up
New Project, Modeling	Control + N
No Directional Snap	Control + Alt + X
No Object Snap	Control + Alt + N
Object	Control + Shift + O
OpenGL Display	Control + =
Open	Control + O
Ortho and Diagonal Snap	Control + Alt + D
Ortho Snap	Control + Alt + O
Outline	Control + Shift + C
Paste	Control + V
Perpendicular Switch	Control + Shift + V
Point	Control + Shift + P
Preferences	Control + ,
Preset View 1	Control + 1
Preset View 2	Control + 2
Preset View 3	Control + 3
Preset View 4	Control + 4
Preset View 5	Control + 5
Previous View	Control + Shift + ,
Print Page	Control + P
Project Info	Control + Shift + /
Quick Paint Display	Control + T
QuickDraw 3D Display	Control + J
Quit/Exit	Control + Q
Radial Snap	Control + Alt + R
Radiosity Options	F5
Redo	Control + B
RenderZone Display	Control + K
Replot Drafting Window	Control + W

Command	Keyboard Shortcut
Reset	Control + \
Right	Control + 8
Rotate Negative X	Control + Alt + Arrow, down
Rotate Negative Y	Control + Shift + Arrow, down
Rotate Negative Z	Control + Alt + Arrow, left
Rotate Positive X	Control + Alt + Arrow, up
Rotate Positive Y	Control + Shift + Arrow, up
Rotate Positive Z	Control + Alt + Arrow, right
Save	Control + S
Screen Digitizer Input	F2
Segment	Control + Shift + S
Select All Unghosted	Control + A
Select Previous	Control + `
Set View	Control + ;
Shaded Rendering Display	Control + L
Snap to Center	Control + Alt + C
Snap to Endpoint	Control + Alt + E
Snap to Face	Control + Alt + F
Snap to Intersection	Control + Alt + I
Snap to Line	Control + Alt + L
Snap to Midpoint	Control + Alt + M
Snap to Point	Control + Alt + P
Snap to Segment	Control + Alt + S
Surface Rendering Display	Control + R
Top	Control + 6
Underlay	Control + U
Undo	Control + Z
Uniform Scale Decrease	Option + Arrow, left
Uniform Scale Increase	Option + Arrow, right
Wire Frame Display	Control + W
World Digitizer Input	F3
XY Reference Plane Active	Control + Shift + X
YZ Reference Plane Active	Control + Shift + Y
Zoom In by Frame	Control + Shift + [

Appendix C: Keyboard Shortcuts

Command	Keyboard Shortcut
Zoom In Incrementally	Control + [
Zoom Out by Frame	Control + Shift +]
Zoom Out Incrementally	Control +]
ZX Reference Plane Active	Control + Shift + Z

Subject Index to Learning Objectives

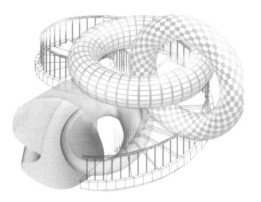

This index is intended to provide a means of accessing the learning objectives contained in chapters 5 through 9 (the tools and exercise chapters) topically. The "Learning Objectives" headings within each of the aforementioned chapters are numbered consecutively within chapters. The reference numbers given here after each entry consist of the chapter number and the learning objective number (followed by the page number). When the material is found on the companion CD-ROM, the page number so indicates.

Numerics

2D contour lines, using to create 3D solid 5-22 *211*
2D Enclosure tool, using 5-18 *204*
2D lines, using to generate parametric objects 5-38 *248*
2D modeling to derive 3D object 8-2 *408*
2D objects, creating with geometry 5-5 *178*
2D path, using for 3D form 7-12 *CD-12*
2D Surface Object tool 5-18 *204*
3D Converged tool, using 5-19 *206*
3D Enclosure tool, using 5-19 *206*
3D extruded forms, creating 5-6 *180*
3D Extrusion tool, using 5-19 *206*
3D form, created from 2D path 7-12 *CD-12*
3D logo, creating 7-5 *378*
3D objects, creating from lines and objects 5-19 *206*
3D solid, creating from 2D contour lines 5-22 *211*
3D surface objects, combining and subtracting 5-42 *260*
3D surfaces, creating with source shapes 5-27 *224*

A

adding and subtracting solid objects 5-41 *256*
advanced form•Z functions, applying 9-2 *451*
aligning and distributing objects 5-67 *305*
analyzing objects by cutting 5-43 *263*
architectural details, creating and presenting 9-10 *CD-189*
architectural modeling 9-11 *CD-198*
architectural monument 7-14 *CD-26*
arcs, creating 5-13 *194*
assigning color to objects 5-70 *312*
Attach tool, using with repetitive elements 7-3 *373*
attaching objects 5-66 *303*
attributes, copying from object to object 5-75 *322*

B

bevels, creating 5-53 *282*
blocks, modeling with single command 7-1 *366*
Bolts/Screw tool 5-24 *216*, 7-9 *CD-3*
Boolean operations, using to create complex forms 8-5 *433*; 5-41 *256*; 7-15 *CD-32*
Break Line tool, using 5-49 *278*

C

Cage tool, using 5-44 *266*
C-Curve tool, using 5-36 *242*
city blocks, creating with cubes 7-4 *375*
cityscape, creating 8-1 *394*
clearing controls 5-58 *287*
color, assigning 5-70 *312*
combining /splitting objects 5-45 *268*
combining and subtracting 3D objects 5-42 *260*
complex curvilinear forms, created with sweep and revolve tools 8-12 *CD-101*
complex model, generated from single curved line 7-2 *370*
modeling simply 8-11 *CD-91*
controlled cubic objects, generating 5-1 *170*
controlled curves, deriving with Vector Line 5-36 *242*
Controlled Mesh tool, using 8-7 *CD-61*
controlled meshes, using to create complex forms 8-5 *433*
controls, clearing 5-58 *287*
converting nonplanar to planar faces 5-33 *236*
converting surface solids and objects 5-57 *286*
copies, using permanent 9-9 *CD-177*
copying objects 5-59 *288*
corners, rounding 5-34 *238*
cursor constraints 6-6 *347*
curved primitive objects, generating and editing 5-2 *171*
curvilinear forms, complex 8-9 *CD-74*

D

deforming meshed objects 5-31 *231*
deleting objects 5-78 *329*; 5-79 *330*
direction reversal for lines and surfaces 5-56 *285*
direction snaps, using 6-5 *344*
displacement maps, using 5-32 *234*
display characteristics, setting for rendering 5-72 *315*
draft angles, creating 5-35 *240*

E

editing controls of parametric objects 5-16 *201*
editing objects 5-60 *291*
extending segments 6-68 *308*

F

fillets and bevels, creating on line segments 5-53 *282*
first points and point markers 5-55 *284*

Subject Index to Learning Objectives

G

geodesic dome, creating 7-10 *CD-6*
geometric projection, using 5-21 *209*
ghosted objects, creating 5-76 *324*
grid lines, snapping to 6-4 *341*
guidelines, setting up 8-6 *CD-51*

H

Helix tool, using 5-24 *216*
hierarchy of objects 5-46 *269*

I

image maps, applying 9-5 *CD-115*
image maps, using 8-3 *418*
inserting points & segments 5-54 *283*
Intersection tool, using 5-42 *260*

J

joining objects 5-45 *268*

L

labels, placing on surface 8-10 *CD-84*
landscape, creating with Terrain tool 7-8 *387*
lathed forms, creating 5-23 *213*
layers, using 5-77 *325*
lighting, advanced uses 9-4 *474*
lighting, setting up for rendering and animations 9-12 *CD-206*
line segment, creating 5-49 *278*
line segments, connecting 5-52 *281*
lines, drawing precisely 6-5 *344*
low-resolution objects 5-44 *266*

M

macros, using to transform geometry 5-65 *301*
material properties, special 8-4 *425*
measuring distances 5-61 *294*
mesh edges, smoothing 5-29 *228*
meshed objects, deforming 5-31 *231*
meshed surfaces with control points, generating 5-37 *245*
meshes, generating 5-28 *226*
Metaballs and organic forms 5-4 *177*
Metaballs, using to create spherical combinations 7-11 *CD-9*
Metaformz and organic forms 5-40 *254*
Mirror tool, using 5-64 *298*
model analysis 9-11 *CD-198*
modeling in 2D 8-2 *408*
modeling using images 9-1 *444*

N

nonplanar to planar faces, converting 5-33 *236*
Nurbz surfaces 5-17 *203*
  advanced modeling 9-3 *465*

O

object hierarchy, creating 5-46 *269*
object snaps, using 6-6 *347*
objects, aligning and distributing 5-67 *305*
objects, attaching 5-66 *303*
objects, copying 5-59 *288*
objects, odd-shaped 7-3 *373*
objects, displaying/editing 5-60 *291*
objects, joining/separating 5-45 *268*
objects, moving 5-62 *295*
objects, removing 5-78 *329*; 5-79 *330*
objects, scaling 5-63 *297*
organic forms, with sweep 5-26 *221*
organic forms, with Metaballs 5-4 *177*
orienting a model in space 6-1 *336*

P

parallel objects, generating 5-20 *207*
parametric objects, generating from 2D lines 5-38 *248*
parametric objects
  clearing controls 5-58 *287*
  editing controls of 5-16 *201*
parametric surfaces, editing 5-17 *203*
patch surfaces, creating and manipulating 5-39 *251*
Patch surfaces, editing 5-17 *203*
Patch tools, creating curved shapes with 9-8 *CD-164*
perpendicularity, working with 6-2 *338*
picking entities 5-15 *199*
placing shapes 5-69 *309*
Plane Reference tool, using 7-16 *CD-38*
point markers, setting 5-55 *284*
points/segments, inserting 5-54 *283*
project information, organizing 5-77 *325*

Q

Query Tool, using 5-60 *291*

R

recording macros 5-65 *301*
Reduce Mesh tool, using 5-30 *230*
reducing object complexity with Cage 5-44 *266*
reference planes, working with 6-3 *339*
reflecting entities 5-64 *298*
Render Attributes tool, using 5-72 *315*
rendering attributes, copying 5-75 *322*
rendering effects, using 9-4 *474*
rendering properties, creating 8-4 *425*
resolution enhancement with Smooth Mesh 5-29 *228*
resolution simplification with Reduce Mesh 5-30 *230*
reversing direction of lines/surfaces 5-56 *285*
Revolved Object tool, using 5-23 *213*; 7-14 *CD-26*
rounding corners 5-34 *238*

S

scaling objects 5-63 *297*
segments, closing open sequence of 5-50 *279*
segments, connecting 5-52 *281*
segments, extending 5-68 *308*
segments, trimming multiple 5-51 *279*
selecting entities 5-15 *199*
separating objects 5-45 *268*
single-command modeling 7-1 *366*
skin operation, used with underlay 9-6 *CD-129*
Skin tool, using 5-27 *224*
smoothing objects 5-71 *313*
snapping to grid lines 6-4 *341*
solid modeling, basics of 7-13 *CD-19*
source objects, using to generate 2D surfaces 5-18 *204*
source sections, using to create a surface 8-7 *CD-61*
spherical combinations 7-11 *CD-9*
spherical objects, generating 5-3 *174*
spiral forms, creating 5-24 *216*
spiral staircases, creating 5-25 *218*
splines, using to create objects and surfaces 5-12 *192*
staircase, easy modeling 7-17 *CD-44*
sstandardizing elements 9-9 *CD-177*
Stitch tool, using 5-42 *260*
subtracting objects 7-15 *CD-32*
surface controls, editing 5-17 *203*
surface objects, converting 5-57 *286*
surfaces, creating with source sections 8-7 *CD-61*
Sweep tool options, using 9-7 *CD-145*
Sweep tool, using 8-6 *CD-51*
sweep, using to create curvilinear forms 5-26 *221*
Symbol Library, using 5-48 *274*

T

text, creating/manipulating 5-47 *271*
texture mapping, basics 7-13 *CD-19*
texture maps, using tactically 7-7 *383*
texture maps, using 8-3 *418*
textures, advanced uses 9-4 *474*; 9-5 *CD-115*
textures, placing 5-74 *320*
textures, positioning and repeating 5-73 *317*
textures, using for realism 9-1 *444*
Topological Levels, using to modify modeling tools 5-14 *196*
Trim Lines, using 5-51 *279*
Trim tool, using 5-42 *260*
two-path sweep 8-8 *CD-67*

U

underlay, with skin operation 9-6 *CD-129*
unfolding 2D/3D shapes 5-21 *209*

V

view tools, using 6-7 *354*
views, adjusting 6-7 *354*
views, changing 6-8 *359*

W

walls, with 3D Enclosure 7-6 *381*

Subject Index to Chapter 13

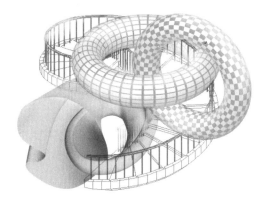

This subject index to Chapter 13, 55 form•Z Tips, is intended to provide a means of locating useful tips about form•Z and its functionality by type. The entries are ordered alphabetically, as in any general index. The reference "Tip" numbers correspond to the tips as consecutively numbered in Chapter 13. These are followed the page numbers where they can be found.

A

Align and Scale Views, using Tip 48, 614
Auto Scroll tool, using Tip 37, 610

B

background, changing Tip 24, 604
Boolean operation, performing multiple Tip 28, 606

C

changing the modeling window background Tip 24, 604
Clear All Ghosted, using Tip 12 599
Clear Rendering Memory Tip 10, 598
Click and Drag, using Tip 39, 610
conserving memory by resetting Undo/Redo Tip 9, 598
cursor, extending Tip 17, 601

D

Display Color, editing Tip 14, 600
dragging in the Layers palette Tip 40, 611

E

editing point types Tip 19, 602

Estimate Memory Usage option Tip 22, 603
extending a grid Tip 41, 611
extending the cursor Tip 17, 601

F

fisheye options Tip 47, 613
flipbooks, making Tip 52, 615

G

Grid Snap, quickly setting Tip 1, 595
grid, extending Tip 41, 611
grid, rendering Tip 27, 605
grouping in the Object palette Tip 18, 601
groups, making Tip 18, 601

H

Hide Palettes, using Tip 21, 603

I

Imager, using Tip 11, 599
inserting a point Tip 16, 600
interactive rendered modeling Tip 55, 616

K

key shortcuts Tip 25, 604

L

Layer Attributes, quickly setting Tip 2, 596
libraries, updating Tip 51, 615
libraries, using Tip 49, 614
lines, closing in Prompts window Tip 5, 597

M

Measure Tool, using Tip 3, 596
memory use in RenderZone Tip 22, 603
modeling interactively in QuickDraw and OpenGL Tip 55, 616
modeling trees Tip 42, 611
mouse tracking Tip 26, 605

N

nonplanar faces, working with Tip 31, 607
nudging, using Tip 15 600

O

object snaps Tip 45, 613
object, separating from a group Tip 36, 609

Subject Index to Chapter 13

P

palettes, hiding Tip 21, 603
pasting using layers Tip 43, 612
performing multiple Boolean operations Tip 28, 606
perspective views, options for Tip 47, 613
Pick Parade, using Tip 38, 610
Pick tool, returning to Tip 8, 598
Plot/Print Setup Tip 50, 615
point types, editing Tip 19, 602
point, inserting Tip 16, 600
polygons, knowing how many Tip 23, 603
preferences, saving Tip 30, 607
previewing your work Tip 29, 606
printing good b/w printouts Tip 7, 597
Project Info tool, using to count objects Tip 23, 603
Prompts window, closing lines in Tip 5, 597

Q

QuickDraw 3D, using Tip 46, 613

R

recording your walkthrough Tip 54, 616
rendering a grid Tip 27, 605
rendering a portion of a window Tip 32, 608
Rendering Memory, clearing Tip 10 598
Replay tool, using Tip 35, 609
rulers, using Tip 44, 612

S

Save Libraries option, using Tip 49, 614
saving as previous version Tip 53, 616
saving preferences Tip 30, 607
separating an object from a group Tip 36, 609
snapping to objects in plan Tip 45, 613
sun, setting the position Tip 20, 602

T

tracking your mouse Tip 26, 605

trees, making Tip 42, 611

U

Undo, adding more for insurance Tip 34, 609
Undo/Redo, resetting to conserve memory Tip 9, 598
updating libraries Tip 51, 615

V

versions, saving in previous Tip 53, 616
viewing relative to a plane Tip 33, 608
views, setting at eye level Tip 13, 599

W

walkthroughs, recording Tip 54, 616
Wire Frame, rendering with Tip 4, 596

Z

Zoom using hot keys Tip 6, 597

Index

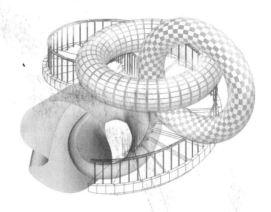

Symbols

Of Faces selection category 57
Of Segments option 186
    setting 189
(+XY) perspective mode 86
(+XY):Bottom command 86
[+XY]:Top command 86
[+YZ]:Right Side command 87
[+ZX]:Back command 88
[-XY]:Bottom command 86
[-YZ]:Left Side command 88
[-ZX]:Front command 88

Numerics

2 Path/Source Sweep option 222
2D contour lines 212
2D drafting module 138
    settings 130
2D drafting window 35
2D drawing CD-76
2D elements, joining 412
2D Enclosure tool 178, 179, 204, 205
2D frame 205
2D layout window, creating 117
2D line editing 278, 279, 281
2D objects
    derived from 3D solid 210
    creating 178
2D path, drawing CD-14
2D profiles, created from 3D object 265
2D Section option 264
2D shapes 128
2D spirals 217
2D Surface Object tool,
    using 371, 387, 422, 426, 434, CD-14, CD-62
2D surface objects 107
    and Boolean tool 179, 415
    editing 408
    generating 179
2D Surface tool 178–179, 204, 205
2D sweep path 223
2D-to-3D conversions 286
2-point rectangle, illustrated 185
3 Points option 175

3D axonometric views 35
    illustrated 181
3D Converged modifier, using 399
3D Converged Solid tool, using 373
3D Converged tool 180–182, 206
3D-first design philosophy 5
3D enclosure solids, illustrated 181
3D Enclosure tool 180–182, 206, 207, 408
    using 366, 381, 384, CD-63
3D extruded solid objects,
    illustrated 180
3D Extruded Solid tool, using 395
3D Extruded/Converged Solid
    options 180
3D Extrusion tool 206, 379, 408
    using CD-32
3D modeling window 4, 6
    pasting 2D elements into 51
3D Object mode, using 435, 436
3D objects
    creating 180, 206
    derived from 2D objects 205
    derived from 2D contour lines 211
    derived from 2D outline 408
    derived from 2D path CD-12, CD-19
    setting heights 182
3D Section option 265
3D shapes, derived from contour
    lines CD-139
3D space, measuring distance in 294
3D Text Editor 273, 379
3D Views option 10, 353, 373
3-Point Arc tools 195
3-point rectangle, illustrated 186
4 Points option 175

A

absolute coordinates, setting 76
absolute units 146
Adaptive Meshing 236
Add button in 23
Add To Pick option 57
Adjust Direction option 300
Adjust Heights At Intervals Of option,
    using 212, 388

Adobe Photoshop, pasting images
    in 52
Adobe Premiere 5
aerial photographs, generating site
    models from 78
Align & Scale Views command 67, 102
    using (Tip 48) 614
Align/Distribute tool 305–308
All Windows option 109, 111
Allow Collinear Points option 119, 191
Allow Intersecting Lines option 119, 191, 196
    warning concerning 192
Alpha Channel option 506, 508
Always Clear Rendering Memory
    command 117
Always Open File Format option 62
ambient illumination 530
analysis tools 515
Angle Display Options dialog
    box 125
Angle option 95
Angle/Slope Snap tool 344, 345, 346
angles, setting accuracy of 125
animation 7, 557, 579–594
    adding background images 578
    adding motion blur 576
    advanced example 579–594
    basics 558–559
    compressing files 566
    exporting files 41, 566–567
    exporting stills from 566
    file types 566, 625–626
    generating 117, 565, 594
    introduced 5
    lighting in 584–589
    path 593
    placing cameras for 589–593
    playing 117
    preview 563–565, 578
    quick guide 559–560
    setting parameters 97
    super sampling in 577
    velocity controls 567, 568–571
    Views palette and 150

animation exercises
  life of a fly 567–572
  walkthrough of a city 561–565
Animation Frame Export Options
    dialog box 566
Animation From Keyframes 82, 97
Animation palette 166, 564
Animation toggle switch 132
antialiasing 111, 503
  defined 498
Apple's MoviePlayer 41
Apply Selected Material button,
    using 430
Apply to Ghosted/Unghosted
    options 57
applying images 444
Arbitrary Plane Active tool 336
Arbitrary Reference Plane tool CD-44
  using CD-51
Arc tools 194–196
  using 428, CD-14, CD-15, CD-142
Arc/Circle/Ellipse tool 120
architectural animation 579–594
architectural modeling exercises CD-
    26–44
architecture, projection suited to 92
Area Lights option 516
Arrange command 66
aspect ratio, preserving 119
Attach Object tool CD-99
Attach Options dialog box 303
Attach tool 303–305
attributes
  editing properties of 293
  querying 291
Attributes tool palette 322–324
  using 430
Auto Save Options dialog box 62
Auto Scroll command 72
  using (Tip 37) 610
auto•des•sys, telephone help
    from 14, 29
Automatic tool 196, 198
Axial Sweep option 222
Axis Marks, showing 73
axis of reflection 298, 299
Axonometric View Parameters dialog
    box 89, 90
axonometric views 35, 54, 83, 84, 85
Axonometric* command 87–91

B

background effects 506
  adding to animation 578
  changing (Tip 24) 604
  setting up 508
backups
  automatic 61
  saving 38
Ball option (Metaballs) 177
Balls tool palette 174
  using 373

barn exercise 366–370
Bevel option 280, 283
  using 404
BitMap fonts, using 63
Blend Colors option, using 516
block and cut operation CD-27
BMP image files 42, 78
bolt exercise, CD-3–CD-5
Bolts/Screws Options dialog box,
    using CD-3
Bolts/Screws tool 213, 216, 217
Boole, George 259
Boolean operations 256–259
  B-Split operation 258
  Difference operation 258, CD-30
  exercise CD-32–CD-37
  Intersection operation 257
  on 2D surface objects 415
  on C-Meshes 247
  performing multiple (Tip 28) 606
  Split tool CD-36
  Union operation 257
  unions 397
  using 372
  working with CD-37
Booleans and Intersections tool
    palette 256–267
  using CD-36
Bottom views
  axonometric mode 87
  perspective mode 87
boundary curves
  drawing CD-76
  smoothing CD-78
Boundary Sweep option 223
Break Distance option 278
Break Line tool 278
  using 416
B-Spline Cubic tool 192, 193
  using CD-104
B-Split tool 256, 258
bulletin boards 30
Bump Map options 582
bump maps, custom 580, 581
Bump option 502
bump textures CD-25

C

Cage tool 266–267
camera
  imaginary 91
  virtual 41, 99
camera path 593
Cameras (Wire Frame) 107
cameras in animation 589–593
candela, defined 528
Capture Digitizer Mesh 131
cardinal points, orientation
    toward 87
Cartesian coordinates 147
cavity, creating 184
C-Curve icon CD-68

C-Curve tool 242–243
  advanced uses 456
  using CD-116, CD-121
Center option 79
  using 421
center of interest (COI) 90, 91, 101
  defined 558
chair modeling exercise CD-51–CD-
    61
chat room, to learn form•Z 14
Choose Site button 98
Circle by Center tool, using 384
Circle tool, using CD-33
Circle, 3 Point tool 188
Circle, Center and Radius tool 188,
    189
Circle, Diameter tool 188
circles 188
city exercise 375–378
Clear All Ghosted command 59, 325
  using (Tip 12) 599
Clear button, warnings about not
    using 162
Clear command 49, 58
Clear Rendering Memory 117
  using (Tip 10) 598
Click and Drag option (Pick) 120,
    199
  using (Tip 39) 610
Clicking On Edges option (Pick) 199
Clip Hither/Yon option 90, 101
Clipboard command 51, 53–54
clipping planes 91
Close All (Windows) command 67
Close At Ends command (Skin) 225
Close box 143
Close command 37, 66, 67
Close Cone Of Vision 103
Close Line tool 279
closing and opening palettes 143
closure settings, primitives 172, 173
C-mesh, fixing twisted surface 286
C-Mesh tool 245–248
  using CD-64
Cog wheel examples 188
COI path, defined 558
coin exercise, 383–387
collapsing palettes 142
color
  adding CD-22
  assigning 122
  setting 322
color depth settings 118
color image maps 447
Color Intensity slider bar 108
Color Palette command 125, 130
  using CD-66
Color Picker dialog box 122, 123
  using CD-120
color printing. See Solid Color
    Printing option
Color selection category 58

Index

color shading, setting intensity 116
color temperature, defined 528
Color tool 313
   using 424, 432
Colors palette 151–153
   using 420, 430
Combination Snap tool 347, 352
Command tools 10
   Copy modifier 51
   modifying the appearance of 31
   Multiple Copy modifier 51
   Self/Copy modifier 50
   through Input Options dialog box 120
commands, grouping families of related 34
   most useful 46
complex forms, modeling CD-91, CD-101
Compress button (Save As) 38
Compression Settings dialog box 566
Cone lights 483
Cone Of Vision 91
   editing CD-31
Cone Of Vision Display Detail dialog box 103
Cone tool 171
   using CD-20
cones 181
   truncated 173
Connect Lines tool 281–282
Connect option (Trim Line) 279
conserving memory by resetting Undo/Redo (Tip 9) 598
Construct Plain Mesh option (C-Mesh) 246
Continuous Copy 288
   using CD-33
Contour tool 263, 265
   using CD-138
contours, deriving 3D shapes from CD-139
control points 201, 249
control vectors 170
Controlled Curve Options dialog box 243
Controlled Curve tool CD-67
controlled curves 192
Controlled Mesh Object dialog box using 435
Controlled Mesh Options dialog box 246
   using CD-64
Controlled Mesh Smoothing options 248
Controlled Mesh tool, using 435, CD-65
Controlled Rounding option 239
   using 403
controls, clearing 287, 331
conventions 33
Converged Solid tool 181
converting between 2D and 3D 286

Coordinate Mode (Ruler) option 76
Coordinates & Prompts palettes 125
Coordinates palette 147, 158–159
Coordinates toggle switch 132
Copy command 49
Copy modifier 51
Copy Options dialog box 289
copying attributes 322
copying tools 288–290, 435
corners, rounding 238
Correct Image Exposure option 537
Create C-Curve tool 245
cross sections, drawing CD-79
cross-pasting 51
Cube Options dialog box 171
Cube tool 170
Cubic Environment option 508
Current Preferences File 60
Cursor Size option 72
cursors
   behavior, constraining 347
   extended 69
   extending (Tip 17) 601
   setting cross hairs 72
curved primitives 171
curved surfaces, inefficiency and CD-102
curves
   generating 192
   smoothing 194, 244
curvilinear forms
   modeling CD-44, CD-74, CD-101
Custom command 81
Custom Cursor Size option 69
Custom Display Scale command 104
Custom Height dialog box 81
custom profiles, creating 461
Custom View Angles command 85
Customize Tools command 132, 133
Cut command 49
CyberSites, Inc., bulletin boards 29–30
Cylinder tool 171
cylindrical objects, modeling 474

D

daylight conditions, simulating 540
Decal tool 320–322
   using CD-84
decals, applying 464
Decals dialog box 321, CD-89
Decompose Non Planar Faces option 113, 536
Default Object Name/Default Group Name fields 160
default values, heights 81
Define Arbitrary Plane tool 89, 339, 340, 402
Define Macro Transformation tool 291
Define Perpendicular Plane tool 339, 340
Define Profile tool 128, 231–234

Definition of Macro dialog box 302
Deform tool 232, 233, 234
   advanced uses 452
Delete Geometry tool 330, 331
Delete Objects tool palette 329
Delete tool, using CD-8
Delete Topology tool 49, 330, 331
Depth Blur option 513
Depth Cue 511
depth effect options 511
Depth Snap options 353
Derivative 2D Enclosure tool 179
Derivative 2D Surface tool 179
Derivative Object tool palette, using CD-7
derivative tools 169
Derivatives tool palette 204–212
Deselect command 58
design environments 4
Desktop button 38
dialog boxes, review of 119
Dialog Preferences dialog box 62
Diameter option 175
Difference tool 256, 258
   using 437, 440, CD-32
digitized images, inputting 53
digitizer options 131
direct illumination 530
Direction Snaps 344–347
Directional light 483
Dismantle One Level option 270
Displacement Map Edit dialog box 235, 449, 450
displacement maps 444
Displacement tool 234–236
   using 444
Display Color, editing (Tip 14) 600
Display Detail command 103
Display menu 103–119
   Always Clear Rendering Memory 117
   Clear Rendering Memory 117
   Custom Display Scale 104
   Display Options command 116
   display types 490
   Draft Layout Mode 117
   Generate Animation 117
   Hidden Line* 109–110
   Image Options 118–119
   metric scale, ratios of 104
   OpenGL* 115
   Play Animation 117
   Quick Paint* 108–109
   QuickDraw 3D 114–115
   Radiosity commands 116
   Redraw Buffers command 118
   RenderZone 113
   scale, ratios of 104
   Shaded Render* 111–113
   Show Surfaces As Double Sided 117
   Surface Render 110–111
   Wire Frame* 105–108

Display On Top Of Grid And Axis option 79
Display Options command 116
distance, measuring 294
Distance From Hither 512
Distance Snap 77
Distribute tool. *See* Align/Distribute tool
Disturb tool 232
dome exercise, CD-6–CD-9
Dot Grid, specifying 74
double clicking 77, 148
Double Parallel option CD-137, 208
Draft Angle tool 240–242
Draft Layout Mode command 117
drafting environment 8
drafting module. *See* 2D drafting module
Drafting Tools, Help window 25
dragging in the Layers palette (Tip 40) 611
dragging palettes 141
Drafting clipboard, shifting to Modeling clipboard 53
drawing, boundary curves CD-76
drawing in 2D CD-14
Drop Controls tool 287
Dump Screen option 44
Dump Window option 44
Duplicate* command 50
Duplication Offset command 51
dust patterns 524
DXF files, importing 37
Dynamic Reflection option 298
Dynamic Scale tool 297–298

E

Edit Color dialog box 152
Edit Cone of Vision command 82, 90, 96, 99–103, 511, CD-31
   controls 100
   reset buttons 102–103
   window 100
Edit Controls tool 170, 201–202, 249
Edit menu 46–65
   Clear All Ghosted command 59
   Clear command 58
   Clipboard command 53–54
   Copy command 49
   Cut command 49
   Delete command 58
   Duplicate* command 50
   Duplication Offset command 51
   Grab Image command 52
   Hide Ghosted command 59
   Key Shortcuts command 59
   Paste command 50
   Paste From Modeling*/Paste From Drafting command 51–52
   Paste Image command 53
   Preferences command 38, 60–65
   Redo command 47
   Replay command 47
   Reset Undo/Redo command 48
   Select All Ghosted command 55
   Select All UnGhosted 54
   Select By command 55–58
   Select Previous command 54
   Undo* command 46–47
Edit Menu command (Heights menu) 81
Edit Surface tool 203–204
Edit Text tool, using 380
Edit View dialog box 97
editing point types (Tip 19) 602
Elevation Oblique View 92, 93
Elevation view, right side 88
Ellipse, Diameter and Radius tool 188–189
Ellipse option 175
Ellipse, Major and Minor Radius tool 188, 189
ellipses 188
Ellipsoid option (Metaballs) 177
enclosed spaces, creating 181
Enclosure Options dialog box 81
Enclosure Wall Width option 181
English units, setting 124
Environment background option 506, 507
Environment Image Limits dialog box 510
environments, in RenderZone 508
Error Messages command 133
Estimate Memory Usage option (Tip 22) 603
evaluation (combining Metaformz) 255
exercises 11
   animation 561–565, 567–572, 579–592
   for advanced users 443–482
   for beginners 366–390, CD-3–CD-48
   for intermediate users 394–440 CD-51–CD-111
   on CD-ROM ix
Exit command 46
Exit Radiosity
   command 116
   menu 535
Export Animation command 41
exposure settings 118
Extend Plane Grid tool 339, 341
Extend Segment tool 308–309
Extended Cursor command 69, 71
   illustrated 70
   using (Tip 17) 601
extending a grid (Tip 41) 611
Extents print type 44
Extrusion/Convergence Options dialog box
   using CD-63
eye path, defined 558
eye point 101
   defined 558
Eye Point View Point 94

F

Face tool 196
faces
   converting nonplanar to planar 236
   defined 197
   generating 183
   tapering 241
facets 168
Facets options 172
figure 165
file compression 566
File Format selection 37
file formats vii, 40, 78
File menu 34–46
   Close command 37
   Export Animation 41
   Import command 42
   New (Draft) command 35
   Open command 37
   Open Recent command 37, 65
   Page Preview command 44
   Page Setup command 43
   Plot/Print Setup command 44
   Print command 45
   Quit/Exit command 46
   Revert To Saved command 41
   Save A Copy As command 40
   Save As command 39–40
   Save command 38–39
   Save QuickTime VR command 41
   View File command 42–43
File Menu, New (Model) command 35
file types 617–626
   .fan 117, 566
   .fmz 39
   3D Studio file format 617
   3DGF 625
   3DMF 625
   3DS 626
   AVI 566, 618, 626
   BMP 617, 626
   CIBSE 622
   CIE 622
   DEM Data 622
   DWG 622
   DXF 37, 622
   EPS 620
   FACT 623
   form•Z 623
   form•Z Animation 625
   form•Z Imager 623
   form•Z Library 623
   form•Z Suspended Render 623
   IES 623
   IGES 37, 624
   Illustrator 624
   JPEG 621
   Lightscape 624
   OBJ 624

Index

PICT 617, 621
PNG 621
QuickTime 566
Quicktime Image 621
Quicktime Movie 618, 626
RIB 624
SAT 624
standard 37
STL 625
Targa 621
TIFF 621
VRML 625
Windows Metafile 617, 626
Windows only 626
files
  finding recently opened 37
  importing 42
  opening 37
  saving 62
  standard (form•Z) 39
  transferring
  viewing 42–43
fillet size CD-55
Fillet/Bevel Lines tool 282–283
  using 412, CD-51
finishing touches, adding CD-100, CD-127, CD-110, 464
First Point, setting 284
fisheye options (Tip 47) 613
Fit All tool 354, 357
Fit Fillet option (Trim Line) 280
Fit Fillet/Bevel Lines dialog box 282
fit operations 121
Five-minute Models 365–391
  barn 366–370
  bolt CD-3–CD-5
  city 375–378
  coin 383–387
  dome CD-6–CD-9
  goblet 370–372
  ice cream cone CD-19–CD-26
  landscape 387–391
  logo 378–381
  mace 373–375
  maze 381–383
  monumental arch CD-26–CD-31
  paper clip CD-12–CD-19
  pear CD-9–CD-12
  portal on a podium CD-38–CD-44
  spiral staircase CD-44–CD-47
  Swiss cheese CD-32–CD-37
flashlight modeling exercise 474–484
Flat option 496
flipbooks, making Tip 52, 615
fly-through animation 589
Fog option 512
Font Options dialog box 63
fonts
  default 63
  using 44
fork modeling exercise CD-67–74
form•Z
  changes between versions ii, 25, 40

interface screen 10, 14
platforms ii
redundancy of 119
standard files 39
version 2.9.5 33
See also learning form•Z
form•Z Imager 537
Frame option (Pick) 200
frame rate 559
Frame selection method 120
frames 7
  in animation 558
From Picked Segment option 228
Front views
  axonometric mode 89
  perspective mode 89
  using 371
fruit bowl modeling exercise CD-115–CD-128
Full Raytrace option 498
Full Z-Buffer option 498
Fuller, Buckminster CD-6

G

General Help dialog box 26
Generate Animation command 117
Generate Exposure Data option 514
Generate Radiosity Solution* command 116, 534
generator tools 168
Geodesic Spheres 176
  using 373, CD-6
Geographic Position dialog box 98
geometric properties 161
Geometric Transformation tool palette 126, 295–303, CD-27
  using 397
geometry, deleting 330
Get Attributes 322–324
Ghost tool 324–325
  using CD-64
ghosted objects
  color in 325
  defined 54, 169, 324
GIF image files 52
Glow options 524
goblet exercise 370–372
Gouraud option 496
Grab Image command 52
Graphic/Keyed option 182
  setting 80
graphical scale, including on drawings 79
grid
  extending (Tip 41) 611
  locking 75
  rendering (Tip 27) 605
  setting line spacing 74
  showing 73
Grid Snap Options dialog box 77, 342
  using 395
Grid Snap Switch 341–344

Grid Snap tool
  using 371, 383, 395, 423, 438, CD-6, CD-13
  using with Graphic/Keyed 80
  quickly setting (Tip 1) 595
grid transmission map 582
Group tool 196, 269
  using CD-10
grouping in the Object palette Tip 18, 601
groups
  defined 198
  making (Tip 18) 601
guidelines
  creating CD-52, CD-62
  for light and shadow 539
  handling image irregularities 536
  saving time 538
  using CD-51, CD-129
  using quick keys 536

H

halo effects 481
hammer modeling exercise CD-91–101
hand modeling exercise 444–451
Hand tool 354, 355
hardware requirements iii
Heights menu 80–82
Heights settings
  default values 81
  storing 81
Helix tool 216
  using 464
Help menu 11, 133–134
  built-in 15–27
  Error Messages 133
  Project Info 134
Hidden Line Options dialog box 109
Hidden Line views, using 397, 402, 492
Hidden Line* command 109–110
Hide Controls option (C-Curve) 245
Hide Ghosted command 59, 325
Hide Palettes option 144
  toggle switch 132
  using (Tip 21) 603
Hither and Yon 511
  clipping planes 91
Hither Plane 101
Hole/Volume tool 196, 197
holes
  holes vs. openings 184
  generating 183
Horizontal Origin field 79
Horizontal Repetition, using 421

I

ice cream cone exercise CD-19–CD-26
icons, modeling, eliminating 17
icons, rows of 15
IGES files, importing 37
illumination 514–516, 541

ambient 530
direct 530
tips for 540
Illumination Analysis Options dialog box, using 515
Image background option 506
Image Color Depth option 119
Image Exposure Options dialog box using 514
image files, loading into form•Z 618
image irregularities, handling 536
Image Map Options dialog box
  using 384, 420, 447, CD-87, CD-118
image maps
  applying 444, CD-115
  color 447
  file types for CD-91
  imported 509
  using for complex models 444
  wrapping 510
Image Options command 118–119, 536
Image Quality option 109
Imager application 537
  using (Tip 11) 599
Import command 37, 42
Inclination Angle option 91, 241
Infinite option, using 421
information, organizing 325
Initialize Radiosity* command 116
Input Options command 119
Insert Face/Outline tool 183
Insert Face/Volume tool, using 398
Insert Hole tool 183, 197
  Heights setting 184
  using 368
Insert Opening tool 183
Insert Outline tool 183
Insert Point tool 283–284
  using (Tip 16) 600
Insert Segment tool 283–284, CD-95
Insertions tool palette 182–184
Inside Boundaries option (Pick) 199
intelligent tools 198
Intensity slider bar 158
interactive rendered modeling (Tip 55) 616
interface screen 8
Internet forum, joining, to learn form•Z 14
Internet help 28, 29, 30
Internet, official form•Z web site 29–30
intersecting shapes 119
Intersection tool 256, 257
Introductory Help window 27
isometric views 83, 84, 86, 91

J

Join and Group tool palette 268–271
  using CD-35
Join Lines tool 281–282

Join option (Close Line) 279
Join tool 268
joining 2D elements 412
joining
  edges 413
  objects CD-35, CD-99
  pieces CD-109
JPEG image files 52
Justification option 179

K

Keep Backup option (Save As) 38
Keep Flat option (Smooth Mesh) 229
Keep Vertical Lines Straight option 94, 362
Key Shortcuts command 59, 60
  using (Tip 25) 604
keyboard commands
  differences between MacOS and Windows 617
Keyboard Layout mode 21
keyboard shortcuts vii, 21–25, 33, 627–634
  Copy 46
  customizing 38
  List Shortcuts button in 23–25
  Manager window 23, 105
  memorizing 86
  Object Snap tools 354
  Sort by Modifier option 24
  Zoom tools 358
keyframes
  defined 558
  setting up 561–563
knife modeling exercise CD-61–CD-66

L

landscape exercise 387–391
landscapes, tool for 211
Lasso option (Pick) 120, 200
lathed forms, creating 213
Layer Attributes dialog box 156
Layer Attributes, quickly setting (Tip 2) 596
Layer dialog box buttons 156
layer groups 157
Layer Name 155
Layer selection category 57
layers
  creating 157
  defined 154
  using 156
Layers command 121, 122, 155–157, 325–329
  toggle switch 132
learning form•Z 13–14
  taking classes 13
  joining an Internet forum or chat room 14
Left Side views
  axonometric mode 88
  perspective mode 88
  using CD-64

lengths, measuring 294
libraries
  of objects 273
  updating (Tip 51) 615
  using (Tip 49) 614
  See also Symbol Libraries command
Light Beam 101
Light Glow option 516
Light Parameters dialog box 500, 585
  using 516
Light Source/Target 101
lighting 113
  adding 480
  advanced use of 474
  effects 481
  options 317
  in QuickDraw 3D 115
  See also illumination
lights
  analyzing 515
  changing parameters 542
  color of 541
  cone 586, 588
  directional 483
  illuminating large models/single objects 541
  positioning 483, 522, 540
  reflective properties 542
  setting intensity 514
  shadowcasting 584
  tips for manipulating 539
  tip for changing status of all 157
  used in animation 584–589
Lights command 126, 157, 158
Lights options (Wire Frame) 107
Lights toggle switch 132
Line Editing tool palette 278–284, 408
Line Grid, specifying 74
Line of Intersection tool 260, 262
Line of Sight 101, 562
Line Style Line Styles 130
  selection category 58
line tools 190
line weights 130
  in wireframes 113
  selection category 58
lines
  2D contours 212
  adding thickness CD-124
  aligning 280
  angled 344
  attaching 304
  closed 178
  closing in Prompts window (Tip 5) 597
  closing segments in 279
  connecting 281
  direction of 282
  generating from underlay file CD-132
  measuring 294

Index

parallel 179
removing redundant 52
reversing direction of 285
segmenting 278
smoothing corners between 282
straight 344
trimming segments 279
Lines, Splines, and Arcs palette 190–196, 408
List Shortcuts button 23–25
Load button in 23
Load Fonts At Launch option 63
Load Preferences button 61
Load Project Lights option 158
Lock Drawing To First Point option 353
logo exercise 378–381
low-resolution "stand-in" objects 266, 267
lumens, defined 528

M

mace exercise 373–375
Macintosh
 commands 38
 hardware requirements iii
 menu 32
MacOS, differences from Windows ii, vi, 617–618
Macro Define tool 301
Macro Transformations command 126–127, 301–303
magnifying glass 42
magnifying glass modeling exercise 425–433
Main Help dialog box 15
Maintain Proportions option 119
Make First Point tool 284–285
Make One/Two Sided tool 168, 286–287
Make Surface Object From Open Source Shape option, using CD-63
Map Options dialog box, using 421
mapped shadows 522, 523
Mapping Type options 236, 449, CD-119
maps, generating site models from 78
Match Grid Module option 74, 342
Match Image Size To Underlay Size option 78
Match View tool 359, 360
 using 362
Maximize Window In Screen option 119, 536, 537
maze exercise 381–383
Measure tool 294
 using (Tip 3) 596
mechanical engineering, projection suited to 91
memory
 clearing 117

use in RenderZone (Tip 22) 603
Menu Commands, Help window 20–21
menus 31–134
 interface for 9, 31, 32
Mesh Depth, using CD-64
Mesh Direction 228
Mesh Length, using CD-64
Mesh Options dialog box 227
Mesh terrain model 211, 212
Mesh tool 226–228
 using with radiosity 534
meshed objects
 changing resolution of 230
 deforming 231
 manipulating 162
 mapping textures onto 318
 meshing recursively 227
Meshed Surface 167
Meshes and Deform tool palette 226–238
Metaballs Evaluation Parameters dialog box
 using CD-10
Metaballs tool 177, 178, 254
 using CD-9
Metaformz 177
 deriving 255
Metaformz tool palette 254–256
metric scale, ratios of 104
metric units
 setting 124
 in Heights command 80
minimizing palettes 142
Mirror tool 298–300
 using 427, CD-136
modeling
 architectural CD-26–3CD-1, CD-38–CD-44
 architectural components CD-44–CD-47, 406
 interactively in QuickDraw and OpenGL (Tip 55) 616
 realistic 444
Modeling clipboard, shifting to Drafting clipboard 53
Modeling Color Palette dialog box 125
Modeling Display Options dialog box 116
modeling exercises CD-19, CD-32–CD-47, 394–440
 chair CD-51–CD-61
 fork CD-67–CD-74
 hammer CD-91–101
 knife CD-61–CD-66
 picture frame 418–425
 screwdriver 433–440
 soda can CD-84–CD-91
 spoon CD-74–CD-84
 tape dispenser 408–418
 teapot CD-101–CD-111
modeling icons, eliminating 17

modeling interface, streamlining 18
modeling tools 9–10, 165–169
 appearance of 31
 customizing 17, 19
 for 3D solids and enclosures 180–182
 for 2D surface object and 2D enclosure 178–179, 204–205
 for 3D extrusion, 3D converged, and 3D enclosure 206–207
 for advanced shape editing 226–238
 for aligning and attaching 303–311
 for arcs 194–196
 for assigning color 312–313
 for attaching objects 303–305
 for attributes 312–323 for Booleans and Intersections 256–267
 for breaking lines 278–278
 for cage 266–267
 for circles and ellipses 188–190
 for clearing controls 287
 for closing lines 279
 for connecting and joining lines 281–282
 for controlled curve 242–245
 for converting surfaces to one or two sides 286–287
 for copying 288–291
 for copying and setting attributes 322–324
 for creating controlled mesh 245–248
 for defining macro transformations 291
 for defining profiles 231–234
 for Deform 231–234
 for deleting objects 329
 for deleting parts 330–331
 for displacement 234–236
 for Disturb 231–234
 for draft angle 238–242
 for edit controls 201–202
 for editing surfaces 203–204
 for extending segments 308–309
 for filleting and beveling lines 282–283
 for geometric transformations 295–303
 for ghosting and layers 324–329
 for graying out objects 324–325
 for grouping and ungrouping 269–271
 for helix and screws 216–218
 for inserting points and segments 283–284
 for join and group 268–271
 for joining and separating 268–269
 for line editing 278–284
 for measuring 294
 for mesh 226–228
 for Metaformz 254–256

for mirroring and reflecting 298–300
for modeling components 196–198
for Move Mesh 231–234
for moving and rotating 295–296
for NURBS and Patches 242–254
for Nurbz 248–250
for organizing information in a model 325–329
for parallel objects 207–209
for parametric derivatives 213–226
for Patches 250–254
for picking 199–201
for placing shapes 309–311
for points and lines 190–192
for polygons 186–188
for positioning textures 317–320
for projection and unfolding 209–211
for querying attributes 291–294
for rectangles 185–186
for reduce mesh 230–231
for rendering attributes 315–317
for reversing direction of lines and surfaces 285–286
for revolved object 213–216
for rounding 238–240
for scaling objects 297–298
for section/contour 263–266
for setting decal attributes 320–322
for setting first points and point markers 284–285
for shading 313–314
for skin operations 224–226
for smooth mesh 228–230
for splines 192–194
for stairs 218–221
for sweeps 221–224
for symbols 273–278
for terrain 211–213
for text 271–273
for text manipulation 271–273
for topologies 284–287
for transforming objects with macros 301–303
for triangulation 236–238
for Trim, Stitch, and Intersection 260–263
for trimming lines 279–280
interface for 165–166
Palettes, customizing 17
Save button in 18
trees (Tip 42) 611
Modeling Tools Help section 15–19
Modeling Tools palettes
 Derivatives 204–212
 Lines, Splines, and Arcs 190–196
 Polygons and Circles 185–190
 Topological Levels 196–198
modifier tools 168
monitor size 138–141

monumental arch exercise CD-26–CD-31
motion blur 567, 576
Mouse Position cursor option 71
mouse tracking
 controlling 147
 using (Tip 26) 605
Move Mesh command 162, 232
 using 462
Move Plane tool 339, 340
Move tool 51, 295–296
 using 397, 407, 429, 435, 460
MoviePlayer 41
Multi-Copy tool 288, 290
Multiple Copy modifier 51

N

Name New Macro Transformation dialog box 301
Name View Dialog Box 97
Navigate View tool 359
Neon Transparency surface style 475
New (Draft) Window 35, 66
New (Model) Window 35, 65
 using 366
No Directional Snap tool 344
No Object Snap tool 347, 348
Noise option 524
nonplanar faces, working with (Tip 31) 607
Normal and Extended Cursor 69
N-Sided Polygon tool 120
nudge keys 120, 200
 in MacOS vs. Windows 618
 using (Tip 15) 600
numeric accuracy, setting 124, CD-12
Numeric Display options 124
NURBS option vs. form•Z Nurbz 243
Nurbz surfaces
 advanced uses 465
 control points on 249
 creating 248
 editing 201–203, 468
 fixing twisted surface 286
 resolution of 250
Nurbz tool 248–250, 438

O

Object Does Not Cast Shadows option 316, 432
Object Duplication Offset dialog box 50, 51
Object Name/Group Name selection category 57
Object Query dialog box 28
Object Snap tools 347–354
 button 77
 keyboard shortcuts 354
 Tip 45, 613
Object Snaps Window Tools palette 347
object status settings 129

Object tool 196
Object Topological Level 197
Object Type tool palette 178–181, 371
 using 373, 379, 384, 434
Object Types selection category 56
objects
 3D extruded 180
 adding and subtracting 256
 aligning 303, 305
 attaching 303
 circular 188
 combining and splitting 268
 complex 213, CD-91
 contours of 265
 controlling behavior of 129
 converting between 2D and 3D 286
 copying 49, 288
 curved 189, 190
 cutting 49
 cylindrical 474
 deforming 233
 distributing 305
 editing 201, 203, 291
 elliptical 188
 generating 178
 graying out 324
 hierarchy of 269
 inserting points into 283
 intersecting 262
 joining CD-35, CD-99
 low-resolution 266
 libraries of 273
 mapping textures onto 317
 meshed 230, 245, 247
 mirroring CD-136
 modeled with line and point tools 191
 modeled with rectangle tools 186
 moving 51, 295
 one-sided 167
 organic-shaped 254
 organizing into groups 269
 parallel 208
 parametric 168, 177, 242
 partially closed 172
 pasting 51
 Patches 251
 picking 199
 placing decals on 320
 polygonal 186
 positioning CD-127
 reflecting 298
 removing 329, 330
 removing from a group 270
 rotating 295
 scaling 297, CD-30, CD-34
 separating from a group (Tip 36) 609
 setting height of 80
 source 169
 spherical 174, 176

Index

spiral-shaped 216
stitching together 263
subtracting CD-32, CD-35, CD-37
text 271
transforming with macros 301
viewing wireframes 106
Objects command 127, 160
Objects palette 159–160
Objects toggle switch 132
Object-to-object attachment,
   illustrated 304
Oblique View Parameters dialog
   box 92
oblique views 83, 84, 86
Oblique* command 92–93
One Copy tool 288, 387, 427
One Way Split (B-Split) 258
One-Sided objects 167
Opacity decal option 322
Open command 37, 66
Open File dialog box 78
   using 421
Open Recent command 37, 65
OpenGL 115, 494, 578
   for Windows only 618
opening palettes 143
openings, generating 183
Options menu 119–131
   Color Palette 125, 130
   digitizer options 131
   Input Options 119
   Layers 121
   Layers command 155
   Lights 126
   Macro Transformations 126–127
   Objects 127
   Pick Options 120–121
   Profiles 128
   Project Colors 122
   Reference Planes 128–129
   Status Of Objects 129
   Surface Styles 130
   Symbol Libraries 123
   Working Units 104, 124–125
   Working Units command 159
   Zoom Options 121
Origin and Handle (Symbols) 275
Ortho + Diagonal Snap tool 344, 345
Orthogonal Snap tool, 344, 423
   using with reflections 300
orthographic drawings,
   creating from 3D object 210
Orthographic Projection tool 209-210
Outline tool 196
outlines
   defined 183, 197
   generating 183
Overlap Pages option 44
over-sampling 504

P

Page Preview command 44

Page Setup command 43
palettes 9, 137–166
   and screen size 138–141
   Animation 166
   appearance of 31
   closing and opening 143
   collapsing 142
   customizing 19
   defined 131
   expanding 144
   hiding (Tip 21) 603
   manipulating 141–166
   ranking 144
   rearranging 141
   transferring 144
Palettes menu 131–132
   Colors palette 151–153
   Coordinates palette 158–159
   Customize Tools 132
   Hide Palettes option 144
   Layers palette 154–157
   Lights palette 157
   Objects palette 159–160
   Planes palette 161
   Profiles palette 162
   Prompts palette 145–148
   Surface Styles palette 151–153
   Symbols palette 162–164
   toggles in 132
   Tool Options 148, 149
   Views palette 150–151
   Window Tools palette 165–166
Palettes tool 15
pan settings 102
Pan tool 42
panning 335
   using Clipboard 54
   using View File command 42
   using Window Tools 354–358
Panorama Cylinder interface 96
Panoramic setting 41
Panoramic View Parameters dialog
   box 95–96
Panoramic* command 94–96
paper clip exercise CD-12–CD-19
Parallel Object dialog box 208
   using CD-82
Parallel tool 207–209
   using CD-137
parametric curves 192
Parametric Derivatives tool
   palette 213–226
   using CD-3, 428
parametric objects 168, 177
   clearing controls on 287, 331
   converting to Nurbz or Patch
      objects 202
   creating 242
   editing 201, 203
   Metaballs 254
Parametric Objects option 108
Parametrics option 505
Partial closure 172

Partial option 172, 173
Paste command 50
Paste From Modeling*/Paste From
   Drafting command 51–52
Paste Image command 53
Paste On Active Layer option 156
pasting 3D object into 2D drafting
   window 51
pasting using layers (Tip 43) 612
Patch objects
   deriving 251
   editing 201–203, 254
   working with 253
Patch tools 250–254
Path Shape button CD-70
patterns, for polygons 187
pear exercise CD-9–CD-12
Per Track editing options pull-down
   menu 569
Perpendicular Switch 338–339, 435
Perpendicular to Surface setting 180
   using 379
perspective mode 86–89
Perspective View Parameters dialog
   box 93, 561
Perspective views 85, 86
   options for (Tip 47) 613
   tips for 362
   using 386
   with Keep Vertical Lines Straight
      option 94
Perspective* command 93–94
Phong option 496
photographic textures 444
photographs, generating site models
   from 78
Pick Crossing option 200
Pick Options command 120–121
Pick Parade, using (Tip 38) 610
Pick tool 58, 159, 199–200
   freeform outlines 200
   MacOS vs. Windows 618
   returning to (Tip 8) 598
   using 388, 398, 435, CD-10, CD-65
PICT image files 78
picture frame modeling
   exercise 418–425
Place On Line tool 310
Place Options dialog box 309
Place Text command 63
Place tool 309–311
Plain Rounding option 239, 240
Plain Text Objects 108
Plan Oblique View 92
Plane Axes options 336
Plane Projection command 89
planes, defining 339
Planes palette 128, 161–162
   using to save custom reference
      plane CD-44
Planes toggle switch 132
Play Animation command 117
Plot Scale field 44

Plot/Print Setup command 44
  using (Tip 50) 615
plotters, using 44
Point light 483
Point Light/Radius option 522
point markers 284
Point tool 190, 191, 196
Point Topological Level 197
  using CD-56
point types, editing (Tip 19) 602
points
  attaching 304
  inserting 283, 600
  measuring 294
  preserving on same line 119
  removing 49
  setting First Point 284
polar coordinates 147
Polygon Options dialog box 187
  using 404
Polygon tool 186–188
  using 404, 439
polygon-based modeling 167
polygons
  drawing with Vector Line tool 191
  illustrated 187
  knowing how many (Tip 23) 603
  patterns for 187, 188
Polygons, Circles and Ellipses
    palette 185–190
  using 374, 439
portal on a podium exercise CD-38–
    CD-44
Postscript fonts, using 44, 63
Predefined button, using 430
Predefined Styles button
  using CD-66
Preference command 60–65
Preferences menu 146
preferences, saving (Tip 30) 607
Preserve Angles option 91
preset viewing angles 82–85
preview mode 108
Preview Raytrace option 498, 538
Preview Z-Buffer option 498, 538
previewing page 44
previewing your work (Tip 29) 606
Previous Session option 61
Previous View tool 354, 357
primitives
  closed 173
  Cone 171
  Cube 170
  Cylinder 171
  editing 201
  editing properties 293
  illustrated 172, 173, 176
  modeling with CD-19
  Sphere 171, 174
  Torus 171
  truncated cone 173
Primitives tool palette 170
  using CD-20, 398

Print command 45
Print Text As Paths option 44
printer setup 44
printing good b/w printouts (Tip
    7) 597
productivity, increasing 18, 141
profiles, custom 461
Profiles command 128
Profiles palette 162
Profiles toggle switch 132
Project Background option 506
Project Colors command 122
Project Environment Settings 506,
    508
Project File Options 61
Project Info command 16, 134
  using to count objects (Tip
    23) 603
Project Option From preferences
    selections 61
Project Options dialog box 40
Project Working Units dialog
    box 124, 159
Projection of View option 209
Projection tool 209–211
Projection Views option 353
projections, preselected 82
projects
  creating new 35
  managing large 328
Prompts palette 81, 145–148
  See also Coordinates palette
Prompts toggle switch 132
Prompts window
  closing lines in (Tip 5) 597
  Grid Snap and 344
"pull-down" menus 32
pyramids 181

Q

QD3D. See QuickDraw 3D command
QTVR. See QuickTime VR movie
    format
Query Attribute tool 291–293
Query Object Attributes dialog
    box 292
Query Object dialog box 291, 292
Query tool 28, 291–294
Quick curves, controls for 243
quick keys, using 536
Quick Paint views, using 491
Quick Paint* command 108–109
QuickDraw 3D command 114–115,
    578
  using (Tip 46) 613
  views 494
QuickDraw 3D renderer 361
QuickTime files 566
QuickTime VR movie format 618, 626
Quit command 46

R

Radial Bend option (Deform) 454

Radial movement (Move Mesh) 462
Radial Shear option, using 478
Radial Snap tool 344, 346
radiosity 525–542
  defined 526
  exiting 535
  generating 532, 534
  getting started with 530
  Mesh tool and 534
  saving 535
  setting intensity 527
  terminating 532, 533
Radiosity button 64
Radiosity Options command 116
Radiosity Options dialog box 529
RadioZity 64
  rendering with 527
Radius option 175
Railings Options 219
RAM requirements iii, 167, 432
raytraced shadows 522, 523
realistic modeling 444
Recent Files button 65
Rectangle tool 185
  using 368, 374, 387, 398, 422
Rectangle, 3 Point tool 185
Redo/Redo All command 47
Redraw Buffers command 118
Reduce Mesh tool 230–231, 267
redundancy 119, 137
Reference Grid Options dialog
    box 74, 337
Reference Plane 6
  3D axonometric view of 35
reference plane axes/grid
  showing 72, 73
Reference Plane Window palette
  using 419
reference planes 339–341
  custom CD-44
  described 335
Reference Planes command 128–129
Reference Planes tool palette 336–
    338
reflection environment, setting
    up 508
Reflection Options dialog 299, 430,
    501
  using 432
reflective properties of light 542
Reflectivity options 481
Relative coordinates, setting 76
Relative Transformations tool
    palette 303–311, 374, CD-99
  using 374
Remove Duplicate & Overlapping
    Lines option 52
Remove From Pick option 57
Remove Object From Group
    option 270
Render As Wire Frame 316, 506
Render Attributes tool 315–317, 506
  using 432

Index

Render Grid And Axes As 3D Lines 113, 506
Render With Fog option 115
Render With Shadows 109, 111, 113
Rendered Environment option 510
Rendered option, using 421
rendering 489, 538
  advanced use of 474
  display types 490–495
  economizing time of 538
  grid (Tip 27) 605
  increasing speed of 329
  portion of a window (Tip 32) 608
  types of 496
Rendering Attributes tool 532
Rendering Engine submenu 115
Rendering Memory, clearing (Tip 10) 598
Rendering Type submenu 114
RenderZone command 64, 113, 495
RenderZone renderer
  using 422, 425, 432
RenderZone version 8, 9, 151
  using 380, 407, 425, CD-11, CD-66
Repeat Copy tool 288, 440
Repeat Last Transformation tool 301–303
Replay command 47
  using (Tip 35) 609
Reset After Saving Project option 49
reset buttons 102
Reset Plane tool 339, 341
Reset tool 354, 358
Reset Undo/Redo command 48
Resize box 144
resolution, low 266, 267
Reverse Direction tool 106, 285–286
  using 415
Revert To Saved command 41
revolution angle 214
Revolve tool
  default settings 213
  using 428, CD-117
Revolve tool, using 371
Revolved Object tool 213–216
  using CD-26, CD-28, 438, CD-85, 476
Revolved Sphere 175
  using CD-119
Right Side views
  axonometric mode 87
  perspective mode 87
  using 439
Rotate Plane tool 339, 340
Rotate tool 295–296
  using CD-27, 440, 460
rotation 54
  snapping CD-108
Rough texture option CD-120
Round Options dialog box 239
Round tool 238–240
  using 403
Rounding Options dialog box 403

rowboat modeling exercise CD-129–CD-145
rulers 75–76
  customizing 76
  showing 75
  using (Tip 44) 612

S

Sampling Box 532
Save A Copy As command 40
Save As command 38–40
Save As Copy option 62
Save As menu 62
Save As options available in Windows only 617
Save button in 23
Save Command 38–39
Save Libraries option, using (Tip 49) 614
Save Preferences button 61
Save Project As dialog box 39, 62
Save Prompts in TEXT File option 61, 146
Save QuickTime VR command 41
Save To Project option 62
Save Undo in Project option 49
Save View 97, 102
saving as previous version (Tip 53) 616
saving preferences (Tip 30) 607
saving time 538
Scale field 79
Scale To Fit Media options 44
Scale With Object option 320
scale, ratios of 104
scaling objects CD-30, CD-34
scanners, inputting from 53
Scratch Disk Preferences dialog box 63
Screen Digitizer Calibration 131
screen size, and palettes 138–141
screwdriver modeling exercise 433–440
scrolling function, setting 72
Section tool 263, 264, 266
Section/Contour of Solids tools 263–266
Segment tool 190, 191, 196
segments
  attaching 304
  defined 197
  extending 308
  inserting 283, CD-95
  measuring 294
Select All Ghosted command 55
Select All UnGhosted command 54
Select By command 55–58
Select Previous command 54, 58
Selection Criteria/Drafting dialog box 58
Selection Criteria/Modeling dialog box 55–58
selection methods 120

Self tool 288
Self/Copy tool palette 50, 288–291
  using CD-33, 435
Separate tool 268
Set Attributes tool 322–324
Set Color tool 312
Set Image Size option 538
  using 407
Set Layer tools, two different 137
Set Range option, using 516
Set View tool 359
Set/Clear Point Marker tool 284–285
setback design, faceted 400
Shaded Render display option
  using 374, 492
Shaded Render* command 111–113
Shader Antialiasing 498, 503
shading, types of 499
Shading Range option 116
shading tool 313–314
shadow leaks 531
shadowcasting point lights 585
shadows
  adding CD-111
  eliminating jagged edges 542
  hard and soft 522
  opaque 523
  setting 99, 316
  tips for manipulating 539
  transparent 523
Shadows option 111, 116, 500
  using 432
Shadows selection category 56
Shape pull-down menu 175
shapes
  placing 309
  transforming 296
Shortcuts 21–25, 33
  bringing up list of 23–25
  to memorize 25
Show Back Faces option 106, 107
Show Color option (Surface Render) 111
Show Cursor At option 71
Show Edges option 109, 111
Show Extended Cursor option 71
Show Face Normals option 106
Show Facets 108
Show First Point option 107
Show Grid* command 73, 100
Show Grid, Axes And Underlay option 113
Show Marked Points 106
Show Objects As Bounding Volume 106
Show Plane Axes* command 72
Show Points option 106
Show Rulers* command 75–76
Show Surfaces As Double Sided command 117
Show Symbol Instances As Bounding Box option 53
Show Transparent Objects 109, 111

Show Underlay option 78, CD-131
Show Wire Edges 108
Show World Axes* command 73
Single Parallel option 208
Skin dialog box 224, 225, CD-81
Skin operation 459
  marking points for 106
  troubleshooting CD-83
  using CD-134
  using underlays in CD-129
Skin tool 213, 224–226, CD-75
  advanced uses 452
Skinning Along Paths option 459
skyscraper towers modeling
  exercise 395–407
  domed tower design 403–406
  faceted setback design 400–402
  setback design 395–397
  ziggurat design 398–399
Slab option 209
Smooth Input Options dialog
  box 194
Smooth Interval 194
  using CD-116
Smooth Line tool 120
Smooth Mesh tool 228–230
  using CD-143
Smooth Shading option 112, 504, 532
  selection category 57
smoothing intervals CD-102, CD-103
Smoothness option 96
  using 380
Snap Options command 76–77
Snap To Angle/Slope tool 77
Snap To Center tool 347, 351
Snap To Endpoint tool 347, 348
Snap To Face tool 347, 352
Snap To Intersection tool 347, 350
Snap To Interval tool 347, 349
Snap To Line tool 347, 350
Snap To Midpoint tool 347, 348
Snap To Perpendicular/Angular
  tool 347, 351
Snap to Point option 347
  using 399
Snap To Segment tool 347, 349
Snap To Tangent tool 347, 351
Snapped Position cursor option 71
snapping to objects in plan (Tip
  45) 613
snaps 341–344
  described 335
  directional 344
soda can modeling exercise CD-84–
  CD-91
Solid Color Printing option 44
solid modeling CD-19, CD-32
Solid/Surface Helix option 217
solids 167–168
Sort by Modifier option 24
sound tracks 5
source objects 204
Source Shape window CD-70

source shapes
  creating CD-69
  using CD-16, CD-57
Sphere tool 171
  using CD-21
spheres
  creating 174
  types of 176
Spherical Environment Image
  Options dialog box 509
Spherical Objects tool 174–176
  using CD-6, 373, 404, 481
Spherical Solid tool, using CD-119
Spin option 91
spiral forms 216–221
  creating 126
Spiral Stair tool 218-219
  using CD-44-45
spiral staircase exercise CD-44–47
spline curves 192
  controls for 243
Spline Sketch tool 192, 193
  using 388
Spline, Cubic Bezier tool 192, 193
Spline, Quadratic Bezier tool 192,
  193, 426
spoon modeling exercise CD-74–84
Stair From Path tool 218
staircase modeling CD-44–47
stairs tools 213, 218–221
statistics, project 134
Status Of Objects command 129, 256
Stepped terrain model 211, 212
stills from animations 566
Stitch Options dialog box 263
Stitch tool 260
  using CD-137
Store New Image Elements in Project
  option 61
Stream Distance (Stream Line) 191
Stream Line tool 120, 190, 191
Stretched Ball option 177
Stretched Ellipsoid option 177
Style Parameters dialog box CD-18
subtracting objects CD-32, CD-35
  colors and CD-37
Summer Solstice button 98
Sun dialog box 98, 541
Sun Position command 98
  setting the position (Tip 20) 602
Super Sampling 504, 567, 577
surface controls 203
Surface Object From Open Source
  option 214
Surface Render command 110–111
  views 492
surface solid, defined 168
Surface Style Color Picker CD-23
surface styles
  applying CD-73
  assigning CD-83
Surface Styles palette 130, 151–153
  toggle switch 132

selection category 57
  using 384, 420, 430, 502, 503, CD-
    66, CD-118
surfaces 167–168
  adding thickness to CD-82
  applying CD-17
  C-Meshed 246
  defined 167
  degrees of 250
  disturbed 233
  measuring 294
  meshed 247
  One-sided 167
  Patches 251
  placing an image on CD-84
  placing textures onto 320
  rendering 178
  reversing direction of 106, 285
  sculpting 203
  Two-Sided 168
sweep, two-path CD-67
Sweep Edit dialog box 223, CD-17,
  CD-70
  using 423
Sweep tool 213, 221—224
  advanced uses 452
  used for two-path sweep CD-67
  using CD-12, CD-16, 423, CD-51,
    CD-101, CD-121, CD-122
Swiss cheese exercise CD-32–CD-37
Symbol Create tool 274
Symbol Creation dialog box 276
Symbol Definition dialog box 275
Symbol Edit tool 274
Symbol Explode tool 274, 277
Symbol Instance Display option 108
Symbol Instance Placement dialog
  box 276
Symbol Instance Selection Criteria
  dialog box 56
Symbol Libraries 123, 278
Symbol Libraries dialog box 164
Symbol Place tool 274, 276
symbols, hierarchical nesting 277
Symbols Edit tool 276
Symbols palette 162–164
Symbols toggle switch 132
Symbols tool palette 124, 273–278
System Option From preferences
  selections 61

T

Tangent Curves option 245
tape dispenser exercise 408–418
tapering faces 241
Targa image files 78
teapot modeling exercise CD-101–111
telephone modeling exercise 451–465
Terrain tool 211–213
  using 388, 389
text
  manipulating 271
  reading each line of 145

Index

Text Edit tool 271–273
Text Line Edit tool 271
text objects 108
   setting smoothness 108
Text Placement dialog box 271, 272
Texture Groups option 319
Texture Map Controls CD-24, CD-118, 448
   using CD-123
Texture map error message CD-89
Texture Map tool 317–320
   using 385, 421, CD-84, CD-118, 119
Texture Mapping Type option 318, 444
texture maps
   defined 317
   importing 317
   using 406
   using for complex models 444
textures 64, 115
   adding 480, CD-22
   advanced use of 474
   applying CD-115, CD-143
   bump maps CD-25
   leather CD-60
   photographic 444
   placing 320
   positioning CD-123
   rough CD-120
   stretched fabric 468
   wood CD-143
   wrapping CD-25
TIFF image files 78
Tile Windows command 66–67
Time Control box 567
time counter, in animation 559
tips
   collection 595–616
   for changing status of all lights 157
   for double clicking 77
   for including graphical scale on drawings 79
   for keyboard shortcuts to memorize 86
   for streamlining modeling interface 18
   for using the Cmd-Z shortcut (Macintosh) 46
   for using Triangulate option in generating mesh 228
tool icons 165
Tool Options palette 148, 149, 165
   using CD-6
   toggle switch 132
tool palettes 15, 16, 166
   Attributes 312–324
   Balls 174
   Booleans and Intersections 256–267
   Delete Objects 329
   Delete Parts 330–331
   Geometric Transformations 295–303

Ghost and Layers 324–329
Insertions 182–184
Join and Group 268–271
Line Editing 278–284
Meshes and Deform 226–238
Metaformz 254–256
NURBS and Patches 242–254
Object Type 178–182
Parametric Derivatives 213–226
Primitives 170
Query 291–294
Relative Transformations 303–311
Rounding and Draft Angles 238–242
Self/Copy 288–291
Symbols 273–278
Text 271–273
Topologies 284–287
tools
   analytical 256, 263, 266, 326
   collections of 443
   customizing 132
   derivative 169
   See also Query tool
toothpaste tube modeling exercise CD-145–163
Top views
   isometric mode 86
   oblique mode 86
   perspective mode 86
   using 397, 437
topological element
   deleting 330
   transforming 296
Topological Level of Group 270
Topological Levels palette 49, 196–198
   using 368, 398, 430
Topologies tool palette 284–287, 415
Topology Count 134
Topology toolbar 106
Torus Edit dialog box 293
Torus tool 171
Total option, using 428
transferring files. *See* Files
transferring palettes 144
Transparencies option 502
   using 432
transparency (Shaded Render) 111
Transparency decal option 321
transparency maps 444, 447
Transparent option (Shading) 500
trees, making (Tip 42) 611
Triangulate tool 213, 236–238
Triangulated terrain model 211
Trim & Stitch tool, using 372
Trim Lines tool 279–280
   using 412
Trim option (Close Line) 279
Trim Segment Options 280
Trim, Stitch, and Intersection tools 260–263
   on C-Meshes 247
Trim/Split tool 260

triple clicking 148, 182, 187, 192, CD-62
TrueType fonts, using 44, 63
Truncated Cone 173
Two Path Sweep option CD-67, CD-122
Two Source Sweep CD-108
Two-Way Split (B-Split) 258
Typological Level of Object 198
Typology Types selection category 56

U

umbrella modeling exercise 465–474
Underlay command 77–80
   using 410
underlay files
   defined CD-129
   importing CD-131
   using 580
   using in skin operation CD-129
   working with CD-130
Undo command 38, 46–49, 192
   adding more for insurance (Tip 34) 609
   turning off 48
   Unlimited Undos 48
Undo/Redo, resetting to conserve memory (Tip 9) 598
Unfold tool 209–211
unfolding, concept of CD-115
Unghost tool 324–325
unghost, defined 54
Ungroup tool 269
Uniform Scale tool 297–298
Union operation 256, 257
   using 397, 402, 406
units, setting 124
updating libraries (Tip 51) 615
Use Accurate Intensity option 514
Use Contour Heights option 212
Use Custom Size option 118, 536, 537
Use Dialog Buffer option 63
Use Face Color option 106
Use Radius option, using 404
Use Window Size option 536, 537

V

Vector Line tool 190, 191, 242
   using 368, 381, 389, 395, 400, 427, 428, CD-15, CD-62
velocity control curves 567, 569
   basics 568–571
   editing 570, 571, 572–576
versions ii, 40, 125
   3.0.2 33
   RenderZone RadioZity 64
   saving in previous (Tip 53) 616
Vertical Origin field 79
Vertical Repetition, using 421
video editing software 5
View Angle 101

View commands
  Edit Cone of Vision menu 102
View Eye Point 90
View File command 42–43
View Image File window 42
View menu 82
  [+XY]:Top command 86
  [+YZ]:Right Side command 87
  [+ZX]:Back command 88
  [-XY]:Bottom command 86
  [-YZ]:Left Side command 88
  [-ZX]:Front command 88
  Axonometric* command 89–91
  Custom command 81
  Custom View Angles command 85
  Edit Cone of Vision 99–103
  Edit Menu 81
  Graphic/Keyed setting 80
  Isometric* command 91
  Oblique* command 92–93
  Panoramic* 94–96
  Perspective* command 93–94
  Plane Projection command 89
  Save View command 97
  Shaded Render* 111–112
  Sun Position command 98
  using 366
  View menu, pull-down 82
  View Parameters command 96
  Views command 97
  z=120° x=20° menu item 84
  z=220° x=45° menu item 84
  z=45° x=45° menu item 83
  z=60° x=30° menu item 85
  z=30° x=60° menu item 82
View Name label 151
View Parameters command 90, 91, 96, 564
View Rotation option 54
View Spin command 101
View tools 359–362
Viewer Position 101
viewing angles, preselected 82–85
views
  axonometric 54, 90
  changing 359
  isometric 91
  perspective 93
  setting at eye level (Tip 13) 599
Views command 97
Views menu, Panoramic setting 41
Views palette 97, 150–151
Views toggle switch 132
virtual camera 41, 99, 150
virtual skyline 394
Visible Entities Only option 40

W

walkthrough options 593

Walkthrough tool 359, 360
  recording (Tip 54) 616
walls in plan or section, creating 179
warnings
  about triple clicking with Vector Line tool 383
  about using Allow Intersecting Lines 192
  about using Clear button 162
Warnings Preferences dialog box 64
wattage, defined 528
Web designers, features benefiting 52
Weight option (Metaballs) 177
Who's Online panel 30
Window Contents print type 44
Window Frames command 68
  using CD-56
Window Level Effects options 499, 504
Window Tools 10, 69, 165–166, 333–362
  appearance of 31
  Direction Snaps 344–347
  Grid Snap Switch 341–344
  Help window 19
  interface for 333–334
  Object Snaps 347–353
  Perpendicular Switch 338–339
  Reference Planes 336–341
  Snap To Angle/Slope tool 77
  toggle switch 132
  View tools 359–362
  Zoom and Pan 358
  zooming and panning in 104, 121, 335
Window Zoom By Corners 355
Windows (Microsoft)
  differences from MacOS vi, 617–618
  hardware requirements iii
Windows menu 65–80
  Auto Scroll 72
  Close/Close All command 67
  Extended Cursor 69
  New (Draft) Window 66
  New (Model) Windows 65
  Show Grid* 73
  Show Plane Axes* 72
  Show Rulers* 75–76
  Show World Axes* 73
  Snap Options 76
  Tile Windows 66–67
  Underlay 77–80
  Window Frames 68
  Windows Setup 73–75
Windows Metafile image files 42, 78
Windows Setup command 73–75, 434
  Show Grid option 100

using 384
windows
  adjustable 68
  opening multiple 65
Windows-only file formats 626
Winter Solstice button 98
Wire And Facets Options dialog box 173
wire density settings 173
Wire Frame views 117
  keyboard shortcut for 38
  using 490
Wire Frame* command 105–108
  rendering with (Tip 4) 596
wire frames 52, 168
Wire Helix options 217
Wireframe Width setting (Shaded Render) 113
Wires and Facets 168, 250
Wires options 172
Working Units command 104, 124–125
World axes
  showing 73
  options 336
World coordinates 147
World Digitizer Calibration 131
Wrapped Textures 449

X

X and Y Justification options 44
XY Reference Plane Active tool 336

Y

Yon Plane 101
YZ Reference Plane Active tool 336

Z

z=120° x=20° viewing angle 84
z=220° x=45° viewing angle 84
z=30° x=60° viewing angle 82
z=30° x=60° viewing angle, using 373
z=45° x=45° viewing angle 83
z=60° x=30° viewing angle 85
zoom commands 104
Zoom In/Out By Frame tools 354
Zoom In/Out Incrementally tools 354
Zoom Options command 102, 121
Zoom tool 354, 355
  keyboard shortcuts 358
zooming 335, 434
  using Clipboard 54
  using hot keys (Tip 6) 597
  using View File command 42
  using Window Tools 358
  using Zoom Options command 121
ZX Reference Plane Active tool 336